D0023264

Encyclopedia of Human Genetics and Disease

Encyclopedia of Human Genetics and Disease

Evelyn B. Kelly

VOLUME 1
A–K

AN IMPRINT OF ABC-CLIO, LLC
Santa Barbara, California • Denver, Colorado • Oxford, England

Library of Congress Cataloging-in-Publication Data

Kelly, Evelyn B.
Encyclopedia of human genetics and disease / Evelyn B. Kelly.
p. cm.
Includes bibliographical references and index.
ISBN 978–0–313–38713–5 (hardback) — ISBN 978–0–313–38714–2 (ebook)
1. Human genetics—Encyclopedias. 2. Genetic disorders—Encyclopedias. I. Title.
QH431.K32 2013
616′.04203—dc23 2012018368

ISBN: 978–0–313–38713–5
EISBN: 978–0–313–38714–2

17 16 15 14 13 2 3 4 5

This book is also available on the World Wide Web as an eBook.
Visit www.abc-clio.com for details.

Greenwood
An Imprint of ABC-CLIO, LLC

ABC-CLIO, LLC
130 Cremona Drive, P.O. Box 1911
Santa Barbara, California 93116-1911

This book is printed on acid-free paper ∞

Manufactured in the United States of America

This book discusses treatments (including types of medication and mental health therapies), diagnostic tests for various symptoms and mental health disorders, and organizations. The author has made every effort to present accurate and up-to-date information. However, the information in this book is not intended to recommend or endorse particular treatments or organizations, or substitute for the care or medical advice of a qualified health professional, or used to alter any medical therapy without a medical doctor's advice. Specific situations may require specific therapeutic approaches not included in this book. For those reasons, we recommend that readers follow the advice of qualified health care professionals directly involved in their care. Readers who suspect they may have specific medical problems should consult a physician about any suggestions made in this book.

Contents

VOLUME 1

Introduction

Did you know that every one of us probably has within our genetic makeup at least 10 to 12 really bad genes? It does not bother most of us because we do not know that we have such genes. These genes have been passed down from generations past to you. You will pass them on to successive generations. You will never suspect that you are harboring these disorders—unless a relative develops something like brittle bone disease or cystic fibrosis. Then a genetic counselor may have you look seriously at a pedigree of family traits.

What Is a Genetic Disorder?

A genetic disorder is an illness caused when something goes wrong in the genes or chromosomes. The disorder can be caused by a different type of a gene, called a variation, or it can cause an alteration of a gene, called a mutation. Some diseases, such as some cancers, are caused by a mutation in a gene or group of genes. When things go wrong in the genes or chromosomes, they can include these types: a small mutation in one single gene; the adding or subtracting of an entire chromosome; or the addition or subtraction of a set of chromosomes. They can also be multifactorial, meaning that many genes interact; and environmental influences are also elements in many diseases, such as heart disease.

Television and other media have allowed genetics to enter our everyday discussions. We watch computer geniuses use genetics to find people with "the criminal mind." Reports of the latest gene, or a possible test for Alzheimer disease, or of a young athlete who has died from an undetected genetic heart disease have led genetics more into the public consciousness.

Purpose

Genetic diseases may be rare, but the number of people with genetic disorders is substantial. There are approximately 7,000 known genetic diseases throughout the world. And, about 4,000 conditions are caused by just one gene gone astray.

Many of these diseases have few or no support groups that encourage or fund research for those who must deal with the disorder. There may be no treatment, and many are easily misdiagnosed. In the United States, a disorder is considered rare if it affects fewer than 200,000 people, but approximately 30 million people are afflicted with rare diseases in the United States. This number emphasizes the point that the disease may be rare, but the number of people is substantial.

The goal of this encyclopedia is to give readers basic information about genetic disorders in general and certain disorders in specific. Although the target audience is high school and college students, it is also an important book for educated lay people who may be interested in the topic.

Occasionally, articles in local newspapers will whet the interest on the public in a disorder. For example, a candidate for president took time off from the campaign to attend to his daughter who has Trisomy 18, or Edwards disease. An article in the *Jacksonville Times-Union* told of one of their staff members whose child has IgM syndrome. A long article discussed the condition and how he had to have $30,000 up front to take the child to a special hospital for the condition. These and other rare diseases are often encountered in daily life. People who are curious can find many of the answers to their questions in the entries in this encyclopedia.

Scope

The *Encyclopedia of Human Genetic Disorders* is presented in two volumes. Following are four types of entries:

- Three general overview essays that lay the scientific foundation for understanding the nature of genetic disorders and that appear at the front of the book, including "Genetics Disorders 101," "Proteomics 101," and "The Genome and the Foundations of Genetics," which includes timelines.
- Eighteen medium-length entries, around 1,000 to 3,000 words, called "Special Topics," which discuss such issues as aging and genetic factors, breast cancer genetics, obesity and genetics, and others.
- Some 355 shorter entries, from around 550 to 1,700 words, on specific disorders.
- Sidebars of related information, biography, or human interest, which accompany some entries.

With the exception of the overview essays at the beginning of Volume 1, all the entries appear in alphabetical order.

In total, there are 376 essays and entries. Many genetic disorders have several names, but the one that is most commonly seen is used as the title for the entry in this encyclopedia. Cross-referencing from other possibly well-known names will assist in finding entries (as will the index). For example, "Presenile and Senile Dementia. *See* Alzheimer Disease."

The information presented here is straightforward and objective. Also included, however, are stories of real people and references to literature, which help to make the concepts come alive. The information presented is solidly based on research. In writing the book, I have usually used a three-pronged approach:

- First, I studied carefully material from peer-reviewed, technical sources that present evidence-based medicine. These sources included publications such as *GeneReviews* from the National Institutes of Health, information from the Centers for Disease Control and Prevention, and research from medical schools.
- Second, I looked at articles written for the general public to see how the information was presented, such as is seen in the NIH publication "Genetics Home Reference."
- Third, I sought to find a support group that included practical information.

I have especially been interested in the history of the disease or disorder and include that information where available.

Structure

The encyclopedia opens with three overview essays, followed by the entries, arranged in alphabetical order. At the back of the book is an annotated list of good websites, books, and dictionaries for the reader to refer to. A comprehensive index concludes the volumes.

Each entry about the *disease or disorder* in this encyclopedia is usually arranged as follows:

- Prevalence of the disorder
- Other names for the disorder
- An introductory paragraph of human interest telling about the discovery of the disorder
- What is the disorder?
- What is the genetic cause of the disorder?
- Treatment for the disorder
- Further reading

The section on treatments is not intended to be complete. Because these disorders are genetic, most have no cure. Treatments mostly are for the symptoms. However, a few tested therapies may be mentioned. Names of drugs for treatment are seldom given because these may vary or be changed over a period of time. Resources, usually three, for further reading are given.

Acknowledgments

I especially acknowledge the encouragement and wise teaching of the staff at the University of Florida. Dr. Sheldon Schuster, Director of the Biotechnology Program at the University of Florida, had the vision to educate teachers and general public about the importance of understanding the basics of biotechnology and molecular genetics.

I especially am indebted to the many editors—such as John Sterling of *Genetic Engineering News*—who assigned articles that honed my investigation skills.

I thank my editor Anne Thompson and the staff of ABC-CLIO, who worked diligently with me through the two years of this project.

Many thanks to the organizers of the American Medical Writers Association and the National Association of Science Writers, who provided field experiences to such places as the research at Massachusetts Institute of Technology, Harvard Medical School, and National Jewish Hospital in Denver. Especially enlightening was a trip to the Lawrence Berkeley National Laboratory, Berkeley, California, that opened my eyes to the great world of protein folding.

Of course, the book would never be made possible without the help and encouragement of my husband Charles L. Kelly, my four children, and four grandchildren.

Evelyn B. Kelly

Genetic Disorders 101

For centuries, people have contemplated the origin of human diseases and disorders. They devised all sorts of answers to their questions. They determined that disabilities were from evil spirits, from moral degeneracy, or from the misalignment of certain humors or the planets. As time progressed, some doctors related disease to small microorganisms, and then others began to note that certain disorders may appear in families. The idea of inheritance and genetics in general was slow to develop. The fact that people looked like other members of the family led to the consideration of the possibility that something related to being a member of a family could be the source of certain health problems.

Likewise, the burgeoning field of genetics has given us the ability to appreciate differences in people and to understand how certain conditions come about. For example, we know today that a person with thick hair all over the body is not a werewolf but has a genetic condition called congenital hypertrichosis (CHG). The genes in the cells that the person inherited were responsible. However, it took some time for the idea of cells, chromosomes, and genes to develop.

It's All Done in the Cells

The science of genetics shows how plants and animals inherit traits from the past generations and then pass on those traits to successive generations. Whether a person has red hair or blue eyes is determined by one's genetic makeup. Similar rules govern all living things. In the seventeenth century, Robert Hooke peered into his crude microscope at the leaf of a plant and saw organized structures that reminded him of jail cells. He called these structures "cells." The name "cell" stuck.

Generations of scientists and researchers determined that the cell is the basic building block of all living things, including human beings. The human body is made up of 100 trillion cells. With the exceptions of red blood cells and a few other types of cells in the bone marrow, each cell has all the genetic information

Changing Attitudes

People who appear different or who have unusual attributes have fascinated society; they were also objects of fear and derision. Going back many centuries ago, people with genetic abnormalities whose differences were clearly visible were often exhibited for entertainment to make money for others.

So called "Freak Shows" were popular in the United States in circuses and carnivals from around 1840 to the 1970s. However, both society and the emergence of scientific knowledge interrupted these circuses and carnivals. In 1954, the landmark decision from the Supreme Court of the United States, *Brown v. Board of Education of Topeka, Kansas,* determined segregation of public schools was unconstitutional and denied black people equal protection under the Fourteenth Amendment. The following decades of the 1960s and 1970s brought periods of great upheaval and societal change for other people seeking equal rights. Among these were individuals with disabilities. The Rehabilitation Act of 1973 and the Americans with Disabilities Act passed in 1975 set the stage for giving people with disabilities rights and opportunities not afforded in the past.

Today defining people by their condition is not acceptable. The individual is not an "epileptic" or "handicapped" or "disabled," but a "person with epilepsy" or a "person with a disability." Thus, we do not define people by their disease or disorder, but as people with the disorder or disability. For the general public in the United States and other countries, the development of an attitude of equality toward people with disabilities has been slowly evolving over the following decades.

that makes up the human being. That total block of information is called the genome.

Most cells have a cell membrane that holds the contents together, a cytoplasm, and a nucleus. Things can pass in and out of the cell membrane. Within the cytoplasm of the human cells, many small structures called organelles or "little organs" perform the work of the cell. Many of the organelles are directly related to human genetic disorders. Following are the important organelles and their functions:

- Mitochondrion: This bean-shaped organelle is the power house of the cell. The mitochondria (plural for mitochondrion) provide energy from the products of digestion. This organ is of special interest to geneticists because, like the nucleus, it has genetic material inherited; however, it is different from the nucleus in that heredity comes only from the mother. A class of genetic diseases is related to mitochondrial DNA.

- Lysosomes: These little organs can be thought of as the cell's "garbage disposals." They contain packets of enzymes that break down bacteria, worn-out parts of other cells, fats, carbohydrates, and other debris. A class of genetic disorders is related to lysosome storage.
- Ribosomes: These organelles provide a scaffold for the synthesis of proteins.
- Endoplasmic reticulum (ER): This membrane network called rough ER is dotted with ribosomes. The ER is the site of protein synthesis and folding, and also lipid or fat synthesis.
- Golgi bodies: Here sugars are made and linked into starches or joined to lipids or proteins. Proteins are finished folding.
- Peroxisome: This sac enables the catalysis or break down of many reactions.
- Vesicle: A sac with a membrane temporarily stores or transports substances.

Within the nucleus are 23 pairs of colored bodies called chromosomes. One chromosome comes from each parent. On the chromosomes are the units called genes that are responsible for heredity.

Deoxyribonucleic Acid (DNA)

The gene is made up of a double-stranded segment of *deoxyribonucleic acid* (DNA). DNA contains the blueprint or recipe for making specific proteins. Four chemical bases are found on the twisted ladder of DNA: adenine (A), thymine (T), guanine (G), and cytosine (C). A is always bound to T; G is always linked with C. These links are called base pairs or nucleotides, and the base pairs make up the genes. Some genes have few base pairs; others have many.

In a process in the nucleus called transcription, the gene sequence is copied from DNA to into ribonucleic acid (RNA). RNA is similar to DNA except it is single-stranded, whereas DNA is double-stranded. RNA also substitutes uracil (U) for thymine and ribose for deoxyribose. RNA becomes the messenger and is referred to as mRNA, taking the genetic information from the nucleus into the cytoplasm where the organelles go to work to make proteins.

Proteins

Genes guide the making of proteins. Certain genes make amino acids. These work-horse amino acids are essential to body function. The 40,000 or so genes in the nucleus program for these proteins. Within the body, millions of chemical reactions are taking place at many different levels.

So, looking into the cell can be pictured as:

> Cell → within the cell is the nucleus → within the nucleus are chromosomes → on the chromosomes are genes → genes are made of DNA → DNA makes amino acids → amino acids form proteins.

Genetic "Trans" Terms That Are Sometimes Confusing

Reverse transcription: The process of making DNA back from RNA mediated by an enzyme reverse transcriptase
Robersonian translocation: A chromosome aberration in which two short arms of non-paired chromosomes break, and the long arms fuse, forming one unusually large chromosome
Transfer RNA: (tRNA): A type of RNA that connects messenger RNA to amino acids during protein synthesis
Transgenic animal: An animal that has been genetically engineered at the embryonic gamete or fertilized ovum stage to have altered cells of another organism
Transcription: The process of manufacturing RNA from DNA
Transcription factor: A protein that activates the transcription of other genes
Translation: Assembly of an amino acid chain according to the sequence of base triplets in a molecule of mRNA
Translocation: Exchange of genetic material between chromosomes that are not paired (nonhomologous chromosomes)
Transposon: A gene or DNA segment that moves to another chromosome

What Can Happen to Genes?

Within the millions of chemical reactions that go on in the human body, so many things can happen. A genetic disorder is caused by abnormalities in the gene or chromosome. The normal working of the gene can be altered or changed. Scientists coined the word "mutation" from the Latin word *muta*, meaning "change." Genetic disorders are the results of mutations.

Following are some of the ways in which mutations occur:

- Deletion: Some misstep in the process leaves out a piece of DNA. The removal of just one DNA base pair may alter the recipe for protein and cause a disease or disorder. Here is an illustration: People in an assembly line are making peanut butter–and–jelly sandwiches for a huge picnic. Each person has a job to do. One opens the bread, another spreads the peanut butter, a third spreads the jelly, and a fourth closes the bread. Suppose that the person spreading the jelly drops the knife and has to go after it. The assembly line continues but without the jelly, and the result would not be a peanut butter–and–jelly sandwich, but a change. This same thing may happen when a gene is deleted.
- Insertion: A piece of DNA that does not belong gets added. The protein that now makes up the gene will not function properly.

- Missense mutation: Here, in one DNA base pair, one amino acid is substituted for another in the protein that makes up the gene.
- Nonsense mutations: The DNA erroneously signals the cell to build the protein early or to stop building the protein altogether. Thus, the shortened protein may not function.
- Frameship mutation: Each gene has a reading frame—a group of three bases that each code for an amino acid. If the grouping of the reading frame changes, the created protein may not function. Framework-reading mutations may cause insertions, deletions, and duplications.
- Repeat expansion: Sometimes a short sequence of DNA is repeated a number of times. One of the most common is the C-A-G repeat found in Huntington Disease.

Finding the Address of Genes

The human genome comes packaged in 23 pairs of chromosomes—22 matched pairs called somatic or body chromosomes, and one pair of sex chromosomes X and Y. The 22 pairs are numbers in approximate order of their sizes, from the largest—chromosome 1—to the smallest—number 22. Women have two large X chromosomes. The size of the X chromosome is between numbers 7 and 8. Men have one X chromosome and one small Y chromosome, which is the smallest of all the chromosomes. The number of 46 chromosomes for human beings is not important. Many species have more, and many have fewer. For example, a cat has 38; a dog, 78; a mouse, 40; and shrimp, 88–92.

Genes are located on the chromosomes. The diagrams of chromosomes are called ideograms. These pictures show a chromosome's size and banding patterns. Just like one may use a roadmap to pinpoint a location, geneticists use chromosome maps to pinpoint the location of certain genes. The word "cytogenetic" is used to describe this map. Cytogenetic comes from two Greek words, *cyto* meaning cell and *gen* meaning "give rise to." In fact, the word genetics comes from the Greek meaning giving rise to or giving birth to. The ideogram is also called a cytogenetic map.

When the cell nucleus is stained, chromosomes appear as colored bands. The position of a certain band is used to describe the location of the gene. The formula for interpreting the address of a gene is as follows:

- Chromosomes are numbered 1 through 22, with the sex chromosomes indicated by X and Y.
- Each chromosome has two arms that are positioned above and below a narrowed or constricted area called the centromere. The shorter part of the chromosome is designated a "p," and the larger arm is called "q." The chromosome arm is the second part of the gene's address. For example, a gene designated 15p would be located on chromosome 15 on the short arm.

If the designation was Xq, the location would be on the long arm of the X sex chromosome.

- Now looking at the short or long arm, distinctive patterns of light and dark bands are seen. Beginning with the centromere, the light and dark bands are counted. The number gets larger as one counts away from the centromere. If the gene number is 13p21, the position will be on the short arm. Band 21 is 21 spaces away from the centromere. The location 13p21 is closer to the centromere than 13p22. Sometimes a decimal point will indicate a sub-band location within a specific band—for example, 14q21.1.
- Three other abbreviations—*cen*, *ter*, and *tel*—may be seen. "Cen" indicates that a gene is close to the centromere. For example, Xqcen would indicate that the gene is located on the long arm of the X chromosome near the centromere. The term "ter" stands for terminus or end; 14pter means the location of the gene is on the short arm of chromosome 14 at the end. The word *telomere* describes the terminal position at the chromosome tip. Occasionally, the shortened term "tel" is used and means the same thing as "ter."
- Some researchers as a result of the Human Genome Project used base pairs to determine the molecular location of a gene. This system is more precise but more complex that use of numbers and letters.

Chromosomes by the Number

Chromosomes are given numbers according to the position in the cell. Genes that determine one's traits, such as eye color and hair color, are located on chromosomes along with those that cause genetic disorders. However, in order to understand genetic disorders, one must know the nature of the chromosomes and basic information about the genes on the chromosome. This section presents a chromosome-by-chromosome overview of the 22 pairs of somatic chromosomes and the sex chromosome X and Y.

1. Chromosome 1: Including about 247 million base pairs, chromosome 1 is the largest of the chromosomes. This chromosome represents about 8% of the DNA in the human body and has about 3,000 genes, which perform a variety of roles in the human body. However, many genetic conditions are related to chromosome 1. Some of the changes include deletions, insertions, and duplications of genes. Following are a few of the conditions related to changes or mutations in certain genes on this chromosome:
   a) Certain forms of cancers are associated with changes in body cells acquired during one's lifetime. Tumors of the brain and kidney have been related to deletions in the short arm (p).
   b) The disorder 1p36 deletion syndrome is caused when a piece of DNA is missing from the short arm (p) of this chromosome. The person has

distinctive facial features, intellectual disability, and problems with various body systems.

c) Mutations can also cause delayed growth, distinctive facial features, and birth defects. Because this chromosome is so large, it may break into pieces and form a ring. Ring chromosomes happen when the chromosome breaks into two pieces and the ends fuse.

2. Chromosome 2: Spanning about 243 million base pairs, this chromosome is the second largest in size and carries about 1,300 to 1,400 genes. Sometimes the number of copies change, and the chromosome is duplicated. When a body is duplicated, it is called a *trisomy*. A condition trisomy 2 may occur when the cells have three copies of the chromosome. A condition myelodysplastic syndrome affects the blood and bone marrow, making the person at increased risk for blood cancer or myeloid anemia. Chromosome 2 has another common abnormality known as ring chromosome 2. Because this chromosome is so large, the breakage forming the ring may cause a syndrome in which the individual has a small head size, slow growth, and heart defects, with distinctive facial features.
3. Chromosome 3: This chromosome has about 200 million base pairs and comprises between 1,100 and 1,500 genes. A number of more common genetic defects are related to this gene, including Charcot-Marie-Tooth disease, essential tremor, Marfan syndrome, and porphyria.
4. Chromosome 4: This chromosome has more than 191 million base pairs and contains between 1,300 and 1,600 genes. Changes in this chromosome have been involved in many types of human cancers. Rearrangements called translocations of the genes between this and several other chromosomes have especially been related to leukemias. Chromosome 4 and chromosome 14 have been known to cross over or translocate, causing a type of cancer known as multiple myeloma.
5. Chromosome 5: This chromosome has about 181 million base pairs and contains about 900 to 1,300 genes. Scientists refer to part of this chromosome as a "gene desert" because no known genes exist in part of the chromosome. Several regions of this chromosome are associated with Crohn disease, a condition in which the intestines harden and do not function. Several genes on this chromosome increase the risk of serious intestinal diseases.
6. Chromosome 6: This chromosome spans about 171 million base pairs and contains about 1,100 to 1,600 genes. Several genetic disorders are related to chromosome 6.
7. Chromosome 7: This chromosome has about 159 million base pairs and contains about 1,150 genes. This chromosome may play a critical role in controlling the growth and division of cells. Without these genes, cells could grow wildly, resulting in cancer or tumors. In addition to relationship to cancer,

several other syndromes such as Russell-Silver syndrome and Willams syndrome are connected with chromosome 7.

8. Chromosome 8: This chromosome spans about 146 million base pairs and contains between 700 and 1,100 genes. Certain cancers such as Burkitt lymphoma can be caused when chromosome 8 is translocated to chromosome 2, 14, or 22. Trisomy 8 can occur when three copies of the chromosome are present. A rearrangement called an inversion duplication may occur when a segment of the chromosome is inverted.
9. Chromosome 9: This chromosome has about 140 million base pairs and consists of about 800 to 1,300 genes. Deletions are commonly found in bladder cancers. Located on this chromosome are genes called tumor suppressors, which help prevent cells from dividing and growing. However, if these genes are missing or if they change, cancer of the bladder may grow.
10. Chromosome 10: This chromosome spans about 135 million base pairs and has about 800 to 1,200 genes. A loss of part of chromosome 10 is involved in fast-growing, aggressive brain cancers, called gliomas.
11. Chromosome 11: This chromosome spans about 134 million base pairs and contains about 1,500 genes. Around 150 of the genes located here appear to give instructions for the sense of smell. This sense of smell is referred to as the olfactory sense; these genes are related to proteins that are used to detect various smells.
12. Chromosome 12: This chromosome has about 132 million base pairs and contains about 1,200 to 1,400 genes. Mutations in this chromosome have been identified with several types of cancers, including leukemias and lymphomas. A condition known as Pallister-Killian syndrome is caused when the arm 12p splits, forming two identical arms. These slits are called isochromosomes.
13. Chromosome 13: This chromosome is made up of about 114 million base pairs and has between 600 and 700 genes. A condition called retinoblastoma is caused by the RBI gene located on the long arm of chromosome 13. Trisomy 13 (three copies of the chromosome) is traced to the translocating to another chromosome during the reproductive process. Translocation refers to the exchange of materials between chromosomes that were not the matching chromosome. This condition disrupts the course of normal development.
14. Chromosome 14: This chromosome has about 106 million base pairs and possesses about 800 to 1,300 genes. Several conditions including cancers are related to this chromosome. Chromosome 14 may form a ring chromosome that causes several genes to be missing. A common symptom of ring chromosomes, including ring 14, is epilepsy.
15. Chromosome 15: This chromosome spans about 100 million base pairs and contains about 650 to 1,000 genes. Many genetic disorders are related

to chromosome 15, including Angelman syndrome, Prader-Willi syndrome, and other changes that cause intellectual disability and characteristic facial changes.

16. Chromosome 16: This chromosome spans about 89 million base pairs and contains between 850 and 1,200 genes. Changes in these genes can cause major health problems. One well-researched condition known as alveolar capillary dysplasia with misalignment of pulmonary veins, a disorder of the lungs, is traced to a deletion at 16q24.1, a region that includes the FOXF1 gene. Changes in this chromosome include several types of cancers.
17. Chromosome 17: This chromosome has about 79 million base pairs and contains about 1,200 to 1,500 genes. Changes in chromosome 17 have a variety of effects including cancers, intellectual disability, weak muscle tone, and short stature.
18. Chromosome 18: This chromosome consists of about 76 million base pairs and approximately 300 to 500 genes. Several disorders are caused when pieces of the long arm or short arm are missing or when extra genetic material is inserted.
19. Chromosome 19: Chromosome 19 spans about 64 million base pairs and has about 1,300 to 1,700 genes. Several conditions including Alzheimer disease are associated with this chromosome.
20. Chromosome 20: This chromosome has about 62 million pairs and contains 700 to 800 genes. One of the most common conditions, Alagille syndrome, involves a gene, JAG1, which presents signaling problems during embryonic development.
21. Chromosome 21: This chromosome spans about 47 million base pairs and contains about 300 to 400 genes. Rearrangements of genetic material have been associated with several types of cancer, including lymphoplastic anemia. Presence of three chromosomes is associated with Down syndrome.
22. Chromosome 22: This chromosome spans about 50 million base pairs and likely has about 500 to 800 genes. Several deletions are common on this chromosome.
23. Sex chromosomes. The sex chromosomes consist of X and Y.
    - X chromosome: This chromosome has about 155 million base pairs and carries about 20,000 to 25,000 genes. Several disorders are related to this chromosome, including Klinefelter syndrome, Turner syndrome, and several abnormal extra chromosome conditions.
    - Y chromosome. This chromosome has only about 58 million base pairs and has only 70 to 200 genes. It is responsible for the determination of sex but has very little other information. The presence of extra Y chromosomes has been involved with infertility and behavioral disorders.

Inheritance Patterns

In the late 1880s a patient monk named Gregor Mendel (1822–1884) worked with peas in his monastery garden in the Austrian Empire. He noted some things happened over and over again. Traits appeared to be passed on through both male and female. Certain traits of the peas would appear in the generations; those traits he called dominant. Other traits might be hidden for several generations and show up in later generations; these traits he called recessive. Years later, Mendel's work became recognized as Mendelian Law. With Mendel and advances in cell biology, researchers opened up the field of genetics, especially studying the nature of abnormal conditions and patterns of inheritance.

In order to understand genetic disorders, comprehension of inheritance patterns is critical. When a person has a genetic disorder, the families begin to look at how the condition may have been passed down from one generation to another and, more importantly, how it might affect unborn generations. Six basic patterns of inheritance are involved: Single-gene dominant, single-cell recessive, X-linked recessive, X-linked dominant, multifactorial, and mitochondrial inheritance.

Single-Gene Traits Dominant

Mendel found that the dominant traits were those that prevailed in the presence of the recessive trait. Dominant disorders happen when an offspring receives a defective gene from either parent, and the presence of the gene expressed a genetic defect. In Mendelian Law, this trait is called dominant.

One affected parent with the dominant gene and a normal recessive, and one unaffected parent with two normal genes, together have offspring in the following pattern: according to chance, two of the offspring will receive both normal genes and will be normal for the condition, but two of the offspring will received the dominant gene and will be affected. Because the gene is located on the body or *autosome*, conditions may affect both sexes. Examples of single-gene dominance inheritance include Huntington disease, neurofibromatosis I, and porphyria.

Single-Gene Recessive

In order to be affected, the person must inherit two copies of the mutated gene. The affected person usually has parents who do not display the characteristic but are carriers of a single copy of the gene. In single-gene traits, the person has a 25% chance of inheriting the condition if both parents are carriers. Examples of single-gene traits are cystic fibrosis, sickle cell disease, and numerous rare conditions.

X-Linked Dominant

In the scheme of chromosomes, females have two X chromosomes, one from the father and one from the mother. Males have an X and a Y chromosome. Thus, it is the male that determines the sex of the child. A gene that is passed on the

X chromosome is called a sex-linked gene. The X chromosome is quite large and carries a lot of genetic information; the Y chromosome is relatively small and has little information. Because males have only one X chromosome, they may inherit the condition because nothing is on the Y chromosome to match the condition. Only a few conditions such as X-linked hypophosphatemia are X-dominant.

X-Linked Recessive

X-linked recessive conditions are quite different. These conditions are carried on the X-chromosome, and males are generally not affected. Examples are hemophilia A and Duchenne muscular dystrophy.

Multifactorial Genes

Multigene traits are those that are caused by several genes. These traits are called polygenic and do not necessarily follow the Mendelian pattern. Many normal characteristics, such as eye color or hair color, do not follow simple Mendelian patterns. The inheritance pattern is much more difficult to understand and interpret because genes may interact. Also, many of the genes may present themselves only in certain environments. Examples of these are legion: breast and ovarian cancer, lung cancer, malignant melanoma, and Alzheimer disease.

Mitochondrial Inheritance Plans

The mitochondria are small, bean-shaped structures found within the cell, sometimes called the powerhouses of the cell. The mitochondria convert molecules into energy. This type of inheritance is also known as maternal inheritance because the genes in mitochondria are passed on only by the mother. The egg cells contribute mitochondria to the developing embryo.

Techniques for Studying Genetics

Three powerful techniques have contributed to the understanding of genetics and locating the position of genes on chromosomes: transgenic animals, twin studies, and quantitative trait loci (QTL). These techniques have helped scientists locate genes on chromosomes, study the nature of the disease, and use the animals in testing for new drugs.

Transgenic Animals

Molecular biology has provided the powerful tool of transgenic animals to study genes. When the root word *trans* is used in a word, it implies that something is moved across, and transgenics means genes are moved from one organism to another. Genetic information from one organism is transferred to another animal, and that animal or organism is called "transgenic." As a result, the transgenic organism has new genetic material that will display different characteristics that the organism did not have before. Scientists are able to "knock out" normal genes and "knock in" a gene that may interest them. For example, knock-out mice are very important in studying Alzheimer disease, Parkinson disease, diabetes, and obesity.

Twin Studies

The genome for individual humans is about 99.9% identical, in contrast to a greater variety among dogs. For example, a Great Dane is very different from a miniature poodle not only in size, but also in basic body functioning. With humans, a relatively narrow range exists in weights and shapes of bodies. For example, no adult is less than 2 feet tall or greater than 10 feet. There is a normal range in which humans fall. Just how much is genetic and how much is environmental is debated.

Studies of identical twins who have been raised apart contributed to the understanding of the effect of genetics and environment. Identical twins, called monozygotic (MZ) twins, share identical heredity because one sperm fertilizes one egg form the mother. After fertilization the egg splits into two cell masses to form two zygotes that develop into two identical individuals. In contrast, fraternal or dizygotic (DZ) twins share only half their genes—no more than any other brother or sister. Comparing MZ and DZ twins indicate a genetic effect. The strength of the effect is called heritability, which is an indication of the proportion of variation within a population that is due to genetics. For example, in studies of twins raised apart, genetics accounted for 50% to 80% of the variation in body mass, with the remaining variation attributed to environment. Studies of twins raised apart and placed in adoptive families showed that adopted children are more similar to their biological parents than to their adoptive parents.

Quantitative Trait Loci

Quantitative trait loci (QTL) use maps of generations to trace and locate various genes. The approach is tedious and limited by the low precision of the DNA area. A famous example of this is the search for the gene that causes Huntington disease. Dr. Nancy Wexler noted that a group of native people living on Lake Maricaibo had a condition that was similar to that of her mother, who died with Huntington disease. She was able to get blood samples and brought them back to Dr. James Gusella, who compared the DNA from the samples and found a gene, later labeled *huntingtin*, located on chromosome 4.

These three methods have afforded a lot of information about genetics. These studies have been made possible through advanced technology, especially high-speed computing. Yet, there are many single gene mutations and multigene disorders to locate.

Many geneticists consider that the study of abnormal traits and search for medical answers has been the foundation of genetics. The fascinating story of the genes is only just beginning.

Naming Genetic Disorders and Genes

When looking at genetic disorders, one may note that several names are given. Many of the conditions are old diseases and names developed through tradition. Others "honored" the doctors who first described and recognized the symptoms of a disease. When researchers investigated the disease and found biochemical pathways, they added the name of the pathway to the mix. The names of genetic conditions are very complex. In this book, the main article is written under the most commonly recognized name for the disorder, with reference from the other names.

Names of Genetic Disorders

The following are some of the ways in which names have developed:

- Person who discovered the disorder: One of the most common ways is using an eponym—that is, naming the disorder after the physician who first described the disorder or who studied the condition and published reports in scientific journals. For example, Alzheimer disease was named after Dr. Alois Alzheimer, the German doctor who first described the plaques and tangles of slides from the brains of patients with dementia. A colleague used his name in a textbook of psychiatry, and the name stuck.
- Symptoms: If a prevalent symptom is noted, the disease may be named for this symptom. For example, sickle cell disease, once called sickle cell anemia, is named for the shape of the red blood cells that are characteristic of the disease.
- Body part: A name may be given for the body part that is affected. For example, retinoblastoma is a condition that affects the retina of the eye.
- A famous person: A famous person who had the disease may contribute the name. For example, Lou Gehrig, a famous baseball player, developed amyelotrophic lateral sclerosis, or ALS. The press began to call the conditions "Lou Gehrig's disease," which is much more recognizable than a long scientific name.
- Region of the world: A condition may be prevalent in a certain region of the country. Familial Mediterranean fever is a disorder found in certain groups of people that live near the Mediterranean Sea.
- Scientific name: Some conditions that have been described more recently use the scientific name of the biochemical defect. The condition ornithine transcarbamylase (OTC) became known when Jesse Gelsinger died during a gene therapy experiment.

When eponymous names were first created, the possessive was always used—for example, "Down's syndrome" or "Parkinson's disease." The medical community now recommends dropping the possessive to make it Down syndrome or

Parkinson disease. The reasons given for dropping the possessive are that the person does not own the disease, and the researcher did not have the disease. In this reference book, eponyms will be referred to without the possessive. The names of genes are always italicized when written.

Further Reading

"Chromosomes." 2012. Genetics Home Reference. National Library of Medicine (U.S.). http://ghr.nlm.nih.gov/chromosomes. Accessed 1/27/12.

Lewis, Ricki. 2009. *Human Genetics: Concepts and Applications*. Dubuque, IA: William Brown and Co.

Kelly, Evelyn B. 2007. *Gene Therapy*. Westport, CT: Greenwood Press.

Ridley, Matt. 2000. *Genome: The Autobiography of a Species in 23 Chapters*. New York: HarperCollins.

Proteomics 101

Proteomics is the study of the structure and function of an organism's complete complement of proteins. With the decoding of DNA in the genome, the next logical step is finding how proteins are made and how they work.

Like the term "genome," the biological name is "proteome." However, unlike the human genome, which has a specific organization and identity, the proteome is dynamic and ever changing. One might think of the decoded genome as the protein parts list; however, missing are the instructions for putting the pieces together.

The Genetic Code: Putting the Pieces Together

The genetic code developed as a set of rules encoded in the DNA or mRNA sequences. The genome is made up of base pairs: Alanine (A) is always paired with thymine (T); and guanine (G) is always paired with cytosine (C). Scientist George Gamow proposed that a set of three base pairs encode the 20 amino acids, which are used by the cells to instruct for proteins. For example, alanine-alanine-alanine, or AAA in sequence, is the codon for the amino acid lysine; the sequence cytocine-cytosine-cytosine, or CCC, codes for the amino acid proline. Translating the codon begins where it starts. The actual reading frame is indicated by a start codon. For example, AUG codon is a start codon. The stop codons include UAG, UGA, and UAA. See Table 1 for a simplification of the codons for the proteins.

To make matters more complex in the genetic code, the number of different protein building blocks (20) exceeds the number of mRNA building blocks (4). So each codon must contain more than one mRNA base. However, this is still inadequate to explain the number of proteins. The idea of beginning at different places in the string of amino acids led scientists to develop the idea of beginning with a second or even a third letter.

In this table, the 20 amino acids found in proteins are listed, along with the single-letter code. A codon is a formula for DNA. The codon is made up of three-letter amino acids that are combinations of T, C, A, and G. Also, a codon exists for stopping the sequence. For example, TAA and TAG indicate when to stop the sequence. To make the issue more complicated, some amino acids have several codons

The proteome is defined as a set of proteins expressed by genes in a specific cell and under a particular set of conditions. Like genes that have a name, proteins do also. Sometimes, the name of the protein is very similar to the name of the gene, and at other times it is not. For each article on genetic disorders in this encyclopedia, the gene(s) will be named according to the nomenclature of the Human Genome Project Organization, and then the protein that the gene instructs for will be named. Within the human proteome, the number of proteins could be as large as 2 million.

The term "proteome" was first coined by the Australian scientist Marc Wilkins in 1995. When he considered the proteome, few envisioned a project similar to the human genome project. However, development of unique tools has enabled scientists to probe the secrets of proteins. The following tools have assisted our understanding of proteins:

- X-ray crystallography: This tool shows pictures of the fold and kinks of proteins; different colors represent different functions. In the grooves and olds, certain molecules fit like a lock to a key.
- Nuclear magnetic resonance (NMR).
- Electrospray ionization: This technique is used in mass spectrometry to produce ions from macromolecules.
- Matrix-assisted laser desorption/ionization (MALDI): This technique is used in mass spectrometry to analyze biomolecules such as proteins, peptides, and sugars.
- FT-ICR mass spectrometry: This type of mass analysis determined mass-to-charge ions based on cyclotron frequency of the ions in a fixed magnetic field
- High-speed computers.

The study of proteomics includes three different efforts:

- Structural proteomics: This study includes the actual study of the makeup and organization of the protein.
- Expression proteomics: This study includes how the proteins express themselves in the organism. For example, a gene normally instructs the protein to do or make certain things. However, a mutation in the gene may program for a protein that causes some malfunction in an organ or system.
- Interaction proteomics: This study includes the totality of human protein-protein interactions. This task is extremely complex and is still a work in progress.

How Are Proteins Organized?

Proteins themselves are very large molecules called macromolecules. The term "macro" comes from a Greek work meaning "large." Amino acids make up these large chains. Actually, the chain is constructed within the cell when the ribosomes translate messenger RNA or mRNA from the DNA in the cell's nucleus. The process of assembling the amino acid chain according to the sequence of base triplets in a molecule of mRNA is called translation. This manufacturing of RNA from DNA is called transcription. The path that is followed is DNA → RNA → Protein. This information transfer is the central dogma of biology.

The structure of proteins may consist of the four following levels:

- Primary: This form is a simple, bead-like chain of usually a 20-letter alphabet of amino acids.
- Secondary: This form is a folded or pleated structure occurring when sequences of amino acids are linked by hydrogen atoms. Loops, coils, and sheets are formed when amino acids attract each other.

Table 1. 20 Amino Acids, Their Single-Letter Database Codes (SLC), and Their Corresponding DNA Codons

Amino Acid	Letter Symbol	DNA Codons
Isoleucine	I	ATT, ATC, ATA
Leucine	L	CTT, CTC, CTA, CTG, TTA, TTG
Valine	V	GTT, GTC, GTA, GTG
Phenylalanine	F	TTT, TTC
Methionine	M	ATG
Cysteine	C	TGT, TGC
Alanine	A	GCT, GCC, GCA, GCG
Glycine	G	GGT, GGC, GGA, GGG
Proline	P	CCT, CCC, CCA, CCG
Threonine	T	ACT, ACC, ACA, ACG
Serine	S	TCT, TCC, TCA, TCG, AGT, AGC
Tyrosine	Y	TAT, TAC
Tryptophan	W	TGG
Glutamine	Q	CAA, CAG
Asparagine	N	AAT, AAC
Histidine	H	CAT, CAC
Glutamic acid	E	GAA, GAG
Aspartic acid	D	GAT, GAC
Lysine	K	AAA, AAG
Arginine	R	CGT, CGC, CGA, CGG, AGA, AGG
Stop codons	Stop	TAA, TAG, TGA

Source: Center for Biological Sequence Analysis, Department of Systems Biology, Technical University of Denmark. n.d. http://www.cbs.dtu.dk/courses/27619/codon.html.

- Tertiary: This form creates a 3D appearance or the entire amino acid sequence.
- Quaternary: This form is a very large unit occurring with the interaction between many protein subunits to form the large structure. There may be more than one polypeptide chain joined in the complex.

Each level is essential to the finished molecule's order. Primary chain determines the order and forms the basis of more complex structures.

Proteomics is a recent and complex field. Because proteins are so diverse, they can have many functions. Analyzing the various proteomes is essential to the understanding of human genetic disorders. Within this analysis may also lie the answer to pharmaceuticals and other interventions to cure serious genetic disorders.

Further Reading

Children's Hospital Boston Introduction to Proteomics. http://www.childrenshospital.org/cfapps/research/data_admin/Site602/mainpageS602P0.html. Accessed 5/1/12.

Proteomics Overview, Agilent. http://www.proteomicworld.org. Accessed 5/1/12.

The Genome and the Foundations of Genetics, with Timelines

The human genome and the Human Genome Project have established the foundation of genetics. The genome is defined as all the heritable traits of an organism, including the full set of chromosomes and genes. Yet, the inquiry into human inheritance began many years ago. A study of human inquiry into inheritance can be divided into five phases:

- Ancient theories or pre-Mendelian thought
- Mendelian inquiry (based on Gregor Mendel's findings)
- The DNA era
- The genomics era ending with the successful completion of the Human Genome Project
- The post-genomic proteomic period

Ancient Theories or Pre-Mendelian Thought

From the time human beings could reason, they realized that there must be a way in which characteristics are passed on from one generation to another. They noted how the chin or nose of a parent was also present in the offspring. The idea that "like begets like" is well known in many cultures.

In the seventeenth and eighteenth centuries in Europe, opposing views of sexual reproduction developed between the ovists and animalculists. The ovists believed that the offspring preexisted in the female egg and only needed to be activated in some way for the baby to form. In 1672, when Regnier de Graaf, an early anatomist, thought he had found the ovum (actually, he saw only the ovarian follicle), this theory seemed to be substantiated. In contrast, the animalculists, also called spermists, believed that the male determined the nature of the offspring. Some of these theorists even thought a tiny preformed man called a "homunculus" was in the sperm. Antoine van Leewenhoek's observation of "animalcules" in semen supported this view.

Also, they began to connect certain conditions as being passed down to offspring. Certain disorders or diseases were noted; for example, ancient Jews recognized a bleeding disease. The Talmud, the primary Jewish religious document of oral and biblical tradition, excused newborn boys from circumcision when a mother had lost several sons due to uncontrolled bleeding, beginning in the eleventh century. The contributors to the Talmud realized that certain disorders were passed from mother to son—long before scientists had discovered Factor VIII and hemophilia. Charles Darwin, who came from a long line of physicians, was fascinated by evidence of disease inheritance.

In the pre-Mendelian period, embryology and genetics were closely related. People could only imagine what happens before birth. Eventually, a few physicians began to connect what happened before birth with physical disorders. In the eleventh century, Trotula, a female physician at Salerno, Italy, wrote a treatise called *The Diseases of Women* that was illustrated with drawings of the child before birth as a miniature adult floating in water. Centuries later, Hieronymus Fabricus (1533–1609) used the same childlike drawings to illustrate his inquiries as to what happened before birth. As sciences, neither embryology nor genetics was given much thought before Mendel and Darwin, although a few thinkers were perplexed about what made people look the way that they did. Tables 2–4 look at a few of these people.

Mendelian Inquiry

Beginning his experiments in 1856, Gregor Mendel, an Austrian monk, noticed that pea plants in his garden defied the dogma of the day of mixing. Crossing peas in his garden, he developed the idea that some traits were dominant and some were recessive and that these traits followed specific laws. When his work was rediscovered, "Mendelianism" became an overnight sensation.

The eugenics propaganda and the ascent of World War II (1941–1945) slowed research into genetics. However, several breakthroughs ushered in a new era—the DNA era. The first great breakthrough was the experiment by Avery, MacLeod, and McCarty with the bacteria pneumococcus. They found that DNA is the material that mediates heredity. However, most scientists were still skeptical of their experiments.

The DNA Era

The Genomics Era Ending in the Human Genome Project

With the announcement of the sequencing efforts in 2000, the genome became the crown jewel of twentieth-century biology. The White House heralded it, and it was plastered in headlines of magazines and newspapers. Yet, from the start it was

Table 2. Timeline, from 5000 BC to 1800s

Year	Scientists	Discovery
5000 BC	Some societies	Selectively bred plants or livestock
400 BC	Aristotle	Observed that some injuries sustained in life could be passed on to offspring; theory of pangenesis attempted to explain how traits are passed through particles called *gemules* to the mother; surmised the form-giving principle was passed through semen and the mother's menstrual blood to interact in the womb to form a new being
Ninth century AD	Al-Jahiz	Animals that survive to breed pass on this successful characteristic to offspring
1000 AD	Arab physician Albucasis	First described hemophilia as a genetic disorder; wrote about an Andalusian family whose males bled to death after simple cuts
1600s	Robert Hooke	Looks at structures under microscope and calls them cells
1600s	Anton van Leeuwenhoek	Dutch craftsman and lens grinder uses crude (by today's standard) microscopes to look at the world of microbes; first to see sperm in semen
Eighteenth century	General acceptance	A hereditary link between hemophilia and color blindness accepted in New England
1814	British physician Joseph Adams	Gives a full summary of the knowledge of inheritance of the time and distinguishes between familial disease that occurs in a single generation and inherited disease passed on from generation to generation
1859	Charles Darwin, Alfred Russel Wallace	Darwin fascinated by human inheritance of disease; theory of natural selection supports that individuals that are better adapted to a situation survive and pass traits on to offspring; Darwin published *Origin of Species*
1800s	General acceptance	Prevailing idea of blending in genetics; if two plants were crossed that were different, the offspring would be a blend of the two

controversial, even called "absurd, dangerous, and impossible" only 15 years earlier. Thousands of people participated, and the intrigue of feuds and disagreements even played a part in the development.

The advance of technology of high-speed analytical machines was instrumental. There are many recognized heroes, but also many unsung heroes who worked in various ways. One of the most important was the refinement of electrophoresis,

Table 3. Timeline, 1800s–1944

Year	Scientist(s)	Discovery
1866	Gregor Mendel	Publishes the results of his experiments of the inheritance factors in pea plants called *Experiments in Plant Hybridization*
1866	J. L. H. Langston-Down	Identifies a cluster of symptoms that now bear his name
1869	Friedrich Miescher	Identifies an acidic substance found in the cell nuclei of white blood cells that is now called DNA; significance is not appreciated for 70 more years
1880	Walther Flemming, Edmund Strasburger, Edouard van Beneden	Describe chromosome behavior during mitosis or division of cells
1883	Francis Galton	Proposes selective breeding as a way to control population; called the proposal "eugenics"
1889	Dutch scientist Hugo deVries	Hypothesizes that specific traits are inherited in small particles; calls these particles (pan)genes
1900	Carl Correns, Hugo deVries, Erich von Tschermak	Independently discover Mendel's work from 1866; age of genetics begins
1900	Englishman William Bateson	Translates Mendel's work in English
1902	Boveri	Recognizes that individual chromosomes are different but does not make connection with inheritance
1902	Walter Sutton	Proposes the chromosome theory of inheritance based on Mendelian patterns
1905	Willam Bateson	Coins the term "genetics" in a letter to Adam Sedgwick and at a meeting in 1906
1905	Nettie Stevens, Edmund Wilson	Independently describe the sex chromosomes XX for females, XY for males
1905	Reginald C. Punnett	Develops squares to illustrate Mendelian inheritance; others begin to suspect that some genes are not linked to Mendelian principles
1903	Archibald Garrod	Proposes that some diseases are lined to an "inborn error of metabolism"
1908	Godfrey Hardy, WilhelmWeinberg	Devise Hardy-Weinberg law of population
1910–1930s	Eugenics movement	Fuels racial resentment leading to involuntary sterilization laws; will eventually establish a negative attitude about genetics

Table 3. (Continued)

Year	Scientist(s)	Discovery
1910	Thomas Hunt Morgan	Confirms the chromosome theory of heredity using studies of eye color of fruit flies; shows genes reside on chromosomes
1911	E. B. Wilson	Identifies X chromosome as the location for gene for color blindness, the first gene to be traced to a specific chromosome
1913	Alfred Sturtevant	Makes the first genetic map of a chromosome; shows genes are arranged in a line on the chromosome
1918	Ronald Fisher	Connects genetics and evolutionary biology; starts population biology
1927	Hermann J. Muller	Uses x-rays to cause artificial mutation in drosophila
1928	Fred Griffith	Shows how heat-killed virulent bacteria can make non-virulent strains to become deadly; proposes some unknown principle is the cause
1931	Barbara McClintock, Harriet Creighton	Observe crossover of some chromosomes
1932	Dutch physician P. J. Waardenburg	Suggests Down syndrome is caused by a chromosome abnormality
1933	Jean Brachet	Finds DNA in chromosomes; RNA in cytoplasm
1934	Norwegian Ivar Asbjorn Folling	Identifies phenylketonuria as a metabolic disease that was ultimately found easy to treat with diet
1941	George Beadle, Edward Tatum	Find one gene encodes for one protein; experiments with red bread mold *Neurospora*; this becomes the central dogma of genetics
1941–1944	Nazi eugenicist Josef Mengele, Russian Trofim Denisovich Lysenko	Genetics comes under attack because of the Nazi eugenics breeding experiments and Russian Lysenko who denied the hereditary properties of chromosomes

a procedure that allows scientists to tear apart the DNA of a specified organism. Many other techniques and procedures were developed in the process. So deciphering those strings of A's, T's, C's and G's was certainly tedious, and the fact that it was done in record time was amazing. Table 5 provides a chronology of the progression of dates, the scientists or agency, and the discovery that brought about the Human Genome Project.

Table 4. Timeline, 1944–1970

Year	Scientist(s)	Discovery
1944	Oswald Avery, Colin MacLeod, Macklyn McCarty	Isolate DNA as the genetic material but at the time called it the transforming principle
1945	Max Delbruck	Organizes a course in molecular biology at Cold Springs Harbor to train students
1946	Ledenberg, Tatum	Find that genetic material can be transferred between bacterial cells
1950	Erwin Cargoff	Shows that four nucleotides are present in DNA—adenine, thymine, guanine, cytosine—are present but not in the same proportions; however, some rules do apply: adenine is present as the same amount as thymine; guanine has the same amount as cytosine
1950	Barbara McClintock	Finds transposons or jumping genes in maize
1951	Rosalind Franklin	Obtains sharp-ray diffraction photographs of DNA
1952	Martha Chase, Alfred Hershey	Uses phages to demonstrate the final proof that DNA is the element of heredity
1953	James D. Watson, Francis Crick	Show the DNA structure is a double helix; get great press acclaim
1956	Joe Hin Tjio, Albert Levan	Establish the correct number of chromosomes in humans to be 46
1958	Arthur Kornberg	Purifies the first enzyme that made DNA in a test tube—DNA polymerase
1958	Matthew Meselson, Frank Stahl	Show DNA replication is semiconservative using centrifuge experiments with isotopes of nitrogen
1961	Sydney Brenner, Francois Jacob, Matthew Meselson	Part of central dogma: Messenger RNA is the intermediate between DNA and protein
1961–1967	Marshall Nirenberg, Har Gobind Khorana, Sydney Brenner, Francis Crick	Combine efforts to crack the genetic code, which consists of three "letters" on the DNA that spell out 20 different proteins
1964	Howard Termin	Shows RNA viruses can reverse the direction of DNA to RNA transcription
1970	Hamilton Smith	Purifies the first restriction enzyme

Post-Genomic Period and the Age of Proteomics

With the completion of 99% of the genome in 2003 with 99.9% accuracy, scientists began thinking about how these letters worked. Now that scientists had the A's, T's, C's, and G's, they set about finding out what all the letters mean. It was

Table 5. Timeline, 1970s–2000s

Date	Scientist(s) or Agency	Discovery or Action
1972–1973	Paul Berg, Herb Boyer	Construct first recombinant DNA
1973	Joseph Sambrook	Led the team at Cold Harbor Laboratory that developed gel electrophoresis using agarose gel and staining with ethidium bromide
1973	Annie Chang, Stanley Cohen	Show a recombinant DNA molecule can be duplicated in *E.coli*
1975	International Meeting	Asilomar Conference in California to adopt guidelines for regulating DNA experiments
1977	Walter Glbert, Allan Maxam, Frederick Sanger	Develop methods for sequencing DNA
1977	Genentech	Becomes the first genetic engineering company designed to make medically important drugs
1981	Three independent teams	Announce the discovery of human oncogenes
1980	David Botstein, Ronald Davis, Mark Skolnick	Propose a method to map the entire human genome using RFLPs (restriction fragment length polymorphisms), a technique that explores variations in like DNA sequences
1982	Japances scientist Akiyoshi Wade	Proposes automated sequencing and gets support from Hitachi to help build robots
1983	James Gusella	Uses blood samples to discover Huntington disease gene on chromosome 4
1984	Charles Cantor, David Schwartz	Develop pulsed field electrophoresis
1985	Robert Sinsheimer at University of California, Santa Cruz (May)	Hosts meeting to discuss the feasibility of sequencing the human genome
1985	Cary Mullis and colleagues at Cetus (December)	Develop a method called polymerase chain reaction (PCR) to replicate large amounts of DNA
1986	Sydney Brenner (February)	Urges European Union to undertake a concentrated program to sequence the human genome
1986	U.S. Department of Energy (DOE) (March)	Hosts a meeting in Santa Fe, New Mexico, to discuss sequencing the human genome
1986	Renato Dulbecco of Salk Institute (March)	Encourages sequencing of the genome in a paper in *Science*
1986	Cold Spring Harbor Laboratory (June)	Scientists hotly debate merits of a human genome project
1986	Leroy Hood, Lloyd Smith	Announce the first automated sequencing machine

(*continued*)

Table 5. (Continued)

Date	Scientist(s) or Agency	Discovery or Action
1986	Charles DeLisi (September)	Begin study of genome at DOE with allocation of $5.3 million from 1987 budget
1987	Walter Gilbert (February)	Resigns from the National Research Council and announces plans to start Genome Corporation to sequence and copyright human genome
1987	Advisory panel of DOE (April)	Promotes spending $1 billion on mapping the genome over the next seven years
1987	David Burke, Maynard Olson, George Carle (May)	Develop YACs for cloning
1987	Helen Donis-Keller and colleagues at Collaborative Research (October)	Publish first genetic map with 403 markers and spark fight over credit
1987	DuPont scientists (October)	Develop a system for rapid DNA sequencing
1987	Applied Scientist, Inc.	Puts first automated sequencing machine on market
1988	NRC (February)	Calls for a phased approach to the Human Genome Project (HGP)
1988	James Wyngaarden, Director of National Institutes of Health (NIH) (March)	Call for the NIH to take over the HGP from the DOE
1988	Cold Spring Harbor Laboratory (June)	Host first annual genome meeting
1988	James Watson (September)	Becomes head of NIH Office of Human Genome Research; wants to includes ethical and social issues as 2% of budget
1988	NIH and DOE (October)	Agree to collaborate on HGP
1989	Norton Zinder (January)	Chairs first panel on HGP at Rockefeller University
1989	Alec Jeffreys	Coins the term DNA fingerprinting
1989	Francis Collins, Lap-Chee Tsui	Traces the gene causing cystic fibrosis to chromosome 7
1989	Olson, Hood Botstein, Cantor (September)	Design a new mapping strategy using STS's (sequence-tagged sites)
1989	NIH and DOE (September)	Appoint a committee to explore social and ethical issues of human genome project
1989	NIH office (October)	Becomes National Center for Human Genome Research (NCHGR); awards grants

Table 5. (Continued)

Date	Scientist(s) or Agency	Discovery or Action
1990	Lloyd Smith, Barry Karger, Norman Dovichi (September)	Develop capillary electrophoresis
1990	W. French Anderson	First gene therapy using the ADA gene
1990	NIH and DOE (April)	Announce a five-year plan
1990	NIH (August)	Gives out sequencing assignments for four model organisms to be completed in three years
1990	NIH and DOE (October)	Reset the clock; official beginning of the Human Genome Project is October 1, 1990
1990	David Lippman, Eugene Myers, at the National Center for Biotechnology Information (NCBI) (October)	Publish the BLAST algorithm system for aligning sequences
1991	J. Craig Venter, NIH biologist (June)	Announces a new strategy using STSs to find expressed genes; fight erupts in Congress over the NIH filing patents of thousands of partial genes
1991	Japanese scientists (October)	Begin sequencing the rice genome
1991	Edward Uberbacher at Oak Ridge Laboratory (December)	Announces GRAIL, the first of many gene finding programs
1992	Bernadette Healy, Director of NIH; James Watson (April)	Resign over disputes about patenting genes
1992	Craig Venter (June); William Haseltine	Leave NIH to establish the Institute for Genomic Research (TIGR); Haseltine heads a sister company to commercialize TIGR's products
1992	Britain's Wellcome Trust (July)	Enters the HGP race with $95 million
1992	Mel Simon of Caltech (September)	Develops BACs for cloning
1992	David Page of Whitehead Institute (October)	Maps the Y chromosome, the smallest chromosome
1992	Daniel Cohen and French teams (October)	Map chromosome 21
1992	NIH and DOE (December)	Release guidelines for sharing data

(*continued*)

Table 5. (Continued)

Date	Scientist(s) or Agency	Discovery or Action
1992	Eric Lander, Jean Weissenbach (December)	Complete genetic maps of mouse
1993	Francis Collins (April)	Named director of NCHGR
1993	NIH and DOE (October)	Revise plan to complete the genome by 2005
1993	John Sulston and Wellcome Trust (October)	Becomes a major sequencing lab in the International consortium
1993	NIH and DOE (October)	Move the GenBank database from Los Alamos to NCBI, ending the struggle over control
1994	Jeffrey Murray of University of Iowa, Cohen of France (September)	Publish a complete linkage map of the genome
1995	Richard Mathies (May through August)	Develop improved sequencing dyes
1995	Venter and Claire Fraser of TIGR, Hamilton Smith of Johns Hopkins (July)	Publish the first sequence of the free living bacteria *Hemophilis influenzae*
1995	Japanese government (September)	Funds several sequencing project for $15.9 million over five years
1995	Patrick Brown of Stanford (October)	Publishes paper using a printed glass microarray of complementary DNA (cDNA)
1995	Lander and Thomas Hudson (December)	Publish a physical map of the human genome with 5,000 markers
1996	Wellcome Trust (February)	Hosts a meeting in Bermuda at which all HGP partners agree to release sequence data to public databases
1996	NIH (April)	Funds six groups to large-scale sequencing of genome
1996	Affymetrix (April)	Makes DNA chips commercially available
1996	DOE (September)	Begins six pilot projects at $5 million to sequence ends of BAC clones
1996	International Consortium (October)	Releases genome sequence of the yeast *S. cerevisiae*
1996	Yoshihide Hayashizaki at RIKEN in Japan (November)	Completes the first set of full length mouse cDNA
1997	Scotland's Roslin Clinic	Announces the birth of Dolly the lamb, the first mammal to be cloned from an adult cell

Table 5. (Continued)

Date	Scientist(s) or Agency	Discovery or Action
1997	NCHGR (January)	Becomes the National Human Genome Research Institute; DOE creates the Joint Genome Institute
1997	Fred Blattner, Guy Plunkett, University of Wisconsin, Madison	Sequence the bacteria *E. coli*
1997	Molecular Dynamics (September)	Introduce MegaBACE, a capillary sequencing machine
1998	NIH (January)	Begins new project to find single nucleotide polymorphisms (SNPs)
1998	Groups from U.S., EU, China and South Korea (February)	Meet in Japan to form international guideline for sequencing rice genome
1998	Phil Green and Brent Ewing of Washington University (March)	Publish a new program called Phred for automatically interpreting sequencing data
1998	PE Biosystems (May)	Introduces the PE Prism 3700 capillary sequencing machine
1998	Venter (May)	Announces a new company named Celera and says it will sequence the entire genome in three years
1998	Wellcome Trust (May)	In response, doubles its investment in HGP to $30 million
1998	NIH and DOE (October)	Speed up action and declare to have a working draft by 2001 and final draft by 2003
1998	Sulston, Robert Waterston (December)	Complete the genome of the roundworm *C. elegans*
1999	NIH (March)	Moves the completion date to spring 2000; large-scale sequencing continues in major centers
1999	Ten companies and Wellcome Trust (April)	Launch the SNP consortium and ledge to release data quarterly
1999	NIH (September)	Launches project to sequence mouse genome in three years
1999	British, Japanese, and U.S. researchers (December)	Announce the complete sequence of human chromosome 22
2000	Celera (March)	Completes the sequence of *Drosophila melanogaster*; validates his whole-genome shotgun approach
2000	HGP and Celera (March)	Disagree over data-release policy

(*continued*)

Table 5. (Continued)

Date	Scientist(s) or Agency	Discovery or Action
2000	HGP consortium led by German and Japanese scientists (May)	Completes sequence of chromosome 21
2000	HGP and Celera (June)	At the White Hose, announce that they are jointly working on project and ending feud
2000	DOE and MRC (October)	Launch project to sequence the genome of the puffer fish
2001	HGP (February)	Publishes its working draft in the magazine *Nature*
2001	Celera (February)	Publishes its draft in the magazine *Science*

like having the ingredients gathered to bake a cake, but no instructions as to how to proceed. The 21,000 different genes encode for millions of different proteins.

In 1994, Marc Wilkins of Australia coined the term "proteome" in his PhD thesis. He used it to describe the complement of proteins expressed in a genome. The Human Proteome Organisation (HUPO) was begun as an international scientific organization representing and promoting proteomics through international cooperation and collaborations. They seek to develop new technologies, techniques, and training. The group is based in Montreal, Canada. The U.S. HUPO also engages in scientific and educational activities regarding the human proteome.

Now the investigations are going from genes to functions. Finding out how they work is essential for medical interventions and the development of a new field called pharmacogenomics.

Further Reading

Arizona State University. 2010. The Embryo Project Encyclopedia. http://embryo.asu.edu. Accessed 2/13/12.

Davies, Kevin. 2001. *Cracking the Genome: Inside the Race to Unlock Human DNA*. New York: Free Press.

Ridley, Matt. 2000. *Genome: The Autobiography of a Species in 23 Chapters*. New York: HarperCollins.

A

Aarskog-Scott Syndrome (AS)

Prevalence Considered a rare disease, but true figures may not reflect mild undiagnosed cases

Other Names Aarskog syndrome; AAS; facio-digito-genital dysplasia; faciogenital dysplasia; shawl scrotum syndrome

In a 1970 article in the *Journal of Pediatrics*, Norwegian pediatrician and human geneticist Dagfinn Aarskog described a syndrome in some families in which individuals were short and had coarse facial appearances and abnormal sex organs. In a 1971 article in *Birth Defects Original Series*, Dr. Charles I. Scott, an American medical geneticist, wrote about families with unusual faces, joint hypermobility, genital anomaly, and short stature; he called the condition "dysmorphic syndrome." The two independently described the same syndrome, and the disorder was given the name Aarskog-Scott syndrome, or AS.

What Is Aarskog-Scott Syndrome (AS)?

AS is a disorder that affects many body parts. The person with AS (usually male) has growth, skeletal, facial, and genital abnormalities and possibly intellectual performance issues. Children with AS may look more like each other than the parents. The following problems may occur:

- Growth problems: The growth of children with AS is delayed, and the short stature is noted by about one to three years of age. Also, growth during adolescence is delayed, but most catch up in later adolescence.
- Skeletal abnormalities
  - Face: The individual has a rounded face and widely spaced eyes, called hypertelorism; droopy eyelids (blepharoptosis); and down-slanting eye slits (palpebral fissures). A very small nose ends with a long area between

the nose and mouth (philtrum) and crease below the lower lip. The mid-part of the face is underdeveloped. The face is capped with a hairline that is shaped like a peak, often called a widow's peak, and the upper part of the ear is slightly folded over. Delayed eruption of teeth is common.

  - Neck: The child has a very short neck, with webbed skin on the side of the neck.
  - Body extremities: Common are very small but broad hands and feet ending with very short, broad fingers and toes, a condition known as brachydactyly. The pinky or fifth finger is curved (clinodactyly), and mild webbing of skin may exist between the fingers and toes. A large crease, known as a simian crease, goes from one side of the hand to the other.
  - Chest and abdomen: The child has a sunken chest, a protruding navel, and often hernias at the navel site or in the groin area.

- Genitalia: The scrotum of most males with AS completely surrounds the penis, a condition known as shawl scrotum; the testicles may not descend.
- Other abnormalities: Often people with AS may have serious heart defects or other problems such as a cleft palate, which is a hole in the roof of the mouth.

The intellectual ability of children with AS varies with the individual. Some may have learning disabilities and hyperactivity, while others are completely normal. In a few rare cases, some children have severe intellectual deficits. However, the children appear to develop very good social skills and get along with other children.

What Is the Genetic Cause of Aarskog-Scott Syndrome (AS)?

Mutations in the gene *FDG1* gene, officially known as the "FYVE RhoGEF and PH domain containing 1" gene, cause Aarskog-Scott syndrome. Normally, *FDG1* instructs for making a protein that turns on or activates the protein Cdc42, which is very important in embryonic development. Changes or mutations in the *FDG1* gene make the abnormal protein that disrupts the Cdc42 signaling and causes the numerous developmental disorders of AS.

This syndrome is transmitted through an X-linked recessive gene. The gene FGD1 is located on the short arm of chromosome X at position 11.21. An X–linked recessive gene means that the sons of the female carriers are at 50% risk of acquiring the gene. Female carriers may have mild symptoms of the condition. Fathers cannot pass the trait to their sons.

What Is the Treatment for Aarskog-Scott Syndrome (AS)?

Although no treatment exists for the disorder, many of the symptoms may be treated. Surgery may correct some of the disorders such as the navel and inguinal hernias, and cosmetic surgery may correct some of the severe facial characteristics. Although trials of growth hormone have been performed on children with growth problems, they have not generally been effective with AS. Children with AS who have learning disabilities qualify under the Americans with Disabilities Act (ADA) and Individuals with Disabilities Educational Act (IDEA).

Further Reading

Aarskog, D. 1970. "A Familial Syndrome of Short Stature Associated with Facial Dysplasia and Genetic Anomalies." *J. Pediatrics*. 77(5): 856–861.

"Aarskog-Scott Syndrome." 2012. Genetics Home Reference. National Library of Medicine (U.S.). http://ghr.nlm.nih.gov/condition/aarskog-scott-syndrome. Accessed 5/1/12.

Scott, C. I. 1971. "Unusual Faces, Joint Hypermobility, Genital Abnormality and Short Stature: A New Dysmorphic Syndrome." *Birth Defects Orig. Artic. Ser*. 7(6): 240–246.

Aase-Smith Syndrome

Prevalence Very rare

Other Names Aase syndrome; congenital anemia; hypoplastic anemia; triphanlangeal thumb syndrome

In 1968, American pediatricians Jon Morton Aase and David Weyhe Smith described a condition in newborns in which the infant had anemia and some unusual joint and skeletal deformities. For their discovery of the new rare syndrome, they were honored with giving the rare condition their names: Aase-Smith syndrome.

What Is Aase-Smith Syndrome?

Anemia and multiple birth defects characterize this rare disease. The physician can see two of the most telling symptoms at birth. He or she cannot fully extend the newborn's joints (contracture deformity) and will see a distinct condition called triphanlangeal thumb syndrome or triple-jointed syndrome. Several other features of this disease include the following:

- Absence of or small knuckles
- Decreased skin creases at joints
- Droopy eyelids
- Deformed ears
- Narrow shoulders
- Pale skin

A cleft palate or hole in the roof of the mouth may also be present. As the child ages, the soft spot or fontanel will close much more slowly than that of a normal child.

A major risk factor of Aase-Smith involves complications that could arise when anemia decreases oxygen supply to the blood cells, causing not only fatigue and weakness but also heart problems. Severe cases of Aase-Smith may be involved in still births or early death.

What Is the Genetic Cause of Aase-Smith Syndrome?

The underdevelopment of blood-forming cells in bone marrow causes the anemia. However, the actual gene related to the syndrome has not been identified. Evidence exists that it is inherited on a somatic chromosome and could be dominant or recessive. Aase and Smith in their first publication about the syndrome in 1968 thought it to be dominant. A single parent may be a carrier of the trait.

What Is the Treatment for Aase-Smith Syndrome?

When the symptoms of Aase-Smith are first noted, the child will be given a complete blood count (CBC) test and a possible bone marrow biopsy. These tests can determine the presence of anemia and low white blood cell count. Frequent blood transfusions may be given during the first year of life, and a bone marrow transplant may be necessary if other treatments fail. The drug prednisone may also be used to treat anemia associated with the disorder, but risk of side effects demand using it with caution and not at all with very young infants.

Certain complications may arise if a heart defect exists. Also, the low white blood count may cause numerous infections. However, with prompt treatment of the anemia and correction of some of the skeletal deformities, the child may show improvement as he or she ages. The syndrome is not associated with mental retardation or neurological problems.

Two well-known clinics specialize in the disorder: Henry Ford Hospital, Detroit; and the Genetic and Rare Disease Information Centre, Gaithersburg, Maryland.

Further Reading

Aase, J. M., and D. W. Smith. 1968. "Dysmorphogenisis of Joints, Brain, and Palate: A New Dominantly Inherited Syndrome." *J Pediatrics* 73(4): 606–609.

"Aase Syndrome." 2003. National Association for Rare Diseases. http://www.rarediseases.org/search/rdbdetail_abstract.html?disname=Aase%20Syndrome. Accessed 5/2/12.

Abetalipoproteinemia. *See* Bassen-Kornzweig Syndrome

Achondroplasia

Prevalence 1 in 15,000 to 40,000 newborns
Other Names dwarfism

Dwarfs or Little People have fascinated those of normal stature throughout history. A fairy tale is told of "Snow White and the Seven Dwarfs." Court dwarfs have been painted as entertainers for royalty. Velasquez, the famous Spanish painter, proudly displayed his work of the court dwarf, Diego Rodriguez de Silva. Dwarfs or "midgets" were featured in circus sideshows and performances. However, the word "midget" is now considered improper and not desirable for use. The proper term is "Little People." There is a Society of Little People that seeks to promote information about the real condition and to help individuals integrate into society. Little People have a condition known as achondroplasia or dwarfism. Recently, dwarfs or Little People have appeared in television and movies, the most recent being Jason Acuña as "wee Man" in *Jackass*.

What Is Achondroplasia?

Recorded in ancient Egyptian art, this disease is one of the oldest on record. The word "achondroplasia" comes from three Greek root words: *a*, meaning "without"; *chondro*, meaning cartilage; and *plas*, meaning form. However, the actual problem is not in the cartilage but in the process of converting cartilage to bone, especially in the long bones of the legs and arms. Cartilage is a hard, rubbery-like material found in joints, ears, and nose. In achondroplasia, something goes wrong in the process with the normal multiplication rate of cells in the growth plates during fetal development and childhood. This leads to short limbs and reduced height. The torso and other features appear normal size.

The person with achondroplasia may have some of the following differences:

- Prominent forehead
- A flat or depressed area at the base of the nose
- Protruding jaw
- Poor dental structure
- Short and stubby hands
- Ligaments in the fingers double-jointed

Prenatal ultrasound can usually detect the condition before birth, and a DNA test can confirm. The condition is present at birth and diagnosis is confirmed at that time. As an adult, the person may reach the average height of about 4′0″.

Health problems may affect people with achondroplasia. Babies may have poor muscle tone and a delay in sitting, standing, and walking. A small hump called kyphosis may develop on the back due to the poor muscle tone. Children may get ear infections and, because of the compression of the chest, may have difficulty

American actor Warwick Davis is an example of a person with achondroplasia. (Hulton Archive / Getty Images)

breathing. For this reason, occasionally, a baby with achondroplasia may die suddenly during sleep.

What Is the Genetic Cause of Achondroplasia?

Dwarfism is due to changes in the *FGFR3* gene, officially known as the Fibroblast Growth Factor Receptor 3 gene. Normally, this gene instructs for making a protein that is part of the family of growth factor receptors and interacts with other growth factors to help regulate the rate of growth in long bones. Two mutations cause more than 99% of cases. Both mutations involve change of one amino acid in the *FGFR3* protein. Here arginine is substituted for glycine at position 380. Just this one substitution causes the receptor to be overly active and causes bone growth disturbances. Now, cartilage is not properly converted to bone, and a shortage of bone exists.

The inheritance pattern involves an autosomal dominant pattern. This means that one parent carrying a mutated gene has a 50% chance or one in two of passing the change to an offspring. Autosomal dominance shows the following pattern:

- It occurs in every generation.
- Both males and females may get the condition.
- Each child of an affected parent has a 50% chance of being affected.
- If both parents are affected, the child will probably have a more severe case and probably die soon after birth.

About 80% of people with achondroplasia are born to average-size people. These cases result from a new mutation in the gene, and the likelihood of having another child with achondroplasia is extremely rare. These cases are called de novo and usually in the sperm of the father. The remainder of the cases occurs when the person inherits an altered gene from the parent. *FGFR3* is located on the short arm (p) of chromosome 4 at position 16.3.

What Is the Treatment for Achondroplasia?

Currently no treatments exist that can reverse the effects of achondroplasia. However, treatment options may improve the quality of life. Growth hormones and/or surgery may be an option for lengthening limbs. Growth hormones may be effective during the first year of life. This hormone targets the pituitary gland but is quite expensive, about $10,000 to $25,000 per year. Limb-lengthening is painful and may cost as much as $100,000. This surgery can increase height up to 14 inches and will affect the alignment of the back.

People with achondroplasia have normal intelligence, and many people with the condition become quite successful. The Little People of America work to help people with dwarfism adjust to life by adapting home, car, and other areas to their short stature.

Further Reading

"Achondroplasia." 2008. Genetics Home Reference. National Library of Medicine (U.S.) http://ghr.nlm.nih.gov/condition/achondroplasia. Accessed 5/3/12.

"Achondroplasia" March of Dimes. http://www.marchofdimes.com/professionals/14332_1204.asp. Accessed 5/3/12.

Little People of America. 2011. http://www.lpaonline.org. Accessed 5/3/12.

ACHOO Syndrome

Prevalence	No studies exist, but informal surveys estimate 10% to 35% of the population are photic sneezers
Other Names	Autosomal dominant compelling helioophthalmic outburst syndrome; photic sneeze reflex

Aristotle (384–322 BC) was puzzled by the question: Why do people sneeze after going out into the bright sunlight? In his *Problems, Book XXXIII*, Aristotle mused about why the heat of the sun on the nose causes the reaction. About 2,000 years later, Sir Francis Bacon in the seventeenth century surmised that if one went out into the sun with the eyes closed, the sneeze would be gone. He guessed that sunlight made the eyes water, and that moisture seeped into the nose and irritated it. Of course, his idea turned out not to be the reason. People are still puzzled by this syndrome dubbed "ACHOO Syndrome."

What Is ACHOO Syndrome?

The ACHOO syndrome, also called photic sneeze reflex, is a reflex that happens when some people are exposed to intense sunlight. A long outburst of sneezing occurs sometimes as much as 30 to 40 times. The sneezing is uncontrollable, and the number of sneezes varies.

According to researcher J. M. Garcia-Moreno, little attention has been paid to this phenomenon, and information about it has been scarcely described in scientific literature. The only references may be in clinical notes or in letters to the editor. Even the sufferers may consider it as an aggravating inconvenience and ignore it with minor complaining.

What Is the Genetic Cause of ACHOO Syndrome?

The fifth cranial nerve, called the trigeminal nerve, is responsible for sneezing. A congenital malfunction in this nerve is probably responsible for ACHOO syndrome. When bright sunlight overstimulates the optic nerve, the malfunctioning trigeminal responds and causes the photic sneeze.

The syndrome is a genetic disorder. In 1978, Dr. Roberta Pagon, a professor at the University of Washington and a sneeze expert, sat at a table with a group of colleagues discussing sternutatory (sneezing) habits. They discovered that of the group of 10, 4 members sneezed in bright sunlight. The number of sneezes went from 3 to 43. The group named the condition and gave it the acronym ACHOO for Autosomal Dominant Compelling Helioophthalmic Outburst Syndrome. Breaking down this name simply says that it is a dominant genetic condition carried on an autosome; helioophthalmic comes from the Greek words *helio*, meaning "sun," and *ophthalmos*, meaning "eye."

ACHOO syndrome appears to be inherited in an autosomal dominant fashion, meaning that if a parent has the condition, there is a 50% chance that the offspring will have the same syndrome. Apparently, the number of sneezes is also inherited in the same way. Because this condition is not life threatening, search for the gene is considered unimportant to serious researchers. However, Dr. Pagon believes that locating any gene that is part of the nervous system is a good thing.

What Is the Treatment for ACHOO Syndrome?

Although most cases are brought by sudden exposure to strong sunlight, other stimuli may also cause the happening. Combing hair, tweezing eyebrows, rubbing the inner corner of the eye, breathing cold air, strong odors or aromas, or even eating too much may trigger the syndrome.

Antihistamines used by people who have seasonal allergies appear to reduce occurrence of sneezes in people who have both allergies and ACHOO syndrome.

Further Reading

Holmes, Hannah. 1997. "The Skinny on Strange Sneezing Situations." 1997. Discovery.com. http://www.discovery.com/area/skinnyon/skinnyon970411/skinny1.html. Accessed 5/3/12.

Garcia-Moreno, J. M. 2006. "Photic Sneeze Reflex or Autosomal Dominant Compelling Helioophthalmic Outburst Syndrome." *Neurologia* Jan–Feb; 21(1): 26–33.

Rosick, Edward R. 2006. "ACHOO Syndrome." http://www.healthline.com/galecontent/achoo-syndrome-1. Accessed 5/3/12.

Achromatopsia

Prevalence 1 in 33,000 in the United States; varies in different parts of the world

Other Names achromatopsia; color blindness; rod monochromatism

In 1996, Dr. Oliver Sacks visited the tiny Pacific atoll of Pingelap and found a population with a large number of people born totally color-blind. He set up a one-room clinic and listened with amazement to how these people described a world where congenital colorblindness is the norm. In a picturesque Pacific island where colorful plants and animals abound, these individuals saw things only in black, white, and shades of gray. These islanders have a condition called achromatopsia.

What Is Achromatopsia?

Achromatopsia is a characterized by color blindness, visual acuity loss, extreme light sensitivity, and erratic movements of the eye called nystagmus. On the retina of the eye, two structures called rods and cones enable individuals to see colors and shades. Cone cells with the 6 million receptors enable people with normal vision to see color. Rod cells with about 100 million photoreceptors enable people with normal vision to see shades but not color. The two systems, the photopic using the cones and the scotopic using the rods, enable people to see in a wide range of lighting conditions. The rod cells function at low levels of light, providing the ability for night vision. However, persons with achromatopsia do not have normal cone vision; they must rely on rod vision, which does not provide color differentiation or good detail vision.

Achromatopsia is present at birth and is frequently noted around the age of six months when the child in uncomfortable in bright light and may have erratic movement of the eyes. The condition becomes more noticeable when the child starts to school and must use their eyes in learning the skills of reading. The congenital disease does not worsen with age.

Two primary forms of complete and incomplete achromatopsia are defined:

- Complete achromatopsia: Persons with this condition have little or no function of cone cells and have no concept of color. They see the world like a black-and-white photograph with varying shades of gray.
- Incomplete achromatopsia (also called dyschromatopsia): These individuals have some function of cones but have color impairment and slightly better visual acuity.

The following five symptoms are associated with the syndrome:

- Loss of color vision
- Amblyopia or reduced visual acuity (sometimes called lazy eye)
- Photophobia or extreme sensitivity to light and glare
- Erratic shaking or wobbling movements of the eye called nystagmus
- Abnormalities in the opening and closing of the iris of the eye

What Is the Genetic Cause for Achromatopsia?

Achromatopsia is recessive condition, meaning that both parents must contribute the gene in order for it to occur in the offspring. All the offspring of a person with achromatopsia carry the gene. In order to pass the gene to the offspring, both parents must carry the recessive gene.

Two groups of isolated peoples have enabled scientists to study this condition. The story of the color-blind people of Pingelap recounts how an isolated population could pass the gene with such a high proportion. Genealogists have traced the condition to 1775, when one man survived a typhoon that killed most of the people of the island. The man had a mutation of *CNGB3* gene, which is responsible for photoreceptors in the eye. Through intermarriage, he passed the gene to offspring who were isolated on this small island. A second group exists in some Western European families. Genetic studies have found another gene *CNBA3* to affect this family.

Cerebral achromatopsia is a form of color blindness acquired from a blow to the head, trauma, illness, or other causes. This form is not inherited and appears to affect only loss of color vision. The condition does not appear to have the severe impaired vision, light sensitivity, as those who have inherited the condition.

What Is the Treatment for Achromatopsia?

At present, no medical or surgical treatments exist. However, diagnosing children early can help with accommodations in environmental modifications such as adapting to the exposure to light.

Further Reading

Achromatopsia Network. http://www.achromat.org/what_is_achromatopsia.html. Accessed 5/3/12.

Sachs, Oliver. 1997. *The Island of the Colorblind*. New York: Vintage Books.

Acoustic Neuroma

Prevalence About 1 in 100,000 worldwide; 300 new cases diagnosed in the United States each year

Other Names acoustic neurinoma; acoustic neurilenoma; vestibular schwannoma

When a person begins to experience hearing loss, dizziness, and ringing in the ears, it may be caused by many factors. If the individual has neurofibromatosis 2, the likelihood is that a tumor may be growing on the nerves in the ear. In the head are 12 cranial nerves. The eighth nerve is the auditory nerve that receives stimuli from the outside and delivers to the part of the brain that permits us to hear.

What Is Acoustic Neuroma?

An acoustic neuroma is a slow-growing, benign tumor growing on the balance and hearing nerves of the inner ear. The tumor is often called a vestibular schwannoma because it involves the vestibular area of the eighth cranial nerve and comes from the Schwann cells, which form the myelin sheath that wraps like an onion skin around the nerves. Vestibular schwannoma is probably a more accurate name for this condition.

The tumor arises from the overproduction of the Schwann cells. As the tumor grows, it presses against hearing and balance nerves, and the person experiences hearing loss, ringing in the ears, and dizziness in that one ear. Noting that problems are present in the one ear is very important to the diagnosis. The growing tumor causes facial numbness by pressing against the trigeminal nerve, which causes facial sensations, and, if left unchecked, may present a life-threatening situation by pressing on the brainstem and cerebellum.

What Are the Genetic Causes of Acoustic Neuroma?

On chromosome 22, a gene normally produces a protein that controls the growth of Schwann cells that cover the myelin sheath. When this gene malfunctions, it causes an overproduction of the growth of these Schwann cells, causing the Schwannoma. The Greek root *oma* means "tumor": hence "schwannoma" is a tumor of the Schwann cells, named from the German scientist Schwann who first described them. The reason for this malfunction is not clear. The neuromas may just occur with no explanation or may arise with exposure to loud noise, low doses of exposure to radiation of the head and neck, or heavy use of cell phones.

The main genetic risk is connected with a disorder called neurofibromatosis 2. Acoustic neuroma is an autosomal dominant disorder, which means it is carried on a nonsex chromosome. A parent carrying the condition has a 50% chance of passing the disorder to offspring. In neurofibromatosis type 1, the schwannoma usually develops in adult life and involves the eighth cranial nerve. In neurofibromatosis II (NFII), the schwannoma may be bilateral appearing on each side. This type is the one that has the strong autosomal dominant pattern, and incidence of the acoustic neuroma occurring with NFII is 5% to 10%.

What Is the Treatment for Acoustic Neuroma?

A CT scan will detect almost all neuromas. Because the tumors are slow growing, the physician usually plans a conservative treatment. First, observation is important especially if the person is 70 or older; the individual may die of other causes before the tumor becomes life-threatening. A second option is microsurgical removal, which may remove the tumor but cause a risk to other areas of the ear and face. A third option is radiation therapy with gamma knife radiosurgery, Linac knife, or proton therapy.

Further Reading

"Acoustic Neuroma." 2010. MayoClinic.com. http://www.mayoclinic.com/health/acoustic-neuroma/DS00803. Accessed 5/3/12.

"Vestibular Schwannoma (Acoustic Neuroma) and Neurofibromatosis." 2004. National Institute on Deafness and other Communication Disorders (U.S.). http://www.nidcd.nih.gov/health/hearing/acoustic_neuroma.aspx. Accessed 5/3/12.

Acute Promyelocytic Leukemia

Prevalence About 1 in 250,000 people in the United States; causes about 10% of acute myeloid leukemia cases

Other Names AML M3; APL; leukemia, acute promyelocytic; M3 ANLL; myeloid leukemia, acute, M3

After the dropping of the atomic bomb over Hiroshima and Nagasaki, Japan, survivors began to show an increased rate of a condition called acute myeloid anemia. Prior to the adoption of safety procedures for X-rays, radiologists and others who worked in the field also developed mysterious symptoms, which were later diagnosed as leukemia.

The condition had been described many years before. In 1827, Alfred Velpeau, French physician, looked at the blood of a 63-year-old patient and described the blood as looking like thick gruel because of the abundance of white blood cells. Later, in 1845, J. H. Bennett, an Edinburgh doctor, found that some patients with enlarged spleens also had changes in the color of the blood. Rudolph Virchow, a famous German doctor, coined the term "leukemia" in 1856. The term "leukemia" comes from two Greek words: *leuko*, meaning "white"; and *heme*, meaning "blood."

What Is Acute Promyelocytic Leukemia?

Acute myeloid leukemia is a type of cancer of the blood-forming cells in the bone marrow. Acute promyelocytic leukemia is one of the types of this cancer. Normally, the bone marrow contains stem cells or precursor cells that produce red blood cells, white blood cells, and thrombocytes or blood platelets. In acute promyelocytic leukemia, precursor white blood cells called promyelocytes gather in the bone marrow and crowd out the normal cells that work for the body.

Following are the symptoms of acute promyelocytic leukemia:

- Very pale skin
- Excessive fatigue
- Infections, because the normal white blood cells are not working properly
- Fever and loss of weight
- Bruising, because the factors for clotting are not working properly
- Excessive bleeding caused by cancerous cells, especially of the gums, and nosebleeds

- Pain in bones and joints when the condition spreads to the areas
- Enlarged spleen

Although the condition may be diagnosed at any age, it usually occurs around the age of 40.

What Are the Genetic Causes of Acute Promyelocytic Leukemia?

This condition is not inherited but is the result of a somatic translocation that occurs in the somatic cells after conception. Random occurrences of the exchange of gene on the chromosomes cause the mutation.

Two genes are involved: *PML* and *RARA*.

PML

Mutations in the *PML* gene, officially called the "promyelocytic leukemia" gene, cause acute promyelocytic leukemia. Normally, *PML* instructs for a protein that is found in the special structures in the nucleus of the cells called PML-NBs or PML nuclear bodies. In the PML-NBs, the PML protein works with other cells that control cell growth, as well as cell destruction or apoptosis. When the PML-NBs are working properly, they keep the cells from growing wildly and therefore acts as a tumor suppressor.

Somatic mutations in *PML* are acquired when certain conditions exist. A translocation occurs where the *PML* located on chromosome 15 fuses with *RARA*, located on chromosome 17. The fused gene does not bind to the PML-NB, and structures do not form properly. The promyelin cells continue to grow abnormally, causing the symptoms of this condition. *PML* is located on the long arm (q) of chromosome 15 at position 22.

RARA

Mutations in the *RARA* gene, officially known as the "retinoic acid receptor, alpha" gene, cause acute promyelocytic leukemia. Normally, *RARA* provides instructions for making a transcription factor called the retinoic acid receptor, alpha (RARα). As a transcription factor, it binds to special areas of DNA to control the action of certain genes. RARα appears to be responsible for maturing the white blood cells beyond the young stage of promyelocytes. The normal action of the protein keeps the white, red, and platelet cells working properly.

Mutations of *RARA* acquired during a person's lifetime can cause acute promyelocytic leukemia. The rearrangement of chromosomes 15 and 17 fuse *RARA* and *PML*. The protein produced from the fused gene does not work normally. It binds to the DNA, but the DNA does not respond to the signal and the genes are repressed. The action is disrupted, causing an abnormal collection of promyelocytes to accumulate in the bone marrow and leading to the symptoms of acute promyelocytic leukemia. *RARA* is located on the long arm (q) of chromosome 17 at position 21.

What Is the Treatment for Acute Promyelocytic (M3) Leukemia?

Early diagnosis of the disorder is imperative. This diagnosis can lead to stopping serious bleeding or clotting problems before they become life-threatening. Chemotherapy drugs are used in addition to non-chemotherapy drugs, such as transretinoic acid. Other treatments may include transfusions of clotting factors or other blood products.

Prognosis is hopeful as about 70% to 90% of people are cured if diagnosed early.

Further Reading

"Acute Promyelocytic Leukemia." 2010. Office of Rare Diseases Research. National Institutes of Health (U.S.). http://rarediseases.info.nih.gov/GARD/Disease.aspx?PageID=4&DiseaseID=538. Accessed 12/24/11.

American Cancer Society. 2011. "Treatment of Acute Promyelocytic (M3) Leukemia." http://www.cancer.org/Cancer/Leukemia-AcuteMyeloidAML/DetailedGuide/leukemia-acute-myeloid-myelogenous-treating-m3-leukemia. Accessed 12/24/11.

Kotiah, Sandy D., MD. "Acute Promyelocytic (M3) Leukemia." Medscape Reference. http://emedicine.medscape.com/article/1495306-overview#showall. Accessed 12/24/11.

AD. *See* Alzheimer Disease (AD)

ADA Deficiency. *See* Adenosine Deaminase Deficiency

ADA-SCID. *See* Adenosine Deaminase Deficiency

Adenosine Deaminase Deficiency

Prevalence Very rare; 1 in 200,000 to 100,000 births

Other Names ADA deficiency; ADA-SCID; "Bubble Boy Disease"

The parents of four-year-old Ashanthi (Ashi) deSilva were frantic. She had a condition in which her immune system did not work and any exposure to the outside world could be deadly for her. The same condition was acknowledged several years earlier by news stories of the "bubble boy" who lived in a large plastic "bubble" to keep germs away; he had died several years earlier. Ashi had taken PEG-ADA, the only medication available, since the age of two; now it was not working, and she was dying. In 1990, the desperate parents understood the risks and signed the papers for the first gene therapy treatment. The treatment was successful. In 2002, Ashi at age 15 was a healthy teenager, playing in the high school band. Ashi had been a victim of the genetic disorder adenosine deaminase deficiency disorder.

What Is Adenine Deaminase Deficiency?

Adenine deaminase deficiency, or ADA deficiency, is a form of genetic disorder that damages the immune system, leaving the person susceptible to all forms of bacteria, viruses, and fungi. The ADA deficiency causes the individual to have severe combined immunodeficiency (SCID) and leaves him or her with absolutely no resistance to these opportunistic infections, which one with a healthy immune system would fight off. The child may have chronic diarrhea, skin rashes, developmental delay, and hearing loss. If not treated, he or she seldom lives past a year or two. However, rarely a case of late-onset ADA deficiency may occur later in life.

What Is the Genetic Cause of Adenosine Deaminase Deficiency?

Mutations in the *ADA* gene, officially known as the "adenosine deaminase" gene, cause adenosine deaminase. The *ADA* gene gives instructions for producing the enzyme adenosine deaminase. Found throughout the body, the enzyme is most active with the lymphocytes, the white blood cells of the immune system that protect the body from outside invaders. The thymus and the lymph nodes produce the lymphocytes. One function of the enzyme is to eliminate a toxic molecule called deoxyadenosine, which is the byproduct of DNA breakdown that destroys lymphocytes.

When the *ADA* gene does not function properly, deoxyadenosine builds up and kills the lymphocytes. Especially vulnerable are the immature lymphocytes in the thymus and other white blood cells in the lymphoid tissues. Thus, the child then develops the fatal symptoms of SCID. *ADA* is inherited in an autosomal recessive pattern. Two copies of the defective gene (one inherited from each parent) must be present to have the disorder. The parents both carry the mutated gene, but neither parent experiences the signs and symptoms of the disorder. The *ADA* gene is located on the q arm of chromosome 20: 20q13.12.

What Is the Treatment for Adenosine Deaminase Deficiency?

There are three basic treatments.

- Enzyme replacement therapy: PEG-ADA consists of bovine (cattle) ADA linked with strands of PEG; the treatment maintains a high level of ADA in the plasma preventing the toxic effects to the lymphocytes.
- Bone marrow transplant from a suitable donor.
- Gene therapy: Although gene therapy was acclaimed in the early 1990s, several trials to cure people with ADA-SCID failed, making the success rate about 1 in 20. The wrinkles in gene therapy still had to be ironed out.

Ashanthi's Story

From the time of her birth, it was obvious that Ashanthi DeSilva was sick. She caught with all types of imaginable viruses, bacteria, and fungi. The physicians told her parents Raj and Van that their daughter was born with a malfunctioning immune system. They would later find out that their child had a rare inherited recessive disease in which her body does not make a protein called adenosine deaminase. Poisons build up in her system that kill the white blood cells or lymphocytes. Without treatment, she would die.

At the age of two, she began taking PEG_ADA, a synthetic form of the enzyme. It worked for a while, but as each new dose was administered, the drug became less and less effective. Another option was a bone marrow transplant, but finding a suitable donor was impossible.

By the age of four, it became obvious that she was dying. Her parents had heard of a doctor in Bethesda, Maryland, who had a new experimental idea called gene therapy and drove 400 miles from their home in North Olmstead, Ohio, to see Dr. W. French Anderson and Michael Blaese. Her parents understood the tremendous risks, but their child was dying; they signed the consent forms.

The time at Bethesda was anxious. Over a period of 12 days, they extracted some of Ashanthi's blood cells, gave those blood cells new working copies of ADA gene, and inserted the healthy gene into a viral vector. On September 14, 1991, they wheeled her into the operating room and injected the cells back into her body. The infusion took about 28 minutes.

The effect was striking. Within six months, her T cells were back to normal level. In the spring of 1991, all members of her family came down with the flu. Ashi was the first one up and about. The child with no immune system was now a healthy, normal child.

As of 2002, at the age of 15, she was in high school developing an interest in music and playing in her high school band.

Further Reading

Hershfield, Michael. 2011. "Adenosine Deaminase Deficiency." *GeneReviews*. http://www.ncbi.nlm.nih.gov/books/NBK1483. Accessed 5/3/12.

Kelly, Evelyn B. 2007. *Gene Therapy*. Westport, CT: Greenwood Press.

Age-Related Macular Degeneration (AMD)

Prevalence Affects about 1 in 2,000 people in the United States and other developed countries; affects people of European descent more frequently than those of African American descent

Other Names age-related maculopathy; AMD; ARMD; macular degeneration, age-related

What Is Age-Related Macular Degeneration (AMD)?

Age-related macular degeneration is a disease of the eye that occurs in older people. The macula is a small area on the retina where central vision occurs. This area is essential for tasks such as reading, driving, and recognizing faces. The loss is gradual and results from the deterioration of the light-sensing rods and cones that detect light and color and are located on the retina. The condition mainly affects central vision; peripheral vision and night vision are not affected.

The condition usually appears in a person's 60s and 70s and is of two types: the dry form and the wet form.

Dry Macular Degeneration

This form accounts for about 85% to 90% of all cases. Yellowish deposits called drusen builds up beneath the retina. It generally occurs in both eyes but not necessarily at the same time.

Wet Macular Degeneration

Although this type is not as common, it is associated with severe, rapidly developing vision loss. Fragile blood vessels under the macula begin to leak blood and fluid. This abnormal blood damages the macula and blurs and distorts vision.

What Are the Genetic Causes of Age-Related Macular Degeneration (AMD)?

Although age-related macular degeneration does not have a clear pattern of inheritance, it does have a genetic connection. About 15% to 20% of all people with the disorder have one sibling or other close relative with the disease. Three genes—*ARMS2*, *CFH*, and *HTRA1*—are associated with age-related macular degeneration.

ARMS2

Mutations in the *ARMS2* gene, officially known as the "age-related maculopathy susceptibility 2" gene are associated with this disorder. Normally, *ARMS2* instructs for a protein with an unknown function. Scientists do know that the *ARMS2* protein is found in two places: the placenta and in the photoreceptors at the back of the eye. The exact role in normal vision is unknown.

Mutations in the *ARMS2* gene are strongly associated with the risk of age-related macular degeneration. Exchange of only one building block may explain the role of the gene on the location of chromosome 10. Another mutation deletes a part of the gene and inserts new material. Chromosome 10 also has another gene,

HTRA1, that is so close it is difficult to tell which gene is involved or if both are. This disease is very complex and may have many factors as well as environmental factors that lead to genetic risk.

HTRA1

Mutations in the *HTRA1* gene, officially known as the "HtrA serine peptidase 1" gene, are associated with macular degeneration. Normally, *HTRA1* instructs for a type of enzyme called serine protease. This enzyme cuts other proteins into smaller pieces. Another action of the enzyme is to bind to the transforming growth factor-β and slow down their ability to help cells grow, move, and even self-destruct. TGF-β also helps form new blood vessels.

The gene, located on chromosome 10 near the *ARMS2* gene, and the two genes are possibly associated. *HTRA1* is located on the long arm of chromosome 19 at position 26.3.

CFH

Mutations in the *CFH* gene, officially called the "complement factor H" gene, is associated with a risk for dry macular degeneration. Normally, *CFH* instructs for making a protein called complement factor H, which is part of the immune system. It is part of the complement system that works to destroy the foreign invader or bacteria and viruses, remove debris from cells, and start the inflammation process. This complement system is highly regulated to prevent it from attacking healthy cells.

Several mutations in the *CFH* gene are related to dry age-related macular degeneration. Scientists surmise that changes in *CFH* make an abnormal complement factor H and disrupt the normal process of fighting inflammation, allowing the yellowish drusen to build up. CFH is located on the long arm of chromosome 1 at position 32.

What Is the Treatment for Age-Related Macular Degeneration (AMD)?

Scientists are working diligently to find a cure for this disorder, but thus far have no cure. However, there are therapies that can slow the progress of the disorder, and some may even restore vision. For dry AMD, a combination of zinc and antioxidants may slow the progression. For wet AMD, certain medical treat or even restore some lost vision, if caught in time. Both forms require a healthy lifestyle and diet to slow the progression.

See also Aging and Genetics: A Special Topic

Further Reading

"Age-Related Macular Degeneration." 2011. Genetics Home Reference. National Library of Medicine (U.S.). http://ghr.nlm.nih.gov/condition/age-related-macular-degeneration. Accessed 12/31/11.

AMD Alliance International. 2010. http://www.amdalliance.org. Accessed 12/31/11.

Haddrill, Marilyn. 2011. "Age-Related Macular Degeneration." All about Vision. http://www.allaboutvision.com/conditions/amd.htm. Accessed 12/31/11.

Aging and Genetics: A Special Topic

In Greek mythology, Eos, the goddess of the dawn, fell in love with Tithonus, a mortal. She was so smitten with love that she begged Zeus to make her mortal lover live forever. Everything was great for a while, until Tithonus began to age. He lost his mind and could not remember anything. Finally, she did not know what to do with him and put in a cave and left his meals at the opening. Eos had neglected to ask Zeus to grant eternal youth to Tithonus. The "Tithonus effect" is applied to the situation in which people live a long time but are in very poor health and have a poor quality of life.

The Tithonus effect may illustrate many lives in the twenty-first century. Scientific advances have increased life expectancy. But with aging and long life come mental and physical problems that may be overwhelming. Are disease and decline inevitable in aging? So, the question arises, why do people age? Is it programmed in the genes of people to die? Genes control the beginning of life; do they also control aging and eventual death?

Gerontologists are scientists who study the process of aging. They seek to answer questions about the nature of the process and study the mechanics of aging. Again, as related to genetics in general, it is all done in the cells and the genes that one inherits. Some gerontologists estimate that genes can explain about 35% of the life span. The life span is the number of expected years of a species. That is not to be confused with "life expectancy," which is an actuarial term used by insurance companies and statisticians to calculate an average of the time a person within a given age may live. For example, the life span for human beings if untouched by disease may be 120 at the maximum. Average life expectancy is calculated according to statistics and may change from time to time. An average life expectancy of 77 means that half of all the people that were born in that person's year of birth are living and half are dead. How the genes, proteins, and chemical pathways work in this process is a subject of intense study.

Theories of Aging

Proponents of various theories of aging use genetic and biochemical terms to explain and justify their positions. They relate various regulatory mechanics that are called pathways. Pathways were first found in small, short-lived organisms such as worms, flies, and yeast. The pathways are regulated by both physiological and environmental signals. Many of the diseases discussed in this encyclopedia discuss the pathways that lead to a genetic condition. Pathways are complex, but molecular biologists have uncovered many of these as well as the genes that instruct for proteins that determine certain pathways.

Although some of the elements of the theories interlink, three basic ideas of aging exist: cumulative random cell damage; programmed cell death theories; and high-level program control theories.

Cumulative Random Cell Damage

This theory states that aging is not part of a planned program but occurs because of random damage to cells. The underlying predicate for this theory is that throughout life, individuals are subject to various cellular accidents or insults that damage cells. This might be considered the "wear and tear" syndrome compared to the natural decay of materials in nature. For example, an apple left outside will slowly rot and decay. A lot depends on what random events happen to the apple. Has it been bruised or knocked around? To what temperatures has it been exposed?

Random theories all point to the damage to which one's cells may be exposed in a lifetime. They point to "free radicals," or chemicals that are released from certain behaviors. As the body takes up oxygen, unstable molecules called "free radicals" bind to cells and cause serious damage. Free radicals come from smoking, pollution, poisons, fried foods, and normal metabolism. When they bond to the cells, they cause damage to the DNA within the cells and are thought to be a major element in the aging process.

Random theory proponents also point to the fact that DNA can just change without a cause. Other random theories include mitochondrial damage and stem cells' inability to multiply. These random changes lead to disorders such as cancer. Also, the body's own waste products can build up, causing random damage, and once the waste buildup damages one part, another one breaks down. They use the analogy of an old car. One thing breaks down, and then another and another.

Those who question this theory ask, if everything related to human birth is so programmed, then how could aging be so random and statistical? Although this theory has been around for a long time and has supporters, others believe that many other factors are necessary for an explanation of aging.

Programmed Cell Death

All cells have genes that control their end or death. This programmed cell death is called "apoptosis." This theory describes a form of cell suicide; genes shut down the cell and tell it when to shrink and disperse. Leonard Hayflick in the 1960s performed a classic experiment, which became known as the "Hayflick limit" or the "Hayflick phenomenon." He showed that cells in a culture would divide and divide up to about 50 divisions and then self-destruct in the process of apoptosis. The self-destruction of normal cells is different from cancerous cells that will continue to multiply within the test tube indefinitely. This theory of aging implies that what happens in aging is that cells divide until they cannot continue. When more are dying than are being replaced, then aging occurs.

Later scientists have tied the Hayflick limit to research into the shortening of telomeres, a bit of DNA located at the end of chromosomes. The research was done on the round worm *Caenorhabditis elegans* or *C. elegans*, whose life span is less than 20 days. Worms that will live longer than the 20 days have been created through manipulating certain pathways.

High Control Theory of Aging

This theory states that some type of master clock determined by genes controls aging. The control could be done in two ways:

1. *Cells act on their own*. They are independent, but they know the age of the DNA program. Each cell knows the age of the person and responds on its own.
2. *The entire body is a system*. The whole body knows its own clock and uses the pathways of cells, enzymes, and organs to accomplish aging. None of the changes happen in isolation but must consider the entire organism.

All the theories do have one thing in common: Genes stored in the DNA control the aging process. Whether it is as independent cells, an entire organism, or random accumulation of damage, genes still instruct for the proteins that form the pathways.

Diseases Related to Aging

Answering the question "What is aging?" is very complex. Obviously, it is not always related to the number of years in which one lives. For example, if one looks at a cohort of 70-year-olds, it is found that they are much more diverse than a cohort of 16-year-olds. Some aging conditions occur in four-year-olds. And the question of what role diseases play in aging has to be considered.

Some diseases appear to be age-dependent, and others appear to be age-related. As people live longer, the chronic diseases related to aging are also on the rise. Research into aging and research into regenerative medicine reflect two sides of the same coin. Regenerative aging focuses on the body's ability to heal itself, and antiaging medicine addresses the process by which the body's cells and tissues regenerate.

Diseases Related to Premature Aging

Some genes program for conditions that cause early cell aging and death. These conditions are not really the same as normal aging, because each disease has other attributes and appearances that are not found in normal aging. Following are some of the conditions in which early aging and death occur:

- Progeria or Hutchinson-Gilford disease, caused by the autosomal recessive gene *LMNA* on chromosome 1
- Werner syndrome, caused by the autosomal recessive gene *WRN* located on chromosome 8

- Ataxia-telangiectasia, or the "fragile chromosome syndrome," caused by the gene ATM on chromosome 11
- Rothmund's syndrome, caused by the gene *RECQL4* on chromosome 8; this condition causes aging syndromes such as osteoporosis, cataracts, skin blotches, and mental retardation
- Progeroid syndrome, a constellation of about six genetic diseases in which the person has the early signs of aging but lives a long life

These early aging diseases are related in some way to the characteristics of aging but are definitely different and specific in their effects.

Diseases that Are Age-Dependent

The following conditions are dependent on and generally occur with age: eye conditions such as cataracts and macular degeneration, osteoarthritis, osteoporosis, bladder and urinary problems in both men and women, brain cell loss, and weak immune system. Generally, it is thought that most of the conditions are preventable with lifestyle changes of exercise and diet, with the exception of the urinary system.

Diseases that Are Age-Related

The following conditions have increasing prevalence with age: atherosclerosis relating to stroke and heart attack, chronic lymphocytic leukemia, hypertension, type II diabetes, Alzheimer disease, Parkinson disease, prostate cancer, skin cancer, breast cancer, colon cancer, Paget's disease of bone, glaucoma, and vulnerability to infections.

Cutting-Edge Research into Aging

Research into aging and regenerative medicine is being conducted at many universities as well as private companies. Following are some of the ideas under investigation that affect aging:

- Reduced calories: Research conducted in the 1930s showed that calorie restriction could extend the lives of rodents. In 2002, scientists confirmed biomarkers that in mice and monkeys were related to aging. Longevity response to dietary restrictions were found to be actively regulated by nutrient-sensing pathways such as kinase target of rapamycin (TOR), AMP kinase, sirtuins, and insulin/insulin-growth factor (IGF-1) signaling.
- Research with the tumor suppressor p53: Research using mice treated with tumor suppressor gene p53 did not have tumors but developed muscle and bone weakness similar to condition in aging. The p53 gene may instruct for the pathway for stem cells to shut down and retard the body's ability to self-renew.
- Telemerase research: Telemerase is an enzyme located at the ends of chromosomes that affect the telomere strands and end cell division. If

overactivated, telomerase can divide wildly and cause cancer, but scientists are creating a three-dimensional structure of the enzyme that would aid in activating the enzyme for regeneration of tissue.

- Sirtuins, anti-aging enzymes: Sirtuins coincide with a marked increase in life span for flies, yeast, and worms. Although this may not translate to longevity in humans, scientists do know that misfiring of certain sirtuins can lead to aging-related disorders such as obesity, diabetes, and cancer.
- Immune system response: As people age, they become less able to content with viral and other illness and less responsive to vaccines. Scientists are studying the pathways of the immune system that are ineffective looking for that missing piece to overcome the inhibitory power.

Until recently, the idea of life extension was considered a fairy tale, not science. It was given the credibility like that of the myth of Eos and Tithonus. Today, aging research is serious. Models of mice and other animals have demonstrated the possibility. However, the U.S. FDA does not recognize aging as a condition to be treated, so now drugs that target age-related mutants will be the target for research and development. Aging as a target will have to wait for the future.

See also Age-Related Macular Degeneration; Alzheimer Disease; Early-Onset Glaucoma; Hutchinson-Gilford Progeria Syndrome; Paget Disease of Bone; Parkinson Disease; Rothmund-Thomson Syndrome (RTS); Werner Syndrome

Further Reading

Kenyon, Cynthia. 2010. "The Genetics of Ageing." *Nature*. March 25. http://www.nature.com/nature/journal/v464/n7288/full/nature08980.html. Accessed 2/10/12.

Melville, Kate. 2000. "The Genetics of Aging." Science a GoGo. March 31. http://www.scienceagogo.com/news/20000230170901data_trunc_sys.shtml. Accessed 2/10/12.

Perls, T.; L. Kunkel; and A. Puca. 2002. "The Genetics of Aging." National Center for Biotechnology Information (U.S.). http://www.ncbi.nlm.nih.gov/pubmed/12076681. Accessed 2/10/12.

Stibich, Mark. 2007. "The Genetic Theory of Aging." About.com. March 28. http://longevity.about.com/od/researchandmedicine/p/age_genetics.htm. Accessed 2/10/12.

Aicardi-Goutieres Syndrome (AGS)

Prevalence	Unknown but has been diagnosed in all continents; about 50 cases are known worldwide, existing mostly in families
Other Names	AGS; Cree encephalitis; encephalopathy with basal ganglia calcification; familial infantile encephalopathy with intracranial calcification and chronic cerebrospinal fluid lymphocytosis; pseudo-TORCH syndrome; pseudotoxoplamosis syndrome

In 1984, French doctors Jean Aicardi and Francois Goutieres described a syndrome present in eight individuals that was quite mysterious in that it appeared to work in phases. In 1988, researchers observed the same syndrome in 11 cases in a Cree community in Canada. Although it is quite rare, the syndrome probably exists worldwide and is mistaken for other rare conditions. The genetic causes are also very mysterious compared to most genetic disorders.

What Is Aicardi-Goutieres Syndrome (AGS)?

Aicardi-Goutieres syndrome affects three systems: the brain and nervous system, the immune system, and skin. At birth, the individual with AGS may not show symptoms, although the condition may be present. However, as the infant gets older, the syndrome does appear. The child may get a fever, but there is no sign of infection. Thus, the condition is referred to as "mimic of congenital infection."

Sometime during the first year of life, the first phase of the condition appears. The child may have severe brain dysfunction or encephalopathy that lasts for several months. Fever may come and go, but there is no infection. He may have seizures and stop developing skills, while losing the skills already acquired. The brain and skull stop growing. In another phase of the disorder, white blood cells connected with inflammation are seen in the spinal fluid, indicating damage throughout the nervous system. Abnormal deposits of calcium at the base of the skull will also show on medical images.

The child with AGS will have serious intellectual and movement difficulty, in addition to skin problems. Large, puffy lesions caused by inflammation of the blood vessel appear on the skin. These lesions, called chilblains, and are worsened by exposure to cold.

Most people with AGS do not live past childhood. However, some affected individuals with late onset or milder neurological problems may live into adulthood.

What Are the Genetic Causes of Aicardi-Goutieres Syndrome (AGS)?

Again, the condition is mysterious. It may be inherited in several ways. Most cases involved an autosomal recessive pattern, meaning that both parents must be carriers of the gene. However, a few cases of an autosomal dominant pattern have been seen. These cases result from a possible new mutation in the gene and occur without previous family history.

Mutation in five genes has been identified in people with AGS. The genes normally give instructions for making nucleases, enzymes that break up molecules of DNA and RNA. Mutations in any of these genes can disrupt the process because the impaired enzyme may cause unneeded DNA or RNA in the cells. The five genes are:

- *TREX1*: Mutations of this gene prevent the production of the 3-prime repair exonuclease 1 enzyme resulting in accumulation of unwanted DNA and

RNA in cells. The unneeded molecules may be mistaken for viral invaders and trigger the immune system to act. Mutations in *TREX1* may also cause lupus. The gene is located on short arm of chromosome 3 at position 21.31.

- *RNASEH2A*: This gene provides instructions for making one part of a group of proteins called the RNaseH2 complex. The enzyme helps break down RNA. There are nine mutations of this gene that may cause the dysfunctional RNA complex and errors in copying. The unneeded cells may be mistaken for viruses causing the immune reaction. The gene is located on the short arm of chromosome 19 at position 13.2.
- *RNASEH2B*: This gene makes instructions for a group of proteins called RNase H2, an enzyme that breaks down RNA. A function is to help prevent inappropriate immune system activation. With a mutation in the gene, the DNA or RNA may generate fragments to which the immune system reacts. This gene is located on the long arm of chromosome 13 at position 14.3.
- *RNASEH2C*: This gene appears to have the same function as the other two RNA genes. Malfunction may cause the unwanted products to be considered viruses, causing an immune reaction. This gene is located on the long arm (q) of chromosome 11 at position 13.1.
- *SAMHD1*: This gene normally provides instructions for making a protein; the function is not well understood. At least 16 mutations are identified in people with AGS. It is located on the very end of the short arm of chromosome 20 at and the long arm (q) at position 12: 20pter-q12.

What Is the Treatment for Aicardi-Goutieres Syndrome (AGS)?

There is no cure for this disorder. However, treatment of some of the symptoms such as skin issues may make the person more comfortable.

Further Reading

"Aicardi-Goutieres Syndrome." Genetics Home Reference. http://ghr.nlm.nih.gov/condition/aicardi-goutieres-syndrome. Accessed 5/4/12.

"TREX1." Genetics Home Reference. http://ghr.nlm.nih.gov/gene/TREX1. Accessed 5/4/12.

Alagille Syndrome

Prevalence 1 in 70,000; may be as high as 1 in 20,000

Other Names Alagille's syndrome; Alagille-Watson syndrome; arteriohepatic dysplasia (AHD); cardiovertebral syndrome; cholestasis with peripheral pulmonary stenosis; hepatic ductular hypoplasia;

hepatofacioneurocardiovertebral syndrome; paucity of interlobular bile ducts; Watson-Miller syndrome

Dr. Daniel Alagille (1925–2006), a French physician, established the first hospital for pediatric hepatology or children with liver disease. He loved children dearly and developed a holistic approach to treating them. In his hospital he established an entire floor into living space dedicated to play and school. The writer of his obituary in the scientific journal *Journal of Pediatric Gastroenterology and Nutrition* wrote that one of his favorite sayings was this: "A physician who attends adults does not need to love his patients, while a pediatrician cannot care for his patients without loving them." Alagille syndrome was named for this caring doctor.

What Is Alagille Syndrome?

This disorder, which generally appears in infancy or early childhood, affects the liver, heart, and other body parts. A major feature is liver damage causing abnormalities, especially narrowing, in the bile ducts that carry bile from the liver to the gall bladder. Bile is essential to the digestion of fats. Because of the narrowed ducts, bile then backs up in the liver, causing serious damage.

Signs of the disorder include yellowing of the skin (jaundice), itchy skin, deposits of cholesterol in the skin, heart problems, and butterfly-shaped vertebrae when viewed by an X-ray. Some patients may have a heart defect called tetralogy, in which a hole exists between the two lower chambers of the heart.

Children with Alagille syndrome have a distinct appearance, which includes a broad, prominent forehead, deep-set eyes, and a small, pointed chin. They may look more like each other than members of their own family. The disorder may also affect the brain, spinal cord, and kidneys. Symptoms may vary from very mild to severe when the person may need a heart or liver transplant.

What Are the Genetic Causes of Alagille Syndrome?

Alagille syndrome is an inherited autosomal disease, which means one copy of the mutated gene in each cell can cause the disorder. In 30% to 50% of cases, the individual acquires the gene from an affected parent. Other cases arise from mutations or new deletions in the gene occurring as random events during embryonic formation. These children may have no history of the disease in their families.

In more than 90% of the cases, mutations in the *JAG1* gene are identified. The other 20% involve mutations in the *NOTCH2* gene. Both of these genes instruct for making proteins that signal how cells are used to build body structures in the developing embryo. People with *JAG1* mutations have one of more of the problems. The lack of *NOTCH2* signaling causes the errors in the bile ducts and liver and heart defects. *JAG1* is located on the short (p) arm of chromosome 20 between positions 12.1 and 11.23. *NOTCH* is located on the short (q) arm of chromosome 1 between positions 13 and 11.

What Is the Treatment for Alagille Syndrome?

Although no cure is known for the syndrome, treatments of the defects of the heart and liver impairments are available. Medications may improve bile flow and reduce the skin itching. Corrective surgery may repair heart defects.

Further Reading

"Alagille's Syndrome." Children's Hospital of Pittsburgh. http://www.chp.edu/CHP/Alagille. Accessed 1/27/12.

"Obituary for Daniel Alagille." 2006. *Journal of Pediatric Gastroenterology and Nutrition*. February. 42(2): 127–128.

Spinner, Nancy B., Anne L. Hutchinson, Ian D. Krantz, and Binita M. Kamath. 2010. "Alagille Syndrome." *GeneReviews*. http://www.ncbi.nlm.nih.gov/bookshelf/br.fcgi?book=gene&part-alagille. Accessed 5/4/12.

Albinism

Prevalence 1 in 17,000

Other Names achromia; achromasia; achromatosis

What Is Albinism?

The word "albinism" comes from the Latin word *albus*, meaning "white." People with albinism do not make sufficient amounts of the pigment melanin, which gives skin its color. They also have very light or platinum-colored hair and light, usually pale blue eyes. Two main categories of albinism exist in humans:

- Ocular albinism: Pigment or eye color is important for normal vision. Ocular albinism is a condition in which the pigment in the iris of the eye and the retina are affected. The iris is the colored part of the eye that acts like a camera shutter to close in strong light and open in darkness; it controls the amount of light entering the eye. The retina is the structure at the back of the eye where visual images are focused. People with this type of albinism will have normal skin pigment and may appear normal. However, vision is severely impaired and permanent, but does not worsen over time.
- Oculocutaneous albinism (OCA): This condition affects both the eyes and the skin. The person appears pale, with light hair and eyes. The condition is combined with serious vision problems and light sensitivity. Four different types of OCA exist: Type 1 is characterized by white hair, very pale skin, and light-colored irises. Type 2 is less severe than type 1; here the skin may be creamy

white and hair a light yellow or blonde. Type 3 usually affects dark-skinned people, who may have reddish-brown skin, red or ginger-colored hair, and hazel or brown irises. Type 4 is associated more with visual problems than skin problems. Because the features of the four types often overlap, a genetic profile of the person will more accurately distinguish the type.

Several other syndromes may present albinism. These include Hermansky-Pudlak syndrome, Chediak-Higasi syndrome, Griscelli syndrome, Waardenbrug syndrome, and Tietz syndrome.

People with albinism are generally as healthy as the rest of the population. Albinism itself does not cause mortality, although exposure to sun does cause the risk of skin cancer, especially the deadly melanoma.

What Are the Genetic Causes of Albinism?

Albinism has several genetic causes.

- *Autosomal recessive*: The genes for OCA are on autosomes, meaning that one gene is inherited from the mother and one from the father. Neither parent has the symptoms.
- *X-linked pattern*: Ocular albinism Type 1 is inherited in an X-linked pattern. The mother is a carrier of the mutated gene *GPR143*, but she does not have

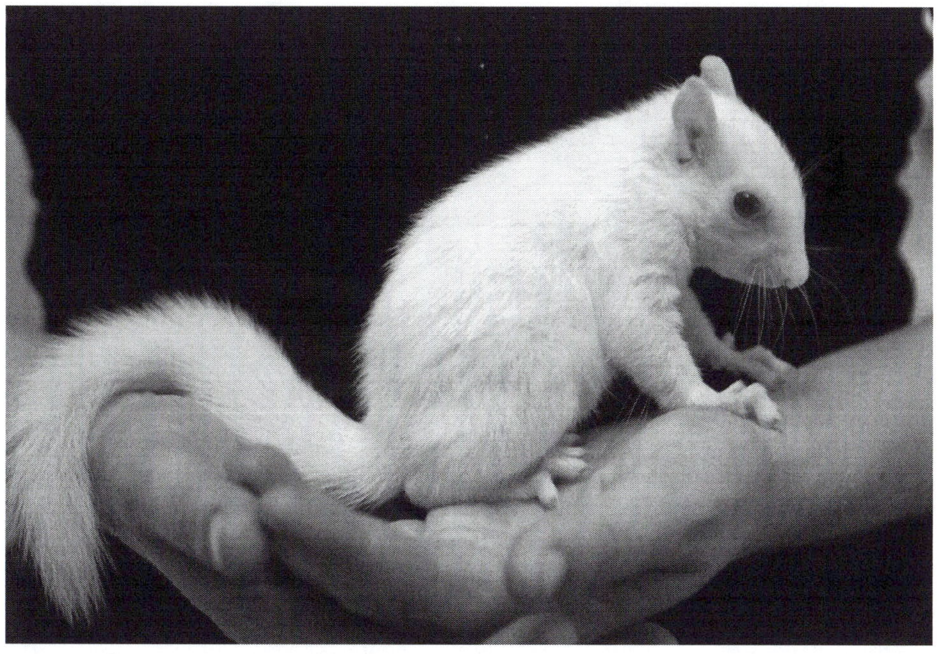

A squirrel in South Africa shows the traits of albinism. (Gallo Images/Getty Images)

the symptoms. Because males do not have a normal copy of the gene, the recessive gene is sufficient to cause the disorder.

In oculocutaneous albinism, mutations in four different genes cause the following types of albinism:

- OCA Type 1 is caused by mutations in the *TYR* gene. Tyrosinase is an enzyme that helps change the amino acid tyrosine into pigment. The more than 100 mutations in the *TYR* gene disrupt the normal production of melanin, causing white hair and very light skin. The *TYR* gene is located on the long arm (q) of chromosome 11, position 14–21: 11q14-q21.
- OCA Type 2 is caused by mutations in the *OCA2* gene. This condition results from a defect in the P protein that helps tyrosinase to function. These individuals make a minimum amount of pigment and can have hair color from light blond to brown. Most affected individuals are of sub-Saharan African American heritage. *OCA2* is located on the long arm (q) of chromosome 15 at position 21.3: 15q21.3.
- OCA Type 3 is caused by mutations in the *TYRP1* gene. This condition is rare and results from a genetic defect in the protein that the enzyme tyrosinase-related protein 1. This enzyme is located in the melanocytes, cells that produce melanin pigment. Individuals with Type 3 can have substantial pigment. The gene is prevalent in very dark-skinned people of southern Africa.
- OCA Type 4 is caused by mutations in the *SLC45A2* gene. *SLC45A2* protein helps tyrosinase to function. People with the type make a minimal amount of pigment. The gene is located on the short arm of chromosome 5 at position 13.2: 5p13.2.

Ocular albinism is caused by mutations in the *GPR143* gene. This is the gene that provides instructions for a protein that gives color to eyes and skin. The gene is located on the short arm (p) of the X chromosome position 22.3; Xp22.3.

What Is the Treatment for Albinism?

Eye problems may arise because of the lack of pigment. Accommodations for most conditions such as strabismus (cross-eyes), astigmatism, and far- or nearsightedness can be made with corrective lenses or surgery. In the United States, most people with albinism live normal life spans.

Further Reading

"Albinism." 2011. National Health Service (UK). http://www.nhs.uk/conditions/albinism/Pages/introduction.aspx. Accessed 1/28/12.

"Oculocutaneous Albinism." 2007. Genetics Home Reference. National Library of Medicine (U.S.). http://ghr.nlm.nih.gov/condition/oculocutaneous-albinism/show/print. Accessed 5/4/12.

"What Is Albinism?" 2002. National Organization for Albinism and Hypopigmentation (NOAH). http://www.albinism.org/publications/What_is_albinism.html. Accessed 5/4/12.

Albinism, Ocular. *See* Ocular Albinism

Alexander Disease

Prevalence Rare—no more than 500 cases worldwide

Other Names Alexander's disease; leukodystrophy with Rosenthal fibers

As interest in genetics became more prominent after World War II, researchers were engaged in investigating several disorders during the late 1940s. In 1949 while studying an infant with hydrocephalus and mental retardation, Dr. W. F. Alexander found that slides taken from spinal fluid of this patient showed that astrocytes, a type of nerve cell, contained deposits of threads associated with myelin disorders. These fibers were later called Rosenthal fibers. Myelin is the white sheath covering nerve fibers, and damage to this structure causes serious motor and neurological symptoms. The condition that he observed later turned out to be the genetic disorder that now bears his name: Alexander disease.

What Is Alexander Disease?

Alexander disease, which affects mostly infants and children, is a rare fatal disorder of the nervous system. It is a member of a family of disorders called leukodystrophies, in which the myelin sheath that covers the nerve cells is destroyed. Myelin is the fatty protective covering of nerve cells that is responsible for aiding the rapid transmission of impulses. Accompanying the myelin sheath damage are abnormal protein deposits called Rosenthal fibers, which are found in certain nerve cells called astroglial cells or astrocytes. Astrocytes are essential for the health of the brain.

Three different clinical forms of Alexander disease are noted:

- Infantile form (birth to 2 years): These children have enlarged heads and brains, seizures, spasticity or stiffness in arms and legs, developmental delay, and intellectual disability. A neonatal form may cause a buildup of fluid in the brain or hydrocephaly. Prognosis for these children is poor, with most of them dying in early childhood.
- Juvenile form: This form starts in school-age children, on average about 9.5 years of age. The main symptom appears to be lack of muscle control or spasticity; however, the patients have none of the cognitive impairments.

- Adult form: This form is sometimes misdiagnosed as multiple sclerosis. The individual may develop speech problems, swallowing difficulties, seizures, and poor coordination.

What Is the Genetic Cause of Alexander Disease?

Alexander disease is inherited in an autosomal dominant pattern, meaning that one copy of the gene must be present in the cells. Most cases develop from mutations and are not present in one of the parents.

Mutations in the gene *GFAP*, officially known as the glial fibrillary acidic protein gene, causes Alexander disease. Normally, *GFAP* gives instructions for making a protein called glial fibrillary acidic fiber, a member of the intermediate filament family of proteins. These filaments form the networks that give strength and support to cells, especially the astroglial cells in the brain and spinal cord. In Alexander disease, mutations change one of the amino acids that make the glial fibrillary acidic protein. The changed protein then accumulates in the astroglial cells, causing the formation of Rosenthal fibers and impairing cell function; the person displays the symptoms of Alexander disease. *GFAP* is located on the long (q) arm of chromosome 17 at position 21.

What Is the Treatment for Alexander Disease?

Because no treatment exists for this condition, making the person comfortable and treating symptoms is the next best thing. Prognosis is usually poor, with death occurring within 10 years after onset. A registered charity called the Stennis Foundation is committed to raising funds and awareness for research into diseases of leukodystrophy.

Further Reading

"Alexander Disease." 2010. Cleveland Clinic. http://myclevelandclinic.org/disorders/alexander_disease/hic_alexander_disease.aspx. Accessed 5/4/12.

"Alexander Disease." 2010. Genetics Home Reference. National Library of Medicine (U.S.). http://ghr.nlm.nih.gov/condition/alexander-disease. Accessed 5/4/12.

Rodriguez, Diana. 2004. "Alexander's Disease." Orphanet. http://www.orph.net/consor/cgi-bin/OC_exp.php?Expert=58. Accessed 5/4/12.

Alkaptonuria

Prevalence Rare—1 in 250,000 to 1 million people worldwide; certain areas of the world such as Slovakia and the Dominican Republic have incidences as high as 1 in 19,000.

Other Names alcaptonuria; homogentisic acid oxidase deficiency; homogentisic aciduria

British physician Dr. Archibald Garrod (1857–1936) was fascinated by the reasons for colors of urine. As a doctor at the Great Osmond Street Hospital in London, he called on Thomas P., a three-month-old child, whose urine was a deep reddish brown. He diagnosed the condition as alkaptonuria, a rare condition in which alcapton is not broken down and builds up in the bloodstream. When talking to the family, he found that five family members had the same symptoms. Garrod surmised that those with the disorder had some "inborn error of metabolism" that was passed on to family members. Garrod was the first to make the connection between inheritance and a genetic mutation. This story of the recognition of the first metabolic disease is of great importance in medical history.

What Is Alkaptonuria?

This condition is noted when the urine turns black when exposed to air. Usually appearing after age 30, a pigment called ochonosis builds up in connective tissues such as cartilage and skin. This pigment is excreted in the urine. People with the condition develop arthritis in large joints of the spine. Also, in early adulthood, individuals may experience prostate stones, kidney stones, and heart problems.

Alkaptonuria is called a metabolic disease. In a normal person, an enzyme, homogentisate oxidase, breaks down the amino acids phenylaniline and tyrosine. These two amino acids are important building blocks in the body. But if the enzyme is not working, a toxin homogentisic acid is deposited in the connective tissue, causing the cartilage and skin to malfunction and darken. The buildup can lead to arthritis and accumulation of stones in the kidney and prostate. It is this homogentisic acid that causes the urine to turn dark when exposed to air.

What Is the Genetic Cause of Alkaptonuria?

Mutations in the *HGD* gene, officially called the "homogentisate 1,2-dioxygenase" gene, causes alkaptonuria. Normally, *HGD* instructs for making the enzyme homogentisate oxidase, which breaks down the amino acids phenylalanine and tyrosine. The breakdown of these amino acids is important to maintain healthy metabolism.

More than 65 mutations in the *HGD* gene have been identified in people with alkaptonuria. Most of these changes are in a single amino acid. For example, substitution of the amino acid valine for the amino acid methionine at position 368 is most common in European populations. The condition is rare and inherited in an autosomal recessive pattern. Both parents must carry the gene. The carriers typically do not have any of the symptoms of the condition. The gene is located on the long arm (q) of chromosome 3 at the 13.33 position.

What Is the Treatment for Alkaptonuria?

The clue to the diagnosis is the dark urine. Symptoms of arthritis and kidney stones can be managed with drugs or other strategies. The *HGD* gene may be a candidate for gene therapy.

Further Reading

"Archibald Garrod Biography (1857–1936)." Faqs.org. http://www.faqs.org/health/bios/42/Archibald-Garrod.html. Accessed 5/4/12.

"HGD." 2010. Genetics Home Reference. National Library of Medicine (U.S.) http://ghr.nlm.nih.gov/gene/HGD. Accessed 5/4/12.

Introne, W.; M. Kayser; and W. Gahl. 2011. "Alkaptonuria." *GeneReviews*. National Library of Medicine (U.S.). http://www.ncbi.nlm.nih.gov/bookshelf/br.fcgi?book=gene&part=alkap. Accessed 5/4/12.

Allan-Herndon-Dudley Syndrome

Prevalence Rare, affecting about 25 families with individuals worldwide

Other Names Allan-Herndon syndrome; MCT8 (SLC16A2)-specific thyroid hormone cell transporter deficiency; mental retardation X-linked, with hypertonia; monocarboxylate transporter 8 (MCT8) deficiency

In 1944, the world was focusing on the effects of World War II. The study of genetics did not get a lot of attention. Yet, in that year, three doctors—William Allan, Nash Herndon, and F. C. Dudley—were studying inheritance patterns. Observing several large families, they determined that a form of mental retardation was passed on by the mother as a sex-linked disorder. This study, published in the *American Journal of Mental Deficiencies*, was entitled "Some Examples of the Inheritance of Mental Deficiency; Apparently Sex-Linked Idiocy and Microcephaly." This was the first of the X-linked mental retardation syndromes to be identified and later the first to be regionally mapped in 1990. The condition was named for the three men: Allan-Herndon-Dudley syndrome.

What Is Allan-Herndon-Dudley Syndrome?

Allan-Herndon-Dudley syndrome is a disorder of the brain. This rare syndrome is caused by delayed brain development that causes severe mental retardation and motor development. Occurring only in males, the disorder is obvious at birth. Following are the symptoms of the disorder:

- Poor muscle tone that appears in early childhood
- Reduced skeletal body mass
- Distinct facial appearance: There may be some elongation of the face, abnormal folding of the ears, and narrowing of the face at the temples.
- Severe cognitive impairment
- Impaired speech development

- Joint disorders: As the person ages, he develops joint deformities and cannot move certain joints
- Muscle weakness: He may have muscle weakness that in adulthood makes him unable to walk independently and causes him to become wheelchair-bound

What Is the Genetic Cause of Allan-Herndon-Dudley Syndrome?

Mutations in the *SLC16A2* gene, officially called the "solute carrier family 16, member 2 (monocarboxylic acid transporter 8)" gene, cause Allan-Herndon-Dudley syndrome. Normally, the *SLC16A2* gene (also known as *MCT8*) instructs for a protein that plays a critical role in the nervous system. The protein transports a hormone called triiodothyronine, or T3, into the developing brain. This hormone is produced by the thyroid gland. When it arrives inside the nerve cells, T3 then interacts with receptors in the nucleus to turn specific genes off or on. T3 may play a critical role in the development of dendrites and synapses, parts of the nerve messaging pathway. T3 is also produced in the liver, kidney, heart and other tissues, and it controls many of the chemical reactions important to body metabolism.

Mutations change the structure of *SLC16A2* protein, making it unable to transport T3 into nerve cells. Because brain development is affected, the person becomes intellectually disabled and has major problems with movement. T3 is not taken up by nerve cells, causing excess amounts to become toxic and damaging to organs. This toxic buildup also produces the other symptoms of AHD syndrome. The X-linked pattern means that a female with the mutated gene will pass it on to male offspring. The female will not have any symptoms. It appears to have a recessive pattern. *SLC16A2* is located on the long arm (q) of chromosome X at the 13.2 position.

What Is the Treatment for Allan-Herndon-Dudley Syndrome?

There is no cure for this condition. Treating some of the symptoms and intervention with physical therapy may help the person have a better quality of life. The person appears to have no additional health risks and may experience a relatively long life span.

Further Reading

"Allan-Herndon-Dudley Syndrome." 2007. Genetics Home Reference. National Library of Medicine (U.S.). http://ghr.nlm.nih.gov/condition/allan-herndon-dudley-syndrome/show/print. Accessed 5/4/12.

Schwartz, Charles. 2004. "Allan-Herndon-Dudley Syndrome." Orphanet. http://www.orpha.net/data/patho/GB/uk-allan.pdf. Accessed 5/4/12.

"SLC Gene Family." 2007. Genetics Home Reference. National Library of Medicine (U.S.). http://ghr/nlm.nih.gov/gene/SLC16A2. Accessed 5/4/12.

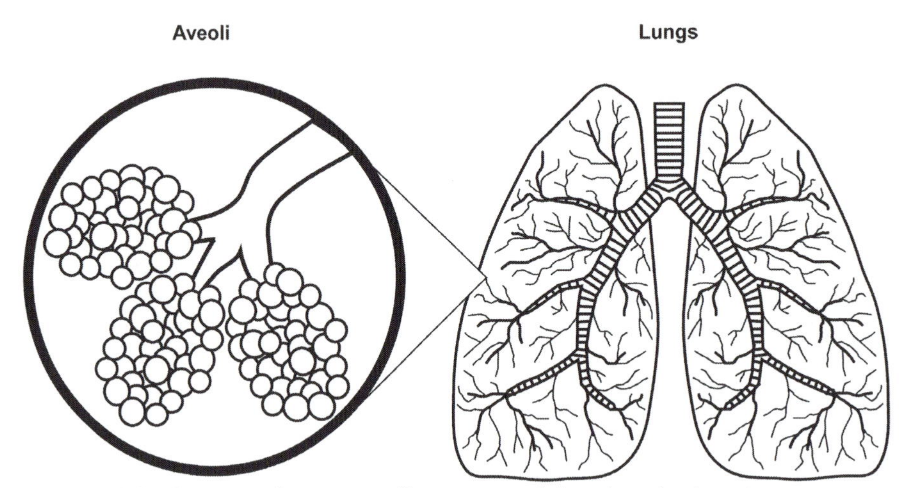

Diagram of the lungs and an inset illustration of the alveoli, the air sacs in the lungs. People with severe alpha-1 antitrypsin deficiency will develop emphysema, from destruction of alveoli and the surrounding tissue. (ABC-CLIO)

Alpha-1 Antitrypsin Deficiency (A1AT)

Prevalence Varies by population; in people of European ancestry, about 1 in 1,500 to 3,000; uncommon in people of Asian descent; many cases are undiagnosed or confused with chronic obstructive pulmonary disease (COPD)

Other Names AATD; alpha-1 protease inhibitor deficiency; alpha-1 related emphysema; ATT; hereditary pulmonary emphysema; inherited emphysema

Physicians in northern Europe noted that some of their very young patients developed a serious breathing disorder similar to emphysema. The condition appeared in many people who lived in the Scandinavian countries, as well as people in Spain and Saudi Arabia. In 1963, Carl-Bertil Laurell (1919–2001) at the University of Lund in Sweden gave the condition a name. Then six years later, another doctor connected the condition with liver disease. The condition that Dr. Laurell discovered was alpha-1 antitrypsin deficiency.

What Is Alpha-1 Antitrypsin Deficiency (A1AT)?

Alpha-1 antitrypsin deficiency is an inherited lung condition in which the patient experiences shortness of breath, coughing, and wheezing. The symptoms resemble asthma attacks that keep recurring but do not respond to treatment. Even with no history of smoking, the person may develop emphysema, a condition in which holes develop in the tiny air sacs called alveoli. It is in the alveoli that the oxygen–carbon dioxide exchange takes place; with holes in the sac, the structures cannot hold air long enough for this exchange to happen. Also different from emphysema, people with A1AT develop the condition between the ages of 20 and 50. Smoking or being around smoking aggravates the condition.

About 10% of infants with A1AT develop liver disease, which will be accompanied by jaundiced skin and yellow eyes. This condition is a leading cause of liver transplants in newborns. About 15% of cases develop cirrhosis of the liver due to scar tissue that has formed. A type of cancer called hepatocellular carcinoma may occur. Rarely, a skin condition called panniculitis appears. This condition, which varies in severity and can occur at any age, is characterized by hardening of the skin and painful lumps or patches.

What Is the Genetic Cause of Alpha-1 Antitrypsin Deficiency (A1AT)?

Mutations in the *SERPINA1* gene, officially called the "serpin peptidase inhibitor, clade A (alpha-1 antiproteinase, antitrypsin), member 1" gene, cause A1AT. Normally, alpha-1 antitrypsin is a protein produced in the liver that protects the lungs from the neutrophil elastase enzyme. This enzyme is released from the white blood cells and plays an important role in fighting infection.

Mutations in *SERPINA1* disrupt the instructions for making the protein alpha-1 antitrypsin protein and allow the neutrophil elastase to destroy the small alveoli air sacs and cause disease. When abnormal amounts of neutrophil elastase accumulate in the liver, the patient experiences liver damage. The common allele of the *SERPINA1* gene is M. Most normal people have two copies of the M gene, one inherited from each parent (MM). Other versions of the gene, such as the S allele, produce low levels of the protective protein. Another allele, Z, produces very little alpha-1 antitrypsin. A person with two copies of the Z allele (ZZ) is likely to have the deficiency. Those with the SZ combination are at increased risk for lung and liver disease. It is estimated that about 161 million people have one copy of S or Z and one copy of M. Their genotype would be MS or MZ. Those with MS or SS produce enough alpha-1 antitrypsin to protect the lungs. The *SERPINA1* gene is inherited in a recessive pattern and is located on the long arm of chromosome 14 at position 32: 14q32.

What Is the Treatment for Alpha-1 Antitrypsin Deficiency (A1AT)?

A1AT deficiency is often undiagnosed or misdiagnosed as COPD without an underlying cause. Patients who have the symptoms without a cause should have a blood test for the A1AT level. Several experimental studies are being tried. The patient may receive intravenous infusions of alpha-1 antitrypsin made from human plasma. A recombinant and inhaled form has been developed and is in the experimental trials.

Further Reading

"Alpha-1 Antitrypsin Deficiency." 2010. Genetics Home Reference. National Library of Medicine (U.S.). http://ghr.nlm.nih.gov/condition/alpha-1-antitrypsin-deficiency. Accessed 5/5/12.

"Alpha-1 Antitrypsin Deficiency." 2010. National Heart, Lung, and Blood Institute (U.S.). http://www.nhlbi.nih.gov/health/health-topics/topics/aat. Accessed 5/5/12.

Silverman, E. K., et al. 2009. "Alpha-1 Antitrypsin Deficiency" *New England Journal of Medicine*. 360(26): 2747–2757.

Alpha Thalassemia

Prevalence A common blood disorder worldwide, particularly in Southeast Asia, the Mediterranean counties, India, and Central Asia

Other Names a-thalassemia

The two thalassemias—alpha and beta—have been frightening disorders throughout history, but with present new therapies, the outlook is encouraging. The quality of life has improved for those individuals with the diseases.

Both of the conditions are blood disorders that involve the decreased production of hemoglobin, the molecule in the red blood cells necessary to carry oxygen throughout the body. The difference in the names—alpha and beta—arises from the part of the hemoglobin molecule that is affected.

What Is Alpha Thalassemia?

Alpha thalassemia is a disorder of the blood. Normal hemoglobin is made of two chains with alpha and beta globins. People with alpha thalassemia have a deficiency in the alpha chain. The individual has a shortage of red blood cells causing a condition known as anemia. This deficiency prevents oxygen from getting to the body's tissues and causes the person to be weak, tired, and pale. Two types of alpha thalassemia are known:

- Hb Bart syndrome: In this syndrome—known as hydrops fetalis—fluid builds up in the fetus before birth. The fetus may have severe anemia, enlarged liver and spleen, heart defects, and abnormalities of the urinary system. These babies are usually stillborn or die in early infancy. During pregnancy, the woman carrying a child with Bart syndrome may have serious complications, such as abnormally high blood pressure, premature delivery, abnormal bleeding, and preeclampsia. Preeclampsia in pregnancy is very serious because it can lead to the more serious eclampsia, a condition in which the mother has seizures, goes into a coma, and possibly dies.

Am I My Brother's Keeper?

The reports from Spain and Chennai, India, told of children who were genetically designed to give blood to a sibling. The experiments used umbilical cord blood for the transplant, and the siblings were not hurt in any way.

However, experiments like these evoke serious ethical questions. What if the child was genetically engineered for "spare parts" for a sibling? A book by Jodi Picoult (and later a film) called *My Sister's Keeper* tells the story of Anna, who is not sick, but she has undergone countless surgeries, transfusions, and shots so that her older sister, Kate, can fight leukemia. Anna was designed with preimplantation genetic characteristics as a bone marrow match for Kate, her older sister. She never questioned her role in life until she became a teenager—and like most teenagers, she began to question who she really was.

In the past, she always defined herself in the terms of her older sister, but she came to the conclusion that she must make a choice. The choice tore her family apart and endangered Kate's life. Anna hired an attorney to have her legally separated from the parents so she would no longer have to undergo the painful operations to save her dying sister.

The book examines what it means to be a good parent, a good sister, and a good person. Should a parent be able to use another child to save one who is ill from a fatal condition? Is it ethical to create "spare parts kids" to save the life of another? Is it morally correct to impinge on the life of a healthy child to save a dying child? Was Anna selfish in not wanted to undergo any more painful bone donor transplants? What about the doctrine of informed consent? Can parents give consent for their children even though it may cause a great deal of pain to the healthy child? Scientists now have the technology to do these things and to create children who will match other children. The question remains: Has science outpaced us as human beings?

- HbH syndrome: This syndrome is a milder form of the disease in which the individual may have moderate anemia, slight enlargement of liver and spleen, and some jaundice or yellowing of the skin and eyes. Bone changes may occur and the child may develop a prominent jaw and forehead in early childhood.

What Is the Genetic Cause of Alpha Thalassemia?

Two genes are involved in alpha thalassemia—*HBA1* and *HBA2*. The genes are inherited in a Mendelian recessive pattern and involve two gene loci or four alleles. In this pattern people have two copies of each of the genes—one copy from the father and one from the mother for each gene. Normally, these two genes give instructions for making alpha-globin, a subunit of hemoglobin. However, the mutated genes disrupt the production of alpha-globin. The four alleles come into play, causing different symptoms. In Bart syndrome, all four alpha-globin alleles are mutated. If three mutated alleles are present, the hemoglobin molecules are so abnormal that they cannot carry oxygen. A loss of two alleles may result in mild anemia and smaller red blood cells. If one allele is lost, the person will carry the trait but basically have no symptoms of thalassemia. *HBA1* is located on the short arm (p) of chromosome 16 at position 13.3. *HBA2* is located on the short arm (p) of chromosome 16 at position 13.3.

What Is the Treatment for Alpha Thalassemia?

There is no cure for the disorder, but modern medicine has made the condition treatable with chelation therapy and blood transfusions. A screening policy in some countries includes prenatal screening and counseling for abortion. In Spain in 2008, a baby was selectively implanted to be a cure for his brother's thalassemia. The cord blood was used for transplantation to his brother, which was successful. In 2009, a group of doctors in Chennai and Coimbatore successfully treated thalassemia with a sibling's umbilical cord blood.

See also Beta Thalassemia

Further Reading

"Alpha Thalassemia." 2010. Genetics Home Reference. National Library of Medicine (U.S.). http://ghr.nlm.nih.gov/condition/alpha-thalassemia. Accessed 5/6/12.

"His Sister's Keeper: Brother's Blood Is Boon of Life." 2009. *Times of India*. September 17.

"What Are Thalassemias?" 2010. Kid's Health from Nemours. http://kidshealth.org/parent/medical/heart/thalassemias.html. Accessed 5/6/12.

Alport Syndrome

Prevalence 1 in 50,000 newborns

Other Names Hematuria-neuropathy-deafness; hemorrhagic familial nephritis; hereditary deafness and nephropathy; hereditary nephritis

William Cecil Alport (1880–1959) was a passionate crusader. Upon the suggestion of Dr. Alexander Fleming (who discovered penicillin), he went to Egypt to be a professor of medicine and work in the King Faud I Hospital in Cairo. At that time Egypt was part of the British Empire. Alport was appalled by the treatment of patients, especially the poor patients, and wrote about it in a book, *One Hour of Justice: The Black Book of Egyptian Hospitals*. Not convinced that he had made a difference in Egypt, he returned to London and devoted himself to poor families. In 1927, he noted that one of the families had several of the members with progressive symptoms of kidney failure ending in death. The condition was also accompanied by hearing and vision losses and affected mostly males. The condition that he observed now bears his name: Alport syndrome.

What Is Alport Syndrome?

Alport syndrome is a disease of the kidneys and urinary system, which also is displayed in several other systems. As Dr. Alport observed, this syndrome is characterized by kidney failure, hearing loss, and eye abnormalities. One of the first signs is blood in the urine, a condition called hematuria; high protein levels will also be in the urine. Because of slow damage to the tiny blood vessels in the kidney (the glomeruli), the kidney's filtering system loses effectiveness, causing toxic wastes to build up. The result is end-stage renal disease (ESRD) at an early age, usually between adolescence and age 40. Although Alport syndrome affects mostly males, females with mild or no symptoms can transmit it to their offspring.

Following are the symptoms of Alport syndrome:

- Abnormal urine color indicating blood and/or protein in the urine
- Swelling in ankles, feet, and possibly overall
- Loss of vision, especially in males
- Loss of hearing especially in males
- Swelling around the eyes

Red flags that indicate the condition include end-stage renal disease in several family members and a history of kidney infections or inflammation.

What Are the Genetic Causes of Alport Syndrome?

Alport syndrome has several patterns of inheritance. The following genetic patterns may be involved in this syndrome:

- X-linked recessive: The gene *COL4A5* is carried on the X chromosome. In males who have only one X chromosome, only one altered gene may cause the kidney failure and the other symptoms. Females may have only mild symptoms such as blood in the urine, or no symptoms at all. About 80% of the cases are inherited in this way.
- Autosomal recessive: This type results from mutations in both copies of *COL4A3* or *COL4A4*. Both parents carry a copy of this mutated gene in their genome and are carriers. Most carriers are unaffected or have less severe symptoms. About 15% of cases are inherited in this way.
- Autosomal dominant pattern: Individuals with this form have one mutation in either *COL4A3* or *COL4A4* in each cell. This form is present in about 5% of cases.

Mutations in the genes *COL4A3*, *COL4A4*, and *COL4A5* cause this syndrome. These genes are collagen biosynthesis genes and are responsible for making type

IV collagen. Collagen is a strong fiber-like material that forms the structure of the underlying membranes of the kidney, inner ear, and eye. In the kidney are structures of specialized blood vessels called glomeruli that remove water and waste from the blood and create urine. A disturbance in the gene can result in abnormalities of the type IV collagen and keep the kidney from working properly. Mutations can also disrupt the type IV collagen function in the ear and eye.

Location of the genes follow: *COL4A3* located on the long arm of chromosome 2, 2q36-q37; *COL4A4* is located at 2q35-q37; *COL4A5* is located on the long arm of the X chromosome at position 22.

What Is the Treatment for Alport Syndrome?

Because this is a genetic condition, no cure exists. However, monitoring and control of symptoms, such as blood pressure, is necessary. Controlling diet and fluid intake is very important. If the kidney condition progresses to end-stage renal disease, dialysis or transplantation may be required. Caring for the eyes and ears can assist in alleviating the symptoms of the disorder.

Further Reading

Alport, A. C. 1927. "Hereditary Familial Congenital Hemorrhagic Nephritis." *British Medical Journal* (London). 1: 504–506.

"Alport Syndrome." 2010. Genetics Home Reference. National Library of Medicine (U.S.). http://ghr.nlm.nih.gov/condition/alport-syndrome. Accessed 5/6/12.

"Alport Syndrome." 2010. Medline Plus. National Library of Medicine (U.S.) http://www.nlm.nih.gov/medlineplus/ency/article/000504.htm. Accessed 5/6/12.

ALS. *See* Amyotrophic Lateral Sclerosis (ALS)

Alzheimer Disease (AD)

Prevalence In persons under 65, about 5%; in persons over 65, about 10.3%; in persons over 85, about 47.2%.

Other Names presenile dementia; senile dementia; senile dementia Alzheimer's type (SDAT)

What Is Alzheimer Disease (AD)?

Alzheimer disease (AD) is a condition in which the brain slowly shrivels and dies. Nerve cells in the brain stop working, and brain signals that are essential for life do

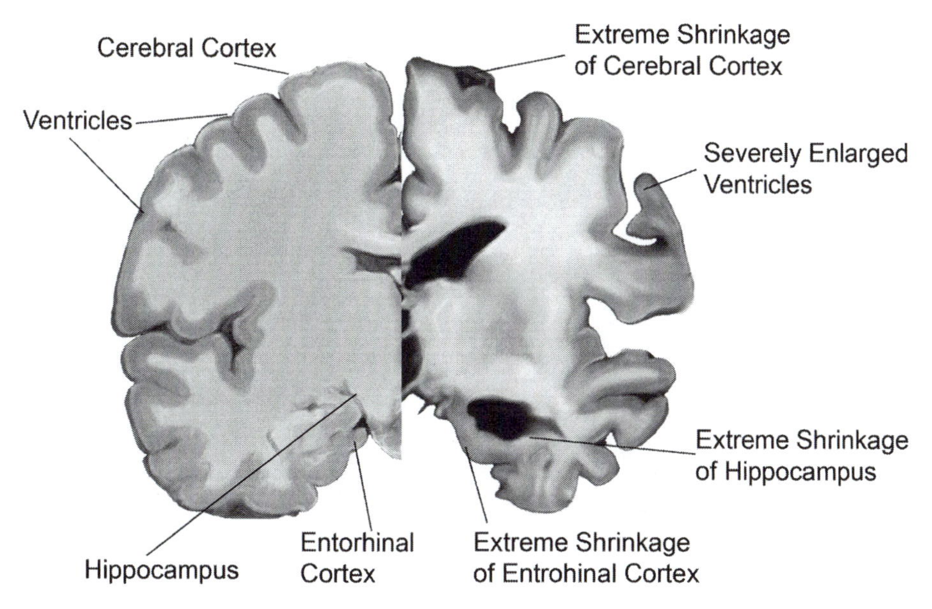

Side-by-side comparison of a normal brain and one with severe Alzheimer disease, showing shrinkages. (National Institute of Health / National Institute on Aging)

not function properly. AD starts with mild memory loss, and people gradually lose judgment, thinking, reasoning ability, and cognition. AD and other dementias are considered illnesses and are not a normal part of aging.

History of Alzheimer Disease (AD)

The disease that is now called Alzheimer disease is not new to the world. The bizarre behavior of people with dementia has been well known for centuries. The philosopher Plato (429–348 BC) considered problems of judgment and irrational behavior that occur in aging were not related to the brain but to moral weakness. He also recognized that some older people were perfectly rational. Years later, the Roman physician Galen (AD 130–200) revived the Greek ideas that bodily "humors" regulated health and that evaporation of bodily heat and moisture caused aging. Frenzy and mania were forms of insanity related to old age. The teachings of Galen prevailed for many centuries. People with mental illness were shunned and later placed in asylums in chains and shackles.

Enter Alois Alzheimer. Alzheimer was born in Markeit, Germany, in 1864 and lived during the second half of the nineteenth century when amazing developments were taking place in the field of medicine. On a gloomy November day in 1901, he was working in an asylum in Frankfurt when he was handed the file of a difficult 51-year-old woman patient named Auguste D. She was rife with hostility and

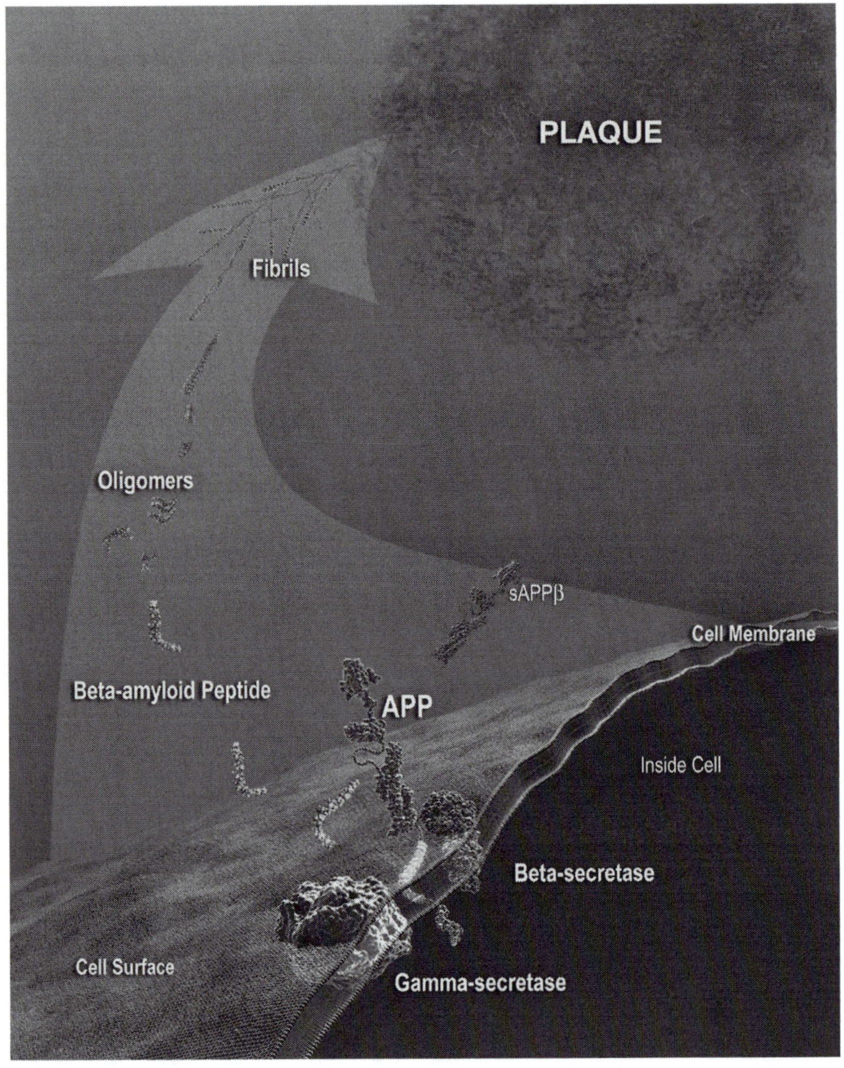

Beta-amyloid plaque associated with Alzheimer disease. (National Institutes of Health)

bizarre behavior, but most of all, she was confused about life and would scream for hours. After her death, he was given permission to study her brain, and what he saw really shocked him. The brain was a mass of tangles; deposits of plaques were spread throughout. He then studied the brains of other people with suspected dementia and found the same plaques and tangles.

He presented his findings at a medical meeting in 1907 and published a paper in a scientific journal a year later. A few years later, the renowned psychiatrist Emil Kraepelin published a famous textbook on psychiatry and included a new condition called "Alzheimer's Disease." The name of the German scientist has stuck and is now one of the most-researched diseases in the world.

Alois Alzheimer discovered the neurological dementia disease that now bears his name. (National Library of Medicine)

Biology of Alzheimer Disease (AD)

The plaques and tangles that Alzheimer saw have been analyzed as plaques of β-amyloid, a sticky protein that appears as tangled threads. β-amyloid is snipped from a larger protein called the amyloid precursor protein, or APP. Surrounding the plaques are bits of dying nerve cells. In addition, twisted nerve cell fibers known as neurofibrillary tangles (NFTs) are present. NFTs are formed in nerve cells when damaged *tau* proteins, which normally stabilize the nerve cell, became entangled with normal *tau* cells.

In advanced Alzheimer disease, the cerebral cortex, the thinking part of the brain, shrivels, and fluid-filled spaced called ventricles have grown larger. Scientists have studied the chemical structure of these proteins, as well as brain structure. Finding the genes that are responsible rivals any detective story.

Rudolph Tanzi: Untying the Gordian Knot

On a cloudy September day in 1980, Rudolph (Rudy) Tanzi applied at the laboratory of a well-known scientist, James Gusella, at Massachusetts General Hospital in Boston. Tanzi, a graduate student, needed the money to help pay for expenses. Gusella was in the midst of researching the gene that causes Huntington disease, the serious nerve disorder. With his background in both microbiology and history, Tanzi became enthralled with the idea that genes causing disease can be located through research. He had the privilege of working with the man who found the gene for Huntington, which became known as the "crown jewel of genetic research."

While working as a graduate student, Tanzi attended a lecture by Dennis Selkoe, an up-and-coming researcher in the field of Alzheimer disease. Selkoe told about how the protein amyloid collected in the brain and blood vessels. Then, the defining moment came that changed Tanzi's life and established his future course. Selkoe showed a slide of Alexander the Great cutting the Gordian knot. In Greek mythology, the Gordian knot was an intricate knot that no one could undo. People from far and near tried to solve the puzzle. But Alexander tried a new way that no one had done before: he cut the knot with his sword. Selkoe then challenged the listeners: Amyloid is like that knot. It would take someone with dedication and patience to think out of the box to solve the puzzle.

The speech challenged Tanzi, and the puzzling protein B-amyloid became his passion. As he recalled Gusella's search for the Huntington gene, he set out to find the gene responsible for Alzheimer disease. His research took him to chromosome 21, and by the end of 1987, he and his colleagues had located *amyloid precursor protein gene (APP)*. He completed his thesis in 1990—and just in time to run in the Boston Marathon. Tanzi became a well-known molecular geneticist and professor of neurology at Harvard Medical School. His journey to untie the Gordian knot of amyloid is detailed in his autobiography, written with Ann B. Parson, *Decoding Darkness: The Search for the Genetic Causes of Alzheimer's Disease* (New York: Basic Books, 2001).

What Are the Genetic Causes of Alzheimer Disease (AD)?

A disease like Alzheimer disease is very complicated and actually has several manifestations. An early form of AD called early-onset Alzheimer disease affects younger adults around age 50; late-onset usually occurs around 60. Scientists have found several genes that are linked with various populations and kinds of AD. Following are some of these genes:

- Chromosome 21: Chromosome 21 is associated with a rare form of the disease called Familial Alzheimer's Disease, or FAD. This rare form affects less than 10% of the persons with AD and progresses faster than common late-onset forms. Scientists use a tool called genetic sequencing to locate suspect genes. Sequencing scans the genomes of people with the condition looking for similar patterns in the genome. An abnormal gene for FAD was found on chromosome 21. The gene *APP* expresses the protein amyloid precursor protein (APP) and differs from the normal gene by only one base pair. In 1991, researchers determined that the mutant gene on chromosome 21 was associated with FAD. Three connections with chromosome 21 were present:

1. A genetic defect of one base pair causes the formation of a defective amyloid protein.
2. The defective protein is deposited in certain places in the brain.
3. The deposit destroys neurons and produces the clinical symptoms of AD.

However, the connection with chromosome 21 was associated with only about 2% to 3% of FAD cases.

- Chromosome 14: Chromosome 14 became a candidate for AD when statistics showed a common marker for the defective amyloid protein (APP) on the chromosome. Investigators named area and the gene *presenilin1* (*PS-1*). Researchers are looking at how presenilins work.
- Chromosome 1: Researchers found a group of Germans living in the Volga Valley of the former Soviet Union and had a very high occurrence of FAD. The studies showed no link between the disease in these people and chromosomes 21 and 14. However, they did find a region on chromosome 1 and named it *presenilin-2* (*PS-2*). Only a very small fraction of early-onset mutations is caused by mutations in the presenilin-2 gene. Mutations in the three genes—APP, *presenilin 1*, and *presenilin2*—account for about 50% of early-onset FAD. Other genes related to FAD have yet to be identified.
- Chromosome 19: The genetics of late-onset AD is much more complicated than FAD. In 1992, researchers at Duke University found an increased risk for late-onset AD with the inheritance of a gene located on chromosome 19 called *apolipoprotein* or *APOE*. This gene makes a protein that sits on the surface of cholesterol molecules. Cholesterol is a form of fat. APOE carries cholesterol in the blood through the body. APOE is also found in glial cells that hold nerve cells in the brain together. Apolipoprotein is found in the plaques of neurofibrillary tangles.

  The gene *APOE* has three different forms called alleles. A person inherits one allele from each parent; everyone has two APOE genes, one inherited from each parent: APOE-ε2, APOE-ε3, APOE-ε4. The Greek symbol for epsilon is ε. APOE-ε3 is the normal form of the protein; APOE-ε2 and APOE-ε4 are the disease-causing forms. The APOE gene has 50,100,878 to 50,104,489 base pairs and is on the long arm (q) of chromosome 19 at 13.23.

  In addition to AD, the APOE-ε4 allele is associated with atherosclerosis or hardening of the arteries. APOE-ε4 has been shown to increase susceptibility to AD. Forty percent to 60% of people with AD have at least one copy of APOE-ε4. People who inherit one copy of the APOE-ε4 allele have an increased chance of getting the disease; those who inherit two copies have an even greater risk. However, now all people with AD have the APOE-ε4 allele, and not all people who have the APOE-ε4 allele develop the disorder.

New Genes

Scientists at Duke scanned the entire genome of a large group of families in which several members had late-onset AD. The strongest pattern was on chromosome 12. In January 2007, scientists from 14 institutions announced a new gene Sorletin-related receptor 1, or *SORL 1*, that may contribute to amyloid plaques. The *PLAU* gene on chromosome 10 may link to late-onset AD. Other novel suspected regions are on chromosomes 3, 4, 5, 6, 8, 9, 13, 16, and X.

What Is the Treatment for Alzheimer Disease (AD)?

Although no cure for AD now exists, the condition is one of the most researched for finding a cure. The U.S. Food and Drug Administration lists more than 75 drugs that are being studied, most of these to alleviate symptoms. About 10 drugs that are on the market have been approved. Most of these drugs help people in the early or middle stages with memory issues.

Drugs work differently in the brain. Following are two types of drugs that are widely used to manage AD:

Acetylcholinesterase Inhibitors

These drugs increase the amount of acetylcholine in the brain. This chemical is important for memory and learning. Three of the acetylcholinesterase inhibitors approved by the FDA to treat Alzheimer's disease are:

- Aricept (donepezil 5 and 10 mg), approved to treat all stages of Alzheimer's disease
- Exelon (rivastigmine tartrate), approved to treat mild to moderate Alzheimer's disease.
- Razadyne (galantamine HBr), approved to treat mild to moderate Alzheimer's disease

Glutamate Pathway Modifiers

This drug works differently from the acetylcholinesterase inhibitors. It affects glutamate, a chemical in the brain that is important for learning and memory. Namenda is currently the only drug of its type approved to treat moderate to severe Alzheimer disease.

Combination Therapy

Because the two types of medications work in different ways, taking them together can be beneficial. Use of the drugs may help some individuals function and maintain a quality of life.

See also Aging and Genetics: A Special Topic

Further Reading

Alzheimer's Association. http://www.alz.org. Accessed 5/6/12.

Alzheimer's Disease Education and Referral Center (ADEAR). http://alzheimers.nia.nih.gov. Accessed 5/6/12.

Alzheimer's Research Forum. http://www.alzforum.org. Accessed 5/6/12.

Kelly, Evelyn B. 2007. *Gene Therapy*. Westport, CT: Greenwood Press.

Kelly, Evelyn B. 2008. *Genes and Disease: Alzheimer's Disease*. New York: Chelsea House.

Amyotrophic Lateral Sclerosis (ALS)

Prevalence	Diagnosed in about 5,000 people in the United States; about 4 to 8 per 100,000 people worldwide; small percentage with known genetic cause
Other Names	ALS; Charcot disease; Lou Gehrig disease; motor neuron disease

A mysterious condition in which healthy young adults suddenly began to lose function of their voluntary muscles has baffled doctors for many years. In 1824, Charles Bell described patients with purely motor disorders. In 1850, autopsies became a way of trying to analyze disorders when English doctor Augustus Waller first described the shriveled nerve fibers of a patient who died with muscle weakness and loss. In 1874, French doctor Jean-Martin Charcot found the relationship between the clinical signs and the findings at autopsy. Charcot was one of the first to make the connection between autopsy and clinical findings, now called the clinicopathologic approach. The approach dominated medical diagnosis in the latter half of the nineteenth century. The disorder received much attention when Lou Gehrig, a famous baseball hero, developed the disease. Amyotrophic Lateral Sclerosis has been commonly known as Lou Gehrig disease.

In the 1950s, a sudden epidemic of ALS occurred among the Chamarro people on Guam, giving researchers the opportunity to study the condition in a contained population. These studies allowed scientists eventually to link a familial form of the disorder to a specific chromosome. ALS continues to be a subject of research with several drugs in clinical trials in the United States.

What Is Amyotrophic Lateral Sclerosis (ALS)?

Amyotrophic lateral sclerosis (ALS) affects the motor neurons. These all-important nerve cells control the movements of the voluntary muscles. The disorder is progressive, meaning that it gets worse over time as the motor neurons do not function and eventually die.

Lou Gehrig Names a Disease

Baseball was America's favorite pastime in the 1930s, relieving many of the discouraging days of the Great Depression. A favorite and hero on the baseball scene was a first baseman whose career with the New York Yankees spanned 17 years, from 1923 to 1939. Lou Gehrig (1903–1941) played in 2,130 consecutive games and held records for the most career grand slams and through his skill as a first baseman.

During the 1938 season, he began to note that he was tired and just could not get going; his statistics were down but still above average. During the next season, his coordination and speed deteriorated. He noted that he was stumbling over curbs, fumbling the baseball, and slipping and falling when running bases. His wife Eleanor took him to Mayo Clinic in Minnesota, and after six days of testing, the diagnosis was ALS. The prognosis was not good. In a few years he would be completely paralyzed.

On July 4, 1939, he was honored as he retired from baseball. He announced to the crowd in a poignant speech that he considered himself the "luckiest man in the world and had a awful lot to live for." He died on June 2, 1941. The disease with the long name "amyotrophic lateral sclerosis" later became known as Lou Gehrig disease.

Source: "Farewell Speech" Lou Gehrig (official website). http://lougehrig.com. Accessed 5/28/12.

The descriptive term "amyotrophic lateral sclerosis" comes from several Greek words: *a*, meaning "no" or "without"; *myo*, meaning "muscle"; and *troph*, meaning "nourishment." The word "lateral" refers to a portion of the spinal cord where parts of the nerve cells are located, and "sclerosis" means scarring or hardening. Thus, ALS is a condition in which the muscle cells do not receive nourishment because of a hardening of the part of the spinal cord where portions of the nerve cells are located.

The structure of motor neurons shows the diverse function. Like other neurons, they have dendrites, cell body, and axons. Two kinds of motor neurons are the upper motor neurons that come from the brain and the lower motor neuron that then go to particular muscles. The upper neurons direct the lower neurons to create such movements as walking, chewing, and many other voluntary functions. The lower axons may be quite long. For example, the axon that connects the spinal cord to the foot may be a yard long.

A disruption of the function of the motor neurons results in motor neuron disease. ALS is one of these diseases. The symptoms of the disease begin very subtly. The person may feel stiffness in the muscles, which is common as one exercises or exerts himself. The following symptoms tell doctors to look for ALS:

Lou Gehrig (1903-1941), the celebrated baseball player who developed the disease known as amyotrophic lateral sclerosis, or ALS, died just before he turned 38. The condition is sometimes called Lou Gehrig's disease. (Harris & Ewing)

- Twitching, cramping, or pain in muscles in one of the legs
- Tripping or stumbling often when walking or running
- Difficulty in doing simple tasks such as buttoning a shirt, writing, or turning a key
- Slurred speech
- Difficulty chewing or swallowing

As the disease progresses, other parts of the body are affected. The weakness in the muscles spread. Now the following problems may arise:

- Problems with moving; person may not be able to get out of bed on their own or use hands and feet.
- Swallowing is more labored, a condition called dysphagia.
- Speaking or forming words is very labored, a condition known as dysarthia.
- Muscles are exceedingly tight and stiff, a condition known as spasticity.
- Exaggerated reflexes that include an overactive gag reflex.
- A common abnormal reflex called Babinski's sign, in which the large toe extends upward from the sole of the foot when stimulated in a certain way. This sign indicates upper motor neuron damage.
- Muscle weakness, cramps, and twitches of muscles observed under the skin.

To have a diagnosis of ALS, people must have the signs of both upper and lower motor neuron damage that cannot be traced to any other conditions.

In later stages of the disease, the person may have difficulty breathing because respiratory muscles are affected. They may require the assistance of a breathing device.

The course of the disorder varies with the individual. Most people do not have cognitive impairment although some may have some memory loss and become depressed.

The average age of onset in those with no family history of ALS is 56; for those with a family history, the average age is 48. Duration is about three years, but this may vary with the individual. Death is usually from the compromised respiratory system.

What Are the Genetic Causes of Amyotrophic Lateral Sclerosis (ALS)?

Mutations in several genes cause ALS: *ALS2*, *SETX*, *SOD1*, and *VAPB*. The following genes increase the risk of developing ALS: *ANG*, *DCTNi*, *NEFR*, *SMN1*, and *SMN2*. Several types of ALS exist relating to genetic cause, ages when symptoms began, and disease progression. Following are the types of ALS and their genetic causes:

- *ALS1*: Onset of this type usually begins in adulthood between the ages of 25 and 44 and progresses slowly. The type is related to the *SOD1* or the "superoxide dismutase1, soluble gene." Normally, *SOD1* instructs for producing an enzyme called superoxide dismutase, which is abundant in the cells of the body. This enzyme neutralizes radical oxygen molecules called superoxides. These superoxide molecules are by-products of energy reactions occurring in the mitochondria and can damage cells if not controlled. In order for the enzyme to function, it must bind to copper and zinc. Of the 100 *SOD1* mutations related to type 1 ALS, most result from a change in only one building block with the most common occurring when valine replaces alanine. The enzyme then becomes harmful, but exactly how the motor neurons become sensitive to the altered enzyme is not clear. Possibly the toxic radicals make if impossible for the abnormal mitochondria to meet the high energy demands of the motor neurons causing cell death. *SOD1* is located on the long arm (q) of chromosome 21 at position 22.1.

- *ALS2*: This type of ALS begins in early childhood or adolescence and slowly worsens for the next 10 to 15 years. The official name of the *ALS2* gene is "amyotrophic lateral sclerosis 2 (juvenile)." Normally, *ALS2* instructs for alsin, a protein abundant in the brain and in the motor neurons. Alsin may play a role in the development of the axons and dendrites, the outgrowths from nerve cells that are involved in the transmission of nerve signals. Two mutations have been identified in which a single building block is substituted for another, causing a disruption in the function of the motor neurons and eventually their death. *ALS2* is located on the long arm of chromosome 2 at position 33.1.
- *ALS3*: This is a rare type found in one large pedigree which began in the legs and onset was age 45 lasting for five years. The gene is not identified but traced to a suspected area of on the long arm of chromosome 18 at position 21.chormosome.
- *ALS4*: This type of ALS typically affects people under 25 and slowly progresses. Mutations in the *SETX* or "senataxin" gene cause ALS4. Normally, *SETX* instructs for senataxin, a protein produced in many tissues, especially the brain, spinal cord, and muscles. Senataxin, one of the classes of helicases, is thought to be involved in DNA repair and production of RNA. Three *SETX* mutations are thought to result from replacement of one amino acid with another, altering the shape of the molecule and disrupting the normal function. *SETX* is located on the long arm of chromosome 9 at position 34.14.
- *ALS5*: The genetic cause of this type is unknown but may be traced to an unnamed gene in the long arm of chromosome 15 at positions 15 through 21.1.
- *ALS6*: Mutations in the *FUS* or "fused in sarcoma" gene cause this type of ALS. Normally, *FUS* instructs for a protein called the heterogeneous nuclear ribonucleoprotein complex (hnRNP), which is involved in the messenger RNA process. Defects in this gene result in ALS type 6. *FUS* is located on the short arm of chromosome 16 at position 11.2.

- *ALS7*: Little is known about this type, which may be traced to the short arm (p) of chromosome 20 at position 13.
- *ALS8*: This type of ALS begins earlier than type 1 and progresses slowly over several years to several decades. *VAPB*, or "vesicle-associated membrane protein-associated protein B and C" gene, causes ALS8. Normally, *VAPB* instructs for a little-known protein found in cells throughout the body. Recent research has found that the protein is related to the membranes that surround the endoplasmic reticulum, a structure within cells that folds proteins for transport to the cell surface. People with ALS8 have a mutation that changes one of the building blocks. Here, the amino acid serine replaces praline and causes ALS8 in some people. *VAPB* is located on the long arm (q) of chromosome 20 at position 13.33.
- *ALS9*: ALS9 has been reported in a small number of French, Italian, North American, and northern European individuals with both sporadic and familial ALS. This type is caused by the *ANG*, or "angiogenin, ribonuclease, RNase A Family, 5" gene. Normally, *ANG* instructs for a protein called angiogenin, which promotes the formation of new blood vessels from existing vessels through a process called angiogenesis. The process is related to a series of reactions that produce ribosomal RNA and also relate to anti-infection process. Mutations in this gene may increase the risk of ALS in people of Irish and Scottish descent. The mutations reduce the activity of angiogenin and impair the transport from the surface of the cells to the nucleus. *ANG* is located on the long arm (q) of chromosome 14 at position 11.1 to 11.2.
- *ALS10*: This type has similar symptoms to sporadic ALS and those with SOD-1. However, mutations in the TARDBP are located on the short arm of chromosome 1 at position 36.2. The gene has been identified in small numbers of northern European, Australian, and Chinese individuals with both sporadic and familial ALS.

Most cases of ALS are sporadic with no family history; about 5% to 10% have a family connection. The disorder can be inherited in an autosomal dominant, autosomal recessive, or X-linked pattern.

What Is the Treatment for Amyotrophic Lateral Sclerosis (ALS)?

The treatment of ALS is palliative, with individuals requiring a multidisciplinary team of neurologist, pulmonologist, speech therapist, occupational therapist, nutritionist, social worker, genetics professional, and other professionals. Depending on the effects of the disorder, different medications to control the symptoms may be used. One medication—Riluzole—is the only medication approved by the FDA for ALS.

Further Reading

"ALS2." 2007. Genetics Home Reference. National Library of Medicine (U.S.). http://ghr.nlm.nih.gov/gene/ALS2. Accessed 5/8/12.

"Amyotrophic Lateral Sclerosis." 2012. Medline Plus. National Library of Medicine. National Institutes of Health. http://www.nlm.nih.gov/medlineplus/amyotrophic lateralsclerosis.html. Accessed 5/8/12.

Donkervoort, Sandra, and Teepu Siddique. 2009. "Amyotrophic Lateral Sclerosis Overview." *GeneReviews*. http://www.ncbi.nlm.nih.gov/books/NBK1450. Accessed 1/26/12.

Androgen Insensitivity Syndrome

Prevalence Affects 2 per 100,000 people who are genetically male

Other Names Aiman's syndrome; AIS; androgen receptor deficiency; androgen resistance syndrome; AR deficiency; DHTR deficiency; dihydrotestosterone receptor deficiency; Gilberg-Dreyfus syndrome; Goldberg-Maxwell syndrome; Lub's symdrome; Morris' syndrome; Reifenstein syndrome; Rosewater syndrome; testicular feminization

The many names of this syndrome indicate the interest generated and the complexity of this disorder. Over the last 60 years, steady progress has been made in determining that all these different symptoms and apparently different disorders were really different presentations of the same condition. This condition is given the name androgen insensitivity syndrome, or AIS.

What Is Androgen Insensitivity Syndrome?

Androgen insensitivity syndrome (AIS) is a condition in which the male hormone androgen is not working properly. Androgen is a hormone that affects the sexual development of the male before and after birth. If a problem exists and these males do not respond to the messages of these hormones, the organs of male child organs may appear feminized. Later at puberty, these children with AIS will develop abnormal secondary sex characteristics and in some types infertility.

Following are the three basic classes of androgen insensitivity syndrome:

- Complete AIS (CAIS): These individuals do not respond to androgens or male hormones at all. They are considered females at birth and are raised as females. However, they do not have a uterus, will not menstruate, and are infertile. External genitalia may be normal, but internally they are different. The sex organs are not ovaries but they may be testicles that are undescended and located in the pelvis or abdominal region. This abnormal structure can cause cancer if not surgically removed. Also the person may have longer limbs, larger hands and feet, larger teeth, minimal or no acne, and other dysfunctions such as dry eye and light sensitivity.

- Partial AIS (PAIS): These individuals can use some androgens. They can have normal female sex characteristics, normal male characteristics, or both male and female characteristics. The gonads are testes, but there may also be a wide number of phenotypes. A scale called the Quigley scale quantifies four degrees of feminized genitalia. These are grades 2 through 5 on this scale. It was assumed that these individuals were always infertile, but others have shown this is not the case.
- Mild AIS (MAIS): The individuals are born with a male phenotype but are often infertile and may have breast enlargement and other female characteristics at puberty.

What Is the Genetic Cause of Androgen Insensitivity Syndrome?

This condition is inherited in an X-linked recessive pattern. This means that the mother has passed the mutation on the X to her sons. Because there is no comparable gene on the Y chromosome, the child has the mutated gene. Two-thirds of the defects are inherited in this way; however, cases may result from a new mutation in the egg or during embryonic development.

Mutations in the *AR* gene, officially known as the "androgen receptor" gene, cause AIS. Normally *AR* instructs for making the protein called androgen receptor. These receptors allow cells to respond properly to androgens, such as the male hormone testosterone that directs male development. Both androgen and androgen receptors also have other functions such as regulating hair growth and sex drive.

When this gene has a mutation, the androgen receptors do not work properly, and the cells become less responsive to androgen or do not respond to the hormone at all. The amount of androgen sensitivity determines the degree of the person's sex characteristics, which can range from mostly female to mostly male. According to a 2008 study by Galani et al., over 400 mutations have been reported in the *AR* mutation database. The number grows with additional research. The *AR* gene is located on the long arm (q) of chromosome X at position 11-q12.

What Is the Treatment or Management for Androgen Insensitivity Syndrome?

Currently, management of AIS is to treat only the symptoms. The technology for correcting the malfunctioning androgen receptor gene or protein is not available. Surgery may be used to remove misplaced parts, along with hormone replacement therapy. Genetic and psychological counseling may assist the individual.

Further Reading

"Androgen Insensitivity Syndrome." 2011. Genetics Home Reference. National Library of Medicine (U.S.). http://ghr.nlm.nih.gov/condition/androgen-insensitivity-syndrome/show. Accessed 5/8/12.

Galani, A.; S. Kitsiou-Tzeli; C. Sofokleous; E. Kanavakis; and A. Kalpini-Mavrou. 2008. "Androgen Insensitivity Syndrome: Clinical Features and Molecular Defects." Hormones (Athens). http://hormones.gr/preview.php?c_id=227. Accessed 5/8/12.

Gottlieb, Bruce; Leonore Beitel; and Mark Trifiro. 2010. "Androgen Insensitivity Syndrome." *GeneReviews*. http://www.ncbi.nlm.nih.gov/books/NBK1429. Accessed 5/8/12.

Androgenetic Alopecia

Prevalence Affects about 35 million men in the United States; in women hair loss is most likely after menopause

Other Names androgenic alopecia; female pattern baldness; male pattern alopecia; male pattern baldness; pattern baldness

Throughout history, men and women have bemoaned the loss of hair and tried all sorts of weird potions for cure. In the ancient Middle East, loss of hair was equated to the loss of manliness. The Egyptians, Romans, and Greeks eagerly sought all sorts of concoctions and salves to stop hair loss or to grow new hair. For example, the Papyrus Papers (1500 BC), which chronicled many cures, gave the recipe for baldness: Grind a dog's toe, rotten dates, and the hoof of an ass, and put in balsam; rub into the head. A story from the Old Testament tells about the prophet Elisha, who was traveling along the road to Bethel when some kids from a village jeered at him calling him a "bald head." He called down a curse of God on the youths, and two bears came out of the woods and mauled 42 of the boys (II Kings 2:23). The condition that has troubled so many throughout history is now in the age of youth culture the subject of serious scientific study: androgenic alopecia.

What Is Androgenetic Alopecia?

Androgenetic alopecia is the loss of hair in both men and women. Men lose hair in a defined pattern: starting above the temples, the hair line recedes, and the hair thins at the crown of the top of the head. This characteristic "M" shape hair around the rim of the head is known as male pattern baldness. In women, the hair simply thins all over but rarely leads to complete baldness.

Scientists have determined that this pattern of baldness is caused by dihydrotestosterone (DHT), one of a group of hormones called androgens. Androgens have

English footballer Bobby Charlton shows typical androgenetic alopecia, or male pattern baldness. (Getty Images)

many important functions including male sexual development and regulation of hair growth and sex drive. The normal pattern for growth of hair follows: Hair growth normally begins with the hair follicle located under the skin. Individual strands of hair go through cycles: grow for two year to six years, rest for several months, then fall out. The cycle repeats itself with a new hair produced in that follicle. If the androgens are increased to the hair follicles, the cycle may shorten. The hair growth is thinner and shorter, and growth of new hair is delayed. Androgenetic alopecia in men is also associated with coronary heart disease; problems with the prostate gland; insulin resistance disorders, such as diabetes and obesity; and high blood pressure. In women, it is associated with polycystic ovary syndrome (PCOS), a condition that can lead to irregular menstruation and weight gain.

What Is the Genetic Cause of Androgenetic Alopecia?

Several genetic and environmental factors play a role, and many of these are still unknown. The gene "androgen receptor," or *AR*, is the one most associated with male pattern baldness. In normal individuals, *AR* gives the instructions for making the protein called the androgen receptor, which allow the body to respond appropriately to dihydrotestosterone and other androgens. Variation in the *AR* gene leads to elevated activity of androgen receptors in the hair follicles. The *AR* gene is located on the long arm of the X chromosome at position 12: Xq12.

A 2007 study suggests another gene on the X chromosome called *EDA2R* gene is also associated. This gene is found on the long arm q position 11-q12. Another gene was found on chromosome 3: 3q26. In addition to genetic factors, some researchers also suggest that lifestyle may play a part. For example, one study suggests that weight training may have a detrimental effect on hair by increasing testosterone levels.

What Is the Treatment for Androgenetic Alopecia?

Most people with male pattern baldness accept the condition. A few treatments have been shown to reduce hair loss, especially in early stages. These treatments include: the drug Finasteride, the drug Minoxidil as a second-line treatment, low-level laser therapy, and topical caffeine.

Further Reading

"Androgenetic Alopecia." 2010. Genetic Home Reference. National Library of Medicine (U.S.). http://ghr.nlm.nih.gov/condition/androgenetic-alopedia. Accessed 5/8/12.

"AR." 2010. Genetics Home Reference. National Library of Medicine (U.S.). http://ghr.nlm.nih.gov/gene/AR. Accessed 9/24/11.

"Male Pattern Baldness Treatments." 2010. WebMD. http://www.webmd.com/a-to-z-guides/hair-loss-treatments. Accessed 5/8/12.

Angelman Syndrome (AS)

Prevalence	1 in every 10,000 to 20,000 births; about 1,000 cases in the United States
Other Names	The Happy Puppet syndrome (obsolete)

Birdie, 16, is happy and affectionate to everyone she meets, but when her parents must still change her diapers, they are aware of her obvious developmental delay. She grunts or points to indicate her needs. Since a tiny baby, she has flailed her hands and arms with jerky motions and appeared happy all the time. Her attention span is very short, and she may have an occasional seizure. Birdie has been diagnosed with Angelman syndrome.

What Is Angelman Syndrome (AS)?

This syndrome is a complex genetic disorder that primarily affects the nervous system. The developmental delay usually is noticeable around 6 to 10 months, when the baby's head does not appear to grow normally. Although the child appears happy, the jerking motions are noticeable—hence the term "the happy puppet syndrome."

In 2005, the Angelman Syndrome Foundation, Inc. issued a Consensus Criteria document that described developmental and physical findings of AS:

- Severe developmental delay—100% of the time
- Movement or balance disorder
- Behavioral uniqueness of frequent laughter for no apparent reason
- Hand flapping or waving movements
- Speech impairments
- A very pronounced, small head in more than 80% of cases
- Seizures beginning about age 3
- A flat, occipital area of the skull in about 20% to 80% of cases
- Protruding tongue
- Sucking and swallowing disorders
- Wide mouth with widely spaced teeth
- Strabismus (crossed eyes)
- Insensitivity to heat
- Obesity
- Scoliosis
- An abnormal encephalogram (EEG)

What Are the Genetic Causes of Angelman Syndrome (AS)?

In 1965, Dr. Harry Angelman, a British physician, noted that some children with severe developmental delay laughed all the time and displayed a severe, stiff, and jerky gait; he called them "happy puppet children." For his studies with these children, he was honored by contributing his name. However, it was not until 1987 that Dr. Ellen Mageris found that children with AS had abnormal deletions in the *UBE3A* gene at position 15q11-q13.

People normally inherit one copy of this gene from each parent, and both genes are turned on or are active in many body tissues. Certain areas of the brain may inherit may inherit only one copy from the mother called genomic imprinting.

If the maternal copy of *UBE3A* gene is lost or mutated, there will be no active copies of the gene in the brain.

Following are five different patterns of genetic changes that are associated with Angelman syndrome:

1. About 70% of the cases happen when the part of the maternal chromosome 15 containing *UBE3A* is missing.
2. About 11% of the cases occur when the gene is changed or mutated.
3. A small percentage occurs with the person inherits two copies of chromosome 15 from the father, instead of one from each parent. This is called uniparental disomy.
4. A rare case may be caused when there is a translocation or mutation in the region of the DNA that controls the activation of the gene.
5. A deletion of a gene called *OCA2* is associated with fair hair and skin. *OCA2* is also on chromosome 15.

In 10% to 15% of cases, the cause may be unknown and other genes or chromosomes may be responsible.

What Is the Treatment for Angelman Syndrome (AS)?

Although no cure exists for AS, a variety of interventions can help the condition. For example, if the person has seizures, antiseizure medication is given. Physical therapy can help improve balance and gait. Speech therapy may assist the person in acquiring acceptable vocabulary.

Individuals with AS may have near-normal life expectancy. However, adults are usually not capable of living by themselves, but they may live successfully in group homes. There are some clinical studies of the effects of vitamin supplements on AS.

Further Reading

"Angelman Syndrome." 2010. Genetics Home Reference. National Library of Medicine (U.S.). http://ghr.nlm.nih.gov/condition/angelman-syndrome. Accessed 5/8/12.

Angelman Syndrome Foundation. http://www.angelman.org. Accessed 5/8/12.

"*UBE3A*." 2010. Genetics Home Reference. National Library of Medicine (U.S.). http://ghr.nlm.nih.gov/gene/UBE3A. Accessed 5/8/12.

Ankylosing Spondylitis (AS)

Prevalence Considered part of a group of diseases known as the spondyloarthropathies, which affect about 3.5 to 13 per thousand people; males are more affected than females

Other Names AS; Bekhterev's disease; Bekhterev syndrome; Marie-Strumpell disease; spondylitis ankylopoietica; spondylitis ankylosing; spondyloarthritis ankylopoietica

Norman Cousins, a noted author and journalist, developed a disease of the connective tissue that caused debilitating back pain. Doctors gave him little hope and told him to prepare for the worse. Cousins decided that he was going to defy the condition by laughing. He began by viewing old films of the Marx Brothers and the Three Stooges, and found that 10 minutes of really hard laughing could give him four hours of pain relief. He wrote about his experiences in a book, *The Anatomy of an Illness as Perceived by a Patient*, in which he proclaimed how positive emotions of hope, happiness, and laughing cured him of an illness that only 1 in 500 recover from. The disease that Cousins laughed at was ankylosing spondylitis.

What Is Ankylosing Spondylitis (AS)?

Ankylosing spondylitis (AS) is a form of arthritis that affects the spine. The term comes from two Greek words: *ankylos*, meaning "bent"; and *spondylos*, meaning "vertebrae." Because of the bent condition of the spine, it is sometimes called "bamboo spine." The word arthritis is also from the Greek word *arthr*, meaning "joint," and the suffix *it is*, meaning "inflammation of."

Beginning in adolescence or early adulthood, the individual develops pain and stiffness in the back, which slowly limits movement as the vertebrae of the spine ankylose or fuse together. The symptoms start with stiffness in the lower back in an area of the sacroiliac joints. Gradually, the pain spreads to joints between the vertebrae and can involve other places such as the shoulders, hips, and knees. When complete fusion occurs, the person has complete rigidity of the spine and the bent appearance.

In 40% of the cases, AS is associated with an eye inflammation called acute iritis, which causes pain and sensitivity to light. If the condition goes to the rib cage, it may affect the lungs and breathing. AS is a member of the group of spondyloarthropathies, which is characterized by chronic inflammation of the connective tissues. The condition has a hereditary component.

AS is an old disease. A 5,000-year-old Egyptian mummy who had evidence of Bamboo spine was unearthed. The physician Galen in the second century recognized that this disease was different from other forms of arthritis. In 1858, David Tucker wrote about Leonard Trask, who had the first documented case of AS in the United States. He subtitled the book "wonderful invalid." In the late nineteenth century, three scientists—Vladimir Bekhterev in Russia, Adolph Strumpell in Germany, and Pierre Marie of France—described the condition. Sometimes the names of these scientists are used when referring to AS.

What Are the Genetic Causes of Ankylosing Spondylitis (AS)?

Many researchers believe that AS is a combination of genetic and environmental causes. The exact pattern is not clear. There are several suspect genes, and the environmental factors have not been identified. One of the major suspects is the *HLA-B* gene, which is present in 90% of the cases; however, people with the "major histocompatibility complex, class I, B" gene, or *HLA-B* gene, do not always develop AS. HLB-A plays an important role in the immune system. This gene is part of a group of genes known as the human leukocyte antigen (HLA) complex, which helps the body distinguish between the foreign substances such as bacteria and viruses. A mutation of the *HLA-B* gene called *HLA-B27* appears to increase the risk of AS. HLA-B gene is located on the short arm of chromosome 6: 6p21.3.

Researchers are also looking at several other genes that have been found in patients with AS. The following genes are also associated with AS:

- *ERAP1*, located on the long arm of chromosome 5 at position 15: 5q15
- *ILIA*, a member of the interleukin family of proteins made in the immune cells; located on the long arm of chromosome 2: 2q14
- *IL23R*, located on the short arm of chromosome 1: 1p31.3

These genes are important to the immune system, but the exact role of the genes is still unknown. Unlike the presence of most genes in genetic disorder, having the gene does not mean the disorder will develop. For example, about 80% of the children who inherit the *HLA-B27* gene from a parent with AS do not develop the disorder.

What Is the Treatment for Ankylosing Spondylitis (AS)?

No cure is known for AS, and the person must rely on medications such as anti-inflammatory drugs, physical therapy, and surgery to relieve the symptoms. The disease can range from mild to debilitating with periods of remission. Other complications to the heart, lungs, eyes, colon, and kidneys may also develop.

Further Reading

"Ankylosing Spondylitis." 2010. Genetics Home Reference. National Library of Medicine (U.S.). http://ghr.nlm.nih.gov/condition/ankylosing-spondylitis. Accessed 5/8/12.

Cousins, Norman. 1979, *Anatomy of an Illness as Perceived by a Patient*. New York: W.W. Norton.

Porter, Robert, et al. 2006. *The Merck Manual of Diagnosis and Therapy*, 290. Rahway, NJ: Merck Research Laboratories.

Apert Syndrome

Prevalence 1 in 65,000 to 88,000 newborns

Other Names acrocephalosyndactyly

In 1906, Eugene Apert, a French physician, described nine people who shared some similar characteristics. He called the condition acrocephalosyndactyly. The long name does not seem as overwhelming when one breaks down the word into its Greek roots: *acro*, meaning "high" or "peaked"; *cephal*, meaning "head"; the prefix *syn*, meaning "together"; and *dactyl*, meaning "digits or fingers and toes." Many long words in medicine can be broken into their root words for easier understanding.

What Is Apert Syndrome?

Apert syndrome is characterized by premature fusion of the bones of the skull. This closure causes the skull or head to have an appearance like a mountain peak. Also present are fused fingers and toes, which appear as webbed digits.

This closure of the skull causes many of the symptoms:

- Unusual sunken appearance to the face
- Bulging and wide-set eyes
- A beaked nose
- Later dental problem because of an underdeveloped jaw
- Webbing of the fingers and toes; this varies—sometimes all of the hands and fingers are fused, but others may have fusion of a minimum of three digits
- Hearing loss
- Heavy sweating
- Very oily skin with severe acne
- Missing hair in the eyebrows
- Ear infections
- Cleft palate

The origin of the condition is in the embryological development of the fetus. At certain stages of development, certain pathways of proteins give signals when to begin the process and when to stop. In the hands and feet are cells that control the stopping process, known as apoptosis. These signals tell certain cells when to stop producing or when to die; therefore, the finger and toes can form separately. In the case of Apert syndrome, cell death does not occur, and the finger and toes remain fused. The cranial bones experience a different pathway of having the facial bones and fetal skull fuse too soon. Apert is one of several conditions in which the skull closes prematurely; this condition is called cranialsynostosis.

What Are the Genetic Causes of Apert Syndrome?

Apert syndrome appears to display an autosomal dominant pattern, and males and females affected equally. Almost all cases result from new mutations in the gene and occur in people with no family history. However, people with Apert can pass the condition on to the next generation.

Apert is caused by mutations in the *FGFR2* gene, officially known as the "fibroblast growth factor receptor 2" gene. Normally, FGFR2 produces a protein called fibroblast growth factor receptor 2, which has many functions. One function is when to signal immature precursor bone cells to become bone. The protein is involved in important process of cell division, regulation of cell growth, forming blood vessels, and embryonic development. If this gene is changed, the signals are altered and cause the overproduction of bone and premature closing of the skull. At least 13 different isoforms of versions of the *FGFR2* protein exist. The *FGFR2* gene belongs to a family of genes called CD molecules. The *FGFR2* gene is inherited in an autosomal recessive pattern and is located on the long arm (q) of chromosome 10 at position 26.

What Is the Treatment for Apert Syndrome?

Plastic surgery may correct the closing of the brain sutures and keep the hard bone from damaging brain development. Aggressive surgery may also separate many fingers and toes. Other conditions may be treated as the symptoms occur.

Further Reading

"Apert Syndrome." 2010. Genetics Home Reference. National Library of Medicine (U.S.). http://ghr.nlm.nih.gov/condition/apert-syndrome. Accessed 5/8/12.

"*FGFR2*." Genetics Home Reference. National Library of Medicine (U.S.). http://ghr.nlm.nih.gov/gene/FGFR2. Accessed 5/8/12.

Teeter's Page. http://www.apert.org. Accessed 5/8/12.

APL. *See* Acute Promyelocytic Leukemia

Arts Syndrome

Prevalence Rare; only a few families with this disorder have been identified

Other Names ataxia-deafness-optic atrophy, lethal; ataxia; fatal X-linked, with deafness and loss of vision

In the Netherlands, physicians noted that several boys from a certain family lacked coordination of movement and were very weak. These boys were profoundly deaf and had vision problems. They usually died early. Likewise, in Australia, doctors noted a family with similar symptoms: weakness, loss of motor control, deafness, vision loss, and intellectual disability. Twelve of the 15 boys from these two families died before the age of six.

Over a period of years, these two families provided blood samples for biochemical and molecular analysis. Using these samples in a technique called sequence analysis, the Dutch scientist W. F. Arts and other researchers were able to pinpoint a new and rare syndrome. The name of the new disorder bore the name of Dr. Arts: Arts syndrome.

What Is Arts Syndrome?

Arts syndrome is a disorder that affects the nervous and immune systems in males. The condition is congenital, from the Greek words *con*, meaning "with," and *gen*, meaning "born"; thus a congenital condition is one that is present at birth. Boys with this disorder have profound hearing loss and are completely or almost completely deaf because of an abnormal inner ear. Other obvious symptoms include: lack of muscle tone called hypotonia, lack of muscle coordination or ataxia, developmental delay, and intellectual disability. The children also have an increased risk for infection, especially involving the respiratory system.

What Is the Genetic Cause of Arts Syndrome?

Arts syndrome is an X-linked disorder, meaning that it is located on the X chromosome. Males have only one copy of the X chromosome, so if the mother is a carrier of the mutated or changed gene, the male son will inherit the condition. Females that are carriers may experience mild symptoms or no symptoms at all.

The condition has been traced to mutations in the *PPRS1* gene, officially called the "phosphoribosylpyrophosphate synthetase 1" gene. Normally, the *PRPS1* gene provides instructions for making an enzyme with a long name: phosphoribosylpyrophosphate synthetase 1, or PRPP synthetase 1. The enzyme produces purines and pyrimidines, the building blocks of DNA. In addition, purines and pyrimidines can be made from smaller molecules that have been recycled in the breakdown of DNA and RNA. This process is the called the salvage pathway. PPRS synthetase 1 and PRPP are involved in making new proteins and are likewise are essential for the salvage pathway.

Mutations in the *PRPS1* involve the replacement of only one building block or amino acid. This one substitution creates an unstable enzyme that is nonfunctional. Thus, both the manufacture of pyrimidines, purines, and the salvage pathway are affected. The result is the impairment of the nervous and immune systems that are characteristic of Arts syndrome. *PRPS1* is inherited in a recessive pattern and is located on the long arm (q) of chromosome X at position 22-q24.

What Is the Treatment for Arts Syndrome?

Because Arts syndrome is a serious genetic disorder, there is no treatment. However, certain losses as that of hearing may be addressed with cochlear implants. One of the major preventions is for the individual with Arts to have routine immunizations and protect against infections.

Further Reading

Arjan, P. M.; J. Duley; and J. Christodoulou. 2008. "Arts Syndrome." *GeneReviews*. http://www.ncbi.nlm.nih.gov/books/NBK2591. Accessed 5/8/12.

Arts, W. F.; M. C. Loonen; R. C. Sengers; et al. 1993. "X-Linked Ataxia, Weakness, and Loss of Vision in Early Childhood with a Fatal Course." *Ann. Neurol.* 33: 535–539.

de Brouwer, A. P.; K. L. William; J. A. Duley; et al. 2007. "Arts Syndrome Is Caused by Loss-of-Function Mutations in PRPS1." *Am J. Hum Genet.* 81: 507–518.

Asphyxiating Thoracic Dystrophy. *See* Jeune Syndrome

Autism/Autism Spectrum Disorders

Prevalence Jumped from 1 in 2,000 about 20 years ago to present prevalence of 1 in 166

Other Names Classical autism; Asperger syndrome, pervasive developmental disorder not otherwise specified (PDD-NOS); high-performing autism

History is rife with anecdotes of encounters with people with very different behaviors. According to his writings, Martin Luther, the fifteenth-century theologian, may have encountered an autistic boy. He tells of meeting a 12-year-old boy who exhibited very strange behavior. He thought that the boy was a soulless mass possessed by the devil. Sometimes such children were referred to as wild ones. Stories of the wild Boy of Aveyron, a province in France, described a feral child who showed many of the signs of autism; a medical student tried to help him improve his behavior and communication. The earliest documented case of autism occurred in 1747 when Hugh Blair of Borgue appeared in a court trial. Because of his strange behavior, his brother-in-law sought to annul Blair's marriage to his sister in order to gain his inheritance.

In 1910, the Swiss psychiatrist Eugen Bleuler first used the word "autism," from the Greek word *autos*, meaning "self." The person with autism was observed to be self-absorbed and lacking communications skills. Another important name in the history of this disorder was Hans Asperger, from Vienna, who used the term "autistic psychopaths" in a 1938 lecture.

What Is Autism?

The single word pictures a monolithic condition in which one diagnosis and one force is at work. In realty, many subtypes of autism exist as well as many diseases in which the child has autistic symptoms. Various degrees of severity exist, so the term autism spectrum or autism spectrum disorder (ASD) sometimes replaces the single word autism. Autism is defined according to the *Diagnostic and Statistic Manual of Mental Disorders-IV-TR* (*DSM-IV-TR*, revised 2000). As of this writing, a new version, *DSM-V*, is being prepared for release in May 2013. The new version, which has been controversial among professional and parents, will have new characteristics for ASD. However, at present, the symptoms as defined in *DSM-IV-TR* are characterized by three widespread symptoms:

- Diverse abnormalities of social interaction
- Communication problems
- Behavioral development that includes very restricted interests and highly repetitive behaviors

ASD has the following five forms:

1. Classical autism: This form is thought of as the silent, mentally disabled, hand-flapping and rocking type. Usually the parent notes the problem early in life. One mother described it: Most children when they are picked up hug and hold on to their mothers; my boy was like a sack of potatoes with no feeling.
2. High-performing autism: This child may be high performing but have odd social interactions, narrowly focused interests, and verbose and pedantic communication styles.
3. Asperger syndrome: Children with Asperger syndrome have normal intelligence and talking communication skills, but they may not be able to interpret nonverbal signals. A common symptom is that the person may focus only on one thing and ignore all other people and objects.
4. Pervasive Developmental Disorder–Not Otherwise Specified (PDD-NOS): This form may be diagnosed when the full criteria for autism or Asperger syndrome are not met.
5. Rett syndrome: Sometimes this is included in ASD spectrum lists because of similar symptoms.

What Are the Genetic Causes of Autism?

Of all the mental disorders, autism is considered the most highly heritable. It exceeds other conditions such as schizophrenia and bipolar in heritability studies with both twin and family. Scientists believe that genetic factors are involved in ASDs, and several genetic disorders have autistic characteristics in addition to their specific symptoms.

About 10% to 15% of cases of ASD have an identifiable Mendelian or single-gene inheritance pattern. Some may also have a chromosome abnormality, or other genetic condition. Thus, the genetics of autism is very complex, and the research continues.

Several scientists are also investigating environmental causes that may contribute to autism. These include viruses, certain foods, and many chemical compounds. The study that blamed autism on the preservative in the MMR vaccine has been shown to be an elaborate fraud.

For classic autism, researchers are probably not looking at just one gene. As many as 12 or more different genes may be involved in various degrees. Following are some of the chromosomes and genes that are highly suspect for ASD:

- Chromosome 2: Chromosome 2 has a group of genes called "homeobox" or *HOX* genes that control growth and development. Expression of these genes is essential to creating the brain stem and cerebellum, two areas that appear to be disrupted in children with ASD. Of special interest is the 2q37 deletion syndrome. This syndrome is caused by a deletion of genetic material from the long arm of chromosome 2 at positions 37. Loss of the multiple genes in this region causes a syndrome in which about 25% of the people have autism.
- Chromosome 7: A region on chromosome 7 has a strong link with autism. Certain areas on chromosome 7 control development of speech and language disorders, one of the several symptoms of ASD. Scientists have called the regions *AUTS1*, which is very likely associated with autism.
- Chromosome 13: Chromosome 13 has been linked with 35% of families with autism in one study. Efforts are continuing to replicate this study in families with autism.
- Chromosome 15: A part of this chromosome may play a part in autism. This chromosome carries two known single genes that, when disrupted, causes Angelman syndrome and Prader-Willi syndrome. Children with each of these conditions have the behaviors of autism.
- Chromosome 16: This chromosome has genes that are vital to many functions. If the genes on this chromosome are disrupted, many problems develop. One known single-gene disorder is tuberous sclerosis, a condition that has the symptoms of autism, including seizures. Another condition has been pinpointed to a deletion of a small part of this chromosome. The deletion is located on the short arm (p) of chromosome 16 at position 11.2. Children with 16p11.2 deletion have serious developmental delay and intellectual disability and most of the features of autism spectrum disorder.
- Chromosome 17: One study of male members in 500 families who were diagnosed with autism found missing or disrupted genes on this chromosome.

Scientists already know that galactosemia, a metabolic disorder, can result in ASD symptoms. Chromosome 17 also carries the gene for the serotonin transporter, which allows nerve cells to respond to serotonin, a neurotransmitter. Disruptions in serotonin is known to cause obsessive-compulsive disorder (OCD), which is marked with recurrent and repetitive patterns of behavior, a hallmark of autism.

- The X chromosome: Indications are that there is a possibility of the X or sex chromosome functions in autism because more males than females are diagnosed with autism. Scientists have already pinpointed Fragile X syndrome and Rett syndrome as related to autism. Both of these disorders result from mutations in genes on the X chromosome.

There are some specific genes that scientists are looking at by focusing studies on hot spots of the chromosome. Following are some of the most promising genes:

- *HOXA1* gene: This gene is a homeobox gene that plays a critical role in brain development, the nerves of the head, the ear, and the bones of the head and neck. In embryonic development, *HOXA1* is very active between the 20th and 24th days after conception. Problems of development during this time period may lead to autism. In one study, 40% of people with autism had a mutation in the *HOXA1* gene.
- Reelin gene (*RELN*) on chromosome 7: Normally, RELN instructs for a protein that plays a critical role in cells of the nervous system. Mutations in *RELN* disrupt the normal function and cause cells of the nervous system not to talk to one another. People with autism have lower levels of the reelin protein.
- Gamma-amino-butyric acid (GABA) pathway genes: The compound GABA is a neurotransmitter, which means it assists the nerve cells in communicating with each other and the transmission of electrical impulses. Normally, early in embryonic growth, GABA genes establish the pathway for nerve communication. However, if there is a disruption in the pathway, the person may have low levels of GABA and show the symptoms of autism and also epilepsy.
- Serotonin transporter gene on chromosome 17: Serotonin is another neurotransmitter. The serotonin transmitter lets nerve cells collect this important compound and use it for cell communications. Disruption in the gene may cause a higher serotonin level and cause the symptoms of depression, alcoholism, obsessive-compulsive disorders, and autism.

According to the Genetics Home Reference published by the National Institutes of Health, people with several conditions exhibit autism spectrum disorders, in addition to specific symptoms of the disease. Following are some of the syndromes and the genes related to them:

- Rett syndrome and related genes *CDKL5* and *MECP2*
- Timothy syndrome and related gene *CACNA1C*

- Cornelia de Lange syndrome and related genes *NIPBL*, *SMC1A*, *SMC3*
- Fragile X syndrome and related gene *FMR1*
- Schindler disease and related gene *NAGA*
- Moebius syndrome
- Angelman syndrome and related chromosome 15 and genes *OCA2*, *UBE3A*
- Smith-Lemli-Opitz syndrome
- Tuberous sclerosis complex and related genes *TSC1* and *TSC2*
- Wilms tumor, aniridia, genitourinary anomalies, mental retardation syndrome and related chromosome 11, and genes *PAX6* and *WT1*
- Guanidinoacetate methyltransferase deficiency and gene *GAMT*

These conditions have separate entries in this encyclopedia.

What Is the Treatment for Autism?

Many children with autism can be helped if interventions are begun early. There is still a social stigma to people with autism because of their communication and other disabilities, and many parents have difficulty dealing with the reality of the disorder. Support groups on the Internet have helped children and families connect with others who have similar problems.

Autism's "First Boy" in Later Years

Donald T. was born in September 1933 to a very wealthy family in Forest, Mississippi. His parents noted that he was different from the very beginning. Instead of being interested in people, he enjoyed things. He could sit for hours twirling pans or playing with objects. He talked little. When he was three, his parents put him in an institution, the typical place for a person with indicated mental disorders. But his parents were not happy with this arrangement and took him back home in 1938.

His parents decided to take him to Baltimore to Dr. Leo Kanner, a psychologist who was interested in children with disorders. He was studying a characteristic that had been described using the word "autistic." At this time it was not diagnosed as a condition, but it was simply an adjective describing people who did not communicate and had repetitive movements. Dr. Kanner published an article in 1943 in which he had studied several children. Donald T. was listed as number 1; he became the first person actually described with the condition "autism." Dr. Kanner dubbed the condition as extremely rare, but he was amazed at Donald's fascination with numbers, music, and things.

Donald could compute large numbers in his head. For example, "What was 86 x 23?" Without blinking, he would answer "2001." In 1957, Donald went to Millsaps College and took French and music. He was a member of an *a capella* choir. It was said that the director did not have to have a pitch pipe because Don would always give the perfect note. As Don got older, he developed a routine in life that involved specific repetitive patterns that included the game of golf. His parents died, but he still lives in their house, which is now in some disrepair.

Fortunately, Don's family had the money to provide for his future and care with trusts. Today with over 500,000 children diagnosed with autism, people may wonder what will happen to them when they become adults. Children with autism will become adults with autism. With education, many of them will learn to contend. Children with autism qualify under the Americans with Disabilities Act and for educational interventions under Individuals with Disabilities Act.

Further Reading

"Autism Research at the NICHD." Eunice Kennedy Shriver National Institutes of Child Health and Development. http://www.nichd.nih.gov/autism. Accessed 2/10/12.

Dana Foundation. 2010. *Brain in the News*. October. 18(8).

Miles, Judith H., MD, PhD; Rebecca B McCathren, PhD; Janine Stichter, PhD; and Marwan Shinawi, MD. "Autism Spectrum Disorders." *GeneReviews*. http://www.ncbi.nlm.nih.gov/books/NBK1442/. Accessed 2/10/12.

Tanguay, Peter E., MD. "Autism Spectrum Disorders." 2010. *Dulcan's Textbook of Child and Adolescent Psychiatry* (Psychiatry Online). http://www.psychiatryonline.com/content.aspx?aID=455994&searchStr=autism+spectrum+disorder#455994 Accessed 5/8/12.

B

Bardet-Biedl Syndrome

Prevalence In North America and Europe, 1 in 140,000 to 160,000 births; more common on island of Newfoundland, with 1 in 17,000 births; among the Bedouin population of Kuwait, 1 in 13,500

Other Names BBS; Laurence-Moon-Bardet-Biedl syndrome; Laurence-Moon-Biedl syndrome; LMBBS; LMS

Practicing in a bleak South London Hospital in 1866, two physicians, Laurence and Moon, observed that four members of the same family had eye problems, difficulty walking, and developmental delays. Later in the 1920s, Georges Bardet in France and Arthur Biedl in Germany independently published papers describing children with extra fingers and toes, childhood obesity, and the same symptoms that Laurence and Moon had described. The name Laurence-Moon-Bardet-Biedl syndrome was applied to this condition. Now it has been shortened to Bardet-Biedl syndrome.

What Is Bardet-Biedl Syndrome?

Bardet-Biedl disorder is complex and difficult to diagnose because so many different parts of the body may be affected. A wide range of symptoms may vary with the individual and among members of the same family.

However, two distinct signs indicate Bardet-Biedl: vision loss and obesity. The loss of vision occurs with the deterioration of the retina, the light-sensing screen at the back of the eye. As the child approaches school age, difficulty with night vision becomes apparent, and blind spots develop in the peripheral vision. The spots begin to merge, and by adolescence of early adulthood, the individual becomes legally blind. A second symptom is obesity and abnormal weight gain. This weight gain then leads to complications of type 2 diabetes, high blood pressure, and high cholesterol levels.

A number of secondary characteristics may present as part of this disorder. Following are some of these features:

- Presence of extra fingers and/or toes, a condition known as polydactyly
- Impaired or delayed speech
- Delayed walking, standing, and other motor skills
- Other eye difficulties, including strabismus or crossed-eyes, cataracts, and astigmatism
- Poor coordination and clumsiness
- Behavioral problems, inappropriate outbursts, and emotional immaturity
- Loss of sense of smell
- Distinctive facial features and dental problems

What Are the Genetic Causes of Bardet-Biedl Syndrome?

Just as the symptoms of Bardet-Biedl are complex, the genetics of the syndrome is also quite complicated. The condition appears to be inherited in an autosomal recessive pattern, meaning that each parent must carry the gene although neither parent shows signs of the condition.

Mutations in a group of genes, called the *BBS* genes, cause Bardet-Biedl syndrome. Normally, the *BBS* gene encodes for the BBS proteins, which are located in the basal body and cilia of the cell. Cilia are tiny microscopic, fingerlike projections that protrude from the surface of several types of cells. Cilia are located especially in the epithelial lining of the bronchi but are also part of the sensory system of sight, hearing, and smell. Genetic mutations cause these created proteins to become ciliopathic, meaning that an altered protein attacks the cilia, leading to the many problems and numerous symptoms. Researchers generally believe that the disturbance of the cilia is responsible for most of the features of Bardet-Biedl.

Several mutations in the *BBS* gene are known. About 25% of the cases are related to the *BBS1* gene, officially known as the "Bardet-Biedl syndrome 1" gene. There are over 30 mutations in the *BBS1* gene, most of which involve replacement of only one amino acid: methionine with arginine at position 390. *BBS1* is located on the long arm (q) of chromosome 11 at position 13.1.

Another 20% of cases are related to *BBS10*, officially known as the "Bardet-Biedl syndrome 10" gene. More than 35 mutations of this gene exist, with mutations either adding or deleting genetic material. Location of *BBS10* is on the long arm (q) of chromosome 12 position 21.2.

A very small percentage of cases have been traced to a number of other genes: ARL6, BBS12, BBS2, BBS4, BBS5, BBS7, BBS9, CEP290, MKKA MKS1, TRIM32, TTC8. The remaining 25% of cases are of unknown origin.

What Is the Treatment for Bardet-Biedl Syndrome?

Because this is a genetic condition that affects the cilia, several areas of research are being done to find drugs that may be effective. The round worm model *C. elegans* enabled researches to know more about the BBS proteins and their involvement in movement of the cells. The model also enabled scientists to study the BBS proteins and determined their role in cell transport. Treatment for BBS is symptomatic and varies with the severity of the case. There are several interventions for the visual disorders.

Further Reading

"Bardet-Biedl Syndrome." 2011. *GeneReviews*. http://www.ncbi.nlm.nih.gov/books/NBK1363. Accessed 2/1/12.

"Bardet-Biedl Syndrome." 2010. Genetics Home Reference. National Library of Medicine (U.S.). http://ghr.nlm.nih.gov/condition/bardet-biedl-syndrome. Accessed 5/7/12.

Ross, Allison, et al. 2008. "The Clinical, Molecular, and Functional Genetics of Bardet-Biedl." In *Genetics of Obesity Syndromes*. New York: Oxford University Press.

Bassen-Kornzweig Syndrome

Prevalence Very, very rare; only about 100 cases known worldwide

Other Names abetalipoproteinemia; abetalipoproteinemia neuropathy; acanthocytosis, Apolipoprotein B deficiency; Betalipoprotein Deficiency Disease; congenital betalipoprotein deficiency syndrome; familial hypobetalipoproteinemia; microsomal triglyceride transfer protein deficiency disease

The Smith family was so happy when their healthy baby Josh was born, but within a few months after taking him home, they began to think something was wrong. First, he did not gain weight normally and had usual bouts of diarrhea with pale, frothy, very foul-smelling stools. But when large chunks of blood and fat were found in the stool, they acted promptly to find the reason. After many visits to physicians and specialists, Josh was diagnosed with Bassen-Kornzweig syndrome, also known as abetalipoproteinemia.

What Is Bassen-Kornzweig Syndrome?

Bassen-Kornzweig syndrome is an inherited condition in which the person has a problem using fat in the body. Fats are essential for healthy nerves, muscles, and digestion, but the fat does not get around the body without help. In order for fats to get into the bloodstream, they must attach to special proteins called lipoproteins. People who have this condition cannot make a group of lipoproteins called betalipoproteins, and thus the ability to absorb dietary fat as well as the fat-soluble vitamins of A, D, E, and K is reduced.

The symptoms that Josh experienced are noted in the first months because pancreatic lipase, which also serves a similar function, is not present. If not diagnosed, the child may suffer serious impairment to the nervous system, including poor muscle coordination, difficulty with balance, and degeneration of the retina of the eye. Many signs of abetalipoproteinemia result from the severe vitamin deficiency because the fat-soluble vitamins are not moving though the system.

The stomach problems and the fat in the stool give the first clues. Blood tests will show abnormal red blood cells and reveal very low levels of cholesterol and triglycerides in the blood. An eye exam may show inflammation in the back of the eye.

What Is the Genetic Cause of Bassen-Kornzweig Syndrome?

Abetalipoproteinemia is inherited in an autosomal recessive pattern, which means both parents must carry the faulty gene for the disorder. The condition appears to affect boys more often than girls—about 70% of the time.

Mutations in the *MTTP* gene, officially known as the "microsomal triglyceride transfer protein" gene, are responsible for this condition. Normally, *MTTP* provides information for making a protein called microsomal triglyceride transfer protein. This MTTP protein is responsible for producing beta-lipoproteins. Beta-lipoproteins are essential for creation of molecules called chylomicrons, which are formed when dietary fats and cholesterol are absorbed from the intestines. Also, these chylomicrons are necessary for the absorption of the fat-soluble vitamins.

Mutations in the genes interrupt the process. More than 30 mutations have been found in the *MTTP* gene. One mutation is especially common in people of Ashkenazi Jewish descent of eastern and central Europe. This lack of lipoprotein causes serious nutritional and neurological problems. The *MTTP* gene is inherited in an autosomal recessive pattern and is located on the long arm (q) of chromosome 4 at position 24.

What Is the Treatment for Bassen-Kornzweig Syndrome?

Treatment for this condition includes a rigorous diet, which includes massive amounts of vitamin E, the protein that helps the body produce lipoproteins. Other symptoms such as muscle weakness and movement are treated with physical therapy.

Further Reading

"Abetalipoproteinemia." 2010. Genetics Home Reference. National Library of Medicine (U.S.) http://ghr.nlm.nih.gov/condition/abetalipoproteinemia. Accessed 5/7/12.

"About Abetalipoproteinemia." About.com: Rare Diseases. http://rarediseases.about.com/cs/abetalipoproteinem/a/072601.htm. Accessed 5/7/12.

"MTTP." 2010. Genetics Home Reference. National Library of Medicine (U.S.). http://ghr.nlm.nih.gov/gene/MTTP. Accessed 5/7/12.

Batten Disease

Prevalence Affects 2 to 4 per 100,000 newborns in the United States; condition has been reported worldwide

Other Names Batten-Mayou disease; Batten Spielmeyer-Vogt disease; classical juvenile NCL; CLN3-related neuronal ceroid; Juvenile cerebroretinal degeneration; Spielmeyer-Vogt disease B

Several physicians from different times and decades began to notice a juvenile form of mental deterioration beginning with vision. Eventually the child becomes bedridden and dies. In 1826, Norwegian doctor Dr. Christian Stengel described a family that had four affected children with mental disturbances and vision loss. Although it is often seen in Jewish families, it may occur in all ethnic groups. In England the Batten brothers, Rayner and Frederick, published papers in 1903 and 1904 describing children with these symptoms. Although several other doctors were publishing and investigating the condition, the name of the Batten brothers stuck.

What Is Batten Disease?

Children with Batten diseases have several symptoms—all of them bad and eventually fatal. Usually first noticed is the loss of vision, beginning between 4 and 8 years of age. Then recurrent seizures occur between ages 5 and 18. At the same time the personality of the child changes, and he or she may show several neurological signs: dementia, clumsiness, difficulty sleeping, speech difficulties, and stumbling. Eventually the child loses ability to walk. Most children with Batten disease do not live past their teens or early twenties.

Batten disease is a member of a group of disorders called neuronal ceroid lipofuscinosis or NCLs. This group of conditions is characterized by buildup of pigments called lipofuscins in body tissues. The name neuronal "ceroid lipofuscinosis" puts together three concepts: "Neuronal," meaning related to the neurons or nerve cells; "ceroid," meaning a fatty pigment in some tissues; and "lipofuscin," meaning the name of a brown fatty pigment. When the suffix "osis" is used on a word, it means "condition of." Therefore, the long name means a buildup of the brown fatty pigment lipofuscin in the nerve cells. Diagnosis of Batten disease includes viewing the presence of these deposits in skin samples under the electron microscope.

What Is the Genetic Cause of Batten Disease?

This rare condition is inherited in an autosomal recessive pattern, which means one gene must come from each parent. The parents do not show symptoms of the disorder. Because it is on an autosome, the condition may affect both boys and girls. The gene connected to this disease is called *CLN3*. Normally, the *CLN3* gene programs for making a protein that appears to play a critical role in the survival of nerve cells or neuron in the brain. If mutations are present in the gene, the normal function of the lysosomes in cell is disturbed. Lysosomes are structures within the cell that break down toxic molecules that might build up. In Batten disease, the brown lipopigment lipofuscin builds up, damaging the neuron and causing it to die. Over time the mutations cause the symptoms associated with Batten disease.

The *CLN3* gene has been related to about 1,000 DNA base pair deletions. With a critical part of the gene missing, it becomes impossible for it to program the needed protein properly. *CLN3* has more than 40 mutations and is located on the short arm (p) of chromosome 16 at position 12.1.

What Is the Treatment for Batten Disease?

Two cutting-edge therapies have been used in Batten disease. In a gene therapy trial at Cornell University, researchers used a gene transfer vector to deliver an experimental drug to the brain of a child with Batten disease. As of 2008, the procedure was shown to be safe, and the disease progression was slowed. In another cutting-edge-experiment Phase I trial in 2006, physicians at Oregon Health and Science Center transplanted fetal stem cells into the brain of a six-year-old child with Batten disease who could not walk or talk. Although both of these Phase I trials were conducted to establish safety, the two efforts did show that potential for treatment of this genetic disorder may be possible. The tests will now progress to Phase II and III.

Further Reading

Klein, Andrew. 2008. "Gene Therapy Trial Offers New Hope for Batten Disease, a Fatal Neurological Disease in Children." *Cornell Chronicle*, May 30. http://www.news.cornell.edu/stories/May08/wcmc.crystal.batten.html. Accessed 5/7/12.

Mole, S., and R. Williams. 2010. "Neuronal Ceroid-Lipofuscinoses." *GeneReviews*. http://www.ncbi.nlm.nih.gov/books/NBK1428. Accessed 5/7/12.

"A Stem Cell First at OHSU." 2006. *Portland Tribune*, November 24. http://www.portlandtribune.com/news/story.php?story_id=116425356905230400. Accessed 5/7/12.

Beare-Stevenson Cutis Gyrata Syndrome

Prevalence Rare; fewer than 20 people with this condition have been reported worldwide

Other Names Cutis gyrate syndrome of Beare and Stevenson; cutis gyrate syndrome of Beare-Stevenson

In 2002, a case was reported in a pediatric medical journal of a baby girl born with eyes that were turned down, red growths of skin tags on the forehead, and an umbilical stump. She soon developed hydrocephalus, or fluid on the brain, as a result of the premature closing of the skull. When her genetic profile was sent for analysis, it confirmed that the child had a very rare condition—Beare-Stevenson cutis gyrate syndrome.

What Is Beare-Stevenson Cutis Gyrata Syndrome?

Both skin abnormalities and fusion of the skull characterize Beare-Stevenson cutis gyrata. The name itself is overwhelming. John Martin Beare was a specialist in pediatric dermatology in England. Drs. Beare and Stevenson had the opportunity to study and pinpoint a new syndrome, which now bears their names. The term "cutis gyrata" comes from the Latin words *cutis*, meaning "skin," and *gyrata*, meaning "folded or convoluted." Thus, the skin abnormality "cutis gyrate" is one of the hallmarks of this disorder. At birth the skin has deep furrows and wrinkles, especially on the face, ears, palms of the hand, and soles of the feet. Also on the skin are velvety dark patches called acanthosis nigricans located primarily on the hands, on the feet, and in the genital region.

The cloverleaf shape of the head and face is caused by the premature closing of the skull. There are eight genetic conditions that have the premature skull closing, or craniosynostosis. The word "craniosynostosis" is comprised of four Greek root words: *cranio*, meaning "skull"; *syn*, meaning "together"; *ost*, meaning "bone"; and *osis*, meaning "condition of." Thus, the word means a condition in which the bones of the skull close together. Many of the facial features of the disorder result from this premature closing of the bones. The head is unable to grow affecting eyes, ears, jaw, and brain development. All children will be mentally retarded.

Some other signs of the condition include blockage of the nasal passages, abnormal genitals, abnormal anus, and overgrowth of the umbilicus or belly button so that it looks like a stump. Medical complications of this disorder are life-threatening.

What Is the Genetic Basis for Beare-Stevenson Cutis Gyrata Syndrome?

This condition is inherited in an autosomal dominant pattern, meaning that one gene in each cell can cause the disorder. All cases have basically been traced to new mutations born to parents who have no history or symptoms of the disease.

Beare-Stevenson cutis gyrata is caused by changes in the *FGFR2* gene, officially known as the "fibroblast growth factor receptor 2" gene. Normally, this gene produces a protein called fibroblast growth factor receptor 2, which plays a critical role in cell signals. The protein tells the cells when to respond to the environment and when to divide or mature. Mutations in the *FGFR2* gene change the protein, interfering with proper skin and skeletal development. The *FGFR2* gene is located on the long arm (q) of chromosome 10 at position 26: 10q26.

What Is the Treatment for Beare-Stevenson Cutis Gyrata Syndrome?

A multidisciplinary approach to management of this condition is necessary. The specialists include plastic surgeons, neurosurgeons, and otolaryngologists (ear, nose, and throat) as well as dentists, speech pathologists, developmental pediatricians, social workers, and medical geneticists. Prevention of secondary complications is essential.

Further Reading

Akai, T., et al. 2002. "A Case of Beare-Stevenson Cutis Gyrata Syndrome Confirmed by Mutation Analysis of Fibroblast Growth Factor Receptor 2 Gene." *Pediatric Neurosurgery*. August. 37(2): 97–99.

"Beare-Stevenson Cutis Gyrata Syndrome." 2010. Genetics Home Reference. National Library of Medicine (U.S.). http://ghr.nlm.nih.gov/condition/beare-stevenson-cutis-gyrata-syndrome. Accessed 5/7/12.

Robin, N. H.; M. J. Falk; and C. R. Haldeman-Englert. 2010. "FGFR-Related Craniosynostosis Syndromes." *GeneReviews*. http://www.ncbi.nlm.nih.gov/books/NBK1455. Accessed 5/7/12.

Beckwith-Wiedemann Syndrome (BWS)

Prevalence 1 in 12,000 newborns worldwide; some people with mild or moderate symptoms never diagnosed

Other Names BWS; EMG syndrome; Exomphalos-Macroglossia-Giantism Syndrome

A condition with the main symptoms of gigantism, macroglossia (large tongue) and a hernia in the abdominal region called an exomphalocele was described in a German family in 1964. Dr. Hans-Rudolf Wiedemann reported the unusual finding at a pediatric meeting and coined the term EMG syndrome, the first letters of exomphalos, macroglossia, and giantism. In 1969, Dr. J. Bruce Beckwith found similar cases in California. Over time, the syndrome was renamed Beckwith-Wiedemann, or BWS. BWS is the most common overgrowth syndrome in infancy.

What Is Beckwith-Wiedemann Syndrome (BWS)?

Beckwith-Wiedemann syndrome (BWS) is an overgrowth condition that has many symptoms and affects many parts of the body. An overgrowth syndrome is one in which the infant is larger than normal at birth and continues to grow very rapidly during childhood. Rapid growth continues until about the age of eight; however, adults with BWS are not unusually tall. Certain body parts may appear abnormally large, a condition known as hemihyperplasia.

The following five features are used to diagnose BWS:

- Birth weight and length are over the 90th percentile for weight and length. This condition is called macrosomia, from the Greek words *macro*, meaning "large," and *some*, meaning "body."
- A very large tongue, which may interfere with breathing, speaking, swallowing, and appearance. This condition is called macroglossia, from the Greek words *macro*, meaning "large," and *gloss*, meaning "tongue."
- Defects in the abdominal wall. The infant may have an opening in the abdominal wall called an omphalocele, an out pooching from the umbilicus or belly button, or other abdominal defects.
- Ear creases or ear pits
- Low blood sugar after birth, called neonatal hypoglycemia

At least two of the five above symptoms indicate BWS; however, other symptoms, such as enlarged heart and kidneys, may be present.

Children with BWS are at a significant risk for cancerous and noncancerous tumors and should receive early screening during childhood. Wilms tumor, a cancer of the kidneys, is a common one of these tumors. About 20% of children with BWS die early in life because of related complications.

What Is the Genetic Cause of Beckwith-Wiedemann Syndrome (BWS)?

The genetics of BWS is quite complicated. It is not like most genetic disorders that can be traced to one genetic mutation, but this condition is caused by a range of defects in the genes. The inheritance pattern is also different. Normally, people get one copy of this chromosome from each parent, and for most genes located on chromosome 11, the genes are active or are turned on. BWS is traced to the short arm of chromosome 11, a region known to contain genes that are imprinted. Imprinting in genes is expressed differently depending of whether the gene came from the father of the mother. In the imprinted gene, only one copy is active and the other is silent.

The cases of BWS are caused therefore in different ways. Most of the cases involve abnormalities in genes on chromosome 11 that undergo genetic imprinting. Half of the cases result from methylation, a process in which small molecules containing methyl groups attach to segments of DNA. Normally, genes of chromosome 11 control for methylation of the specific genes involved in growth. This group of genes is referred to as "imprinting control genes or ICRs." However if abnormal methylation interferes with the proper regulation of these genes, it can lead to the characteristic overgrowth of BWS and other characteristics. Following are the genes involved in ICR: *CDKN1C*, *H19*, *IGF2*, and *KCNQ1OT1*. About 1% of the children with BWS have a chromosomal aberration in chromosome 11 caused by translocation or rearrangement or by abnormal copying in the duplication process.

What Is the Treatment for Beckwith-Wiedemann Syndrome (BWS)?

Because the BWS is a genetic disorder, there is no cure. Many of the symptoms will have to be corrected. Defects in the abdominal wall can be corrected with surgery. For hypoglycemia at birth, it must be corrected with medication or brain damage may occur. If the large tongue is causing breathing or eating problems, a craniofacial surgeon may have to correct. Several other problems can be corrected with surgery.

In general, the prognosis of children with BWS is very good. The overgrowth of tallness usually subsides with age so that as adults they are no taller than average. Although some children may develop cancer, usually it can be treated successfully. Advances in neonatal complications and premature infants in the last 20 years have greatly improved the mortality rate of children with BWS.

Further Reading

"Beckwith-Wiedemann Syndrome." 2010. Genetics Home Reference. National Library of Medicine (U.S.). http://ghr.nlm.nih.gov/condition/beckwith-wiedemann-syndrome. Accessed 5/7/12.

"Beckwith-Wiedemann Syndrome." 2010. Whonamedit? http://www.whonamedit.com/synd.cfm/1198.html. Accessed 5/7/12.

Ferry, Robert J. 2010. "Beckwith-Wiedemann Syndrome." eMedicine Pediatrics. http://emedicine.medscape.com/article/919477-overview. Accessed 5/7/12.

Behçet Disease

Prevalence Common in Mediterranean areas, especially in Turkey where about 420 in 100,000 people are affected; not common in northern Europe or the United States, where it affects fewer than 1 in 100,000 people

Other Names Adamantiades-Behçet disease; Behçet's syndrome; Behçet syndrome; Behçet triple symptom complex; Old Silk Route disease; triple symptom complex; Morbus Behçet

In the 1930s, Hulusi Behçet (pronounced BUH-set), a Turkish dermatologist, began studying a condition that was prevalent in his country and in many surrounding countries. The condition was called the old Silk Road Route disease because it followed that area that was once deemed the Silk Road. This condition has many facets in that it has recurrent ulcers and other conditions. He described the disorder in a medical journal in 1937. The disorder was named after the doctor—Behçet disease.

What Is Behçet Disease?

Behçet disease is a condition that affects the mucous membranes, the eyes, and other parts of the body. Widespread inflammation of the blood vessels, a condition called vasculitis, presents itself in many areas. Following are the symptoms of Behçet disease:

- Sores in the mouth: These sores, which look like canker sores, are the first sign of the disorder. They are called apthous ulcers, meaning "without saliva," and are very painful. The sores appear on the tongues, lips, and especially inside the cheek.
- Sores on the sex organs: Appearing on the scrotum in men and on the labia in women, these sores are similar to those that appear in the mouth.
- Sores on the skin: These sores are filled with pus and resemble acne. Red tender nodules can appear on the legs, face, neck, and arms.
- Swelling of the eyes, called uveitis: This condition can later cause blindness.
- Pain and swelling of the joints
- Later spread to other parts of the body: It may affect the gastrointestinal tract, the lungs, and muscles. The vascular inflammation along the blood vessels can rupture and be life-threatening.
- Central nervous system: Related symptoms include meningitis, headaches, confusion, personality changes, memory impairment, speech loss, and loss of balance.

The disease affects people in their 20s and 30s, and it may take a long time to develop all the symptoms, which appear to come and go.

What Is the Genetic Cause of Behçet Disease?

Mutations in the *HLA-B* gene, officially known as the "major histocompatibility complex, class I, B" gene, cause Behçet disease. Normally, *HLA-B* provides instructions for a protein that plays an important role in the immune system and is a member of a complex of three basic groups that form the major histocompatibility complex called MHC. *HLA-B* has many variations, which closely work with the immune system to react to a wide range of foreign invaders.

A specific mutation in the *HLA-B* called *HLA-B51* is associated with the risk of Behçet disease. Researchers are not sure of the exact relationship between the gene and the disease and think possibly that other factors, such as bacterial or viral infections, influence the development of the complex disorder. *HLA-B* is located on the short arm (p) of chromosome 6 at position 21.3.

What Is the Treatment for Behçet Disease?

This very complex disorder with lots of symptoms cannot be cured. However, the symptoms can be managed. The prime goals are reducing inflammation and controlling the immune system, which can be done with high-dose corticosteroid therapy. Some biologics, such as infliximab and etanercept, have been helpful with controlling skin and mucus outbreaks. The drug thalidomide has also been used.

Further Reading

American Behçet Disease Association. http://www.behcets.com/site/pp.asp?c=bhJIJSOCJrH&b=260521. Accessed 1/1/12.

"Behçet Disease." 2010. Genetics Home Reference. National Library of Medicine (U.S.). http://ghr.nlm.nih.gov/condition/Behçet -disease. Accessed 1/1/12.

"Behcet's Syndrome." 2012. Medline Plus. National Library of Medicine (U.S.). http://www.nlm.nih.gov/medlineplus/behcetssyndrome.html. Accessed 1/27/12.

Berardinelli-Seip Congenital Lipodystrophy

Prevalence About 1 in 10,000 people worldwide; 1 per million in Norway, 1 per 200,000 in Lebanon, 1 per 500,000 in Portugal.

Other Names Berardinelli-Seip syndrome; Brunzell syndrome; BSCL; congenital generalized lipodystrophy; generalized lipodystrophy; lipodystrophy, congenital generalized; Seip syndrome; total lipodystrophy

In Brazil in 1954, Dr. W. Berardinelli described two children born with very little fatty tissue. The children appeared muscular, but the doctor knew that if fat was missing from the tissues, it must be stored in other parts of the body such as the liver and muscles. Other serious medical problems were also noted. Then in 1959, Dr. M. Seip observed a new group of patients originating in the county of Rogaland, Norway, who had similar symptoms. Over the years, the names of the two men were combined to create Berardinelli-Seip congenital lipodystrophy.

What Is Berardinelli-Seip Congenital Lipodystrophy?

Berardinelli-Seip congenital lipodystrophy is a rare condition associated with some distinct symptoms and serious medical problems. The condition is characterized by a lack of fatty tissue, which then causes the fats to go to the muscles and liver. Signs and symptoms are obvious at birth. The child has a muscular look because of the absence of fatty tissue. The chin and bones around the eyes are prominent, and the hands and feet are large. Both males and females have enlarged external genitalia. On the folds and creases of the skin are very dark patches called acanthosis nigricans.

Blood analysis reveals high levels of fats or triglycerides circulating in the bloodstream and obvious resistance to the hormone insulin. As the child grows into adolescence, this resistance may develop into diabetes mellitus. Abnormal storage of fat in the liver, called hepatic steatosis, damages this organ. Likewise, abnormal accumulation of fat in the heart can lead to heart disorders such as cardiomyopathy and the risk for sudden death.

There are two types of Berardinelli-Seip: type 1 and type 2. The two have similar symptoms but different genetic causes. People with type 1 may develop bone cysts in the long bones of the arms and legs. Type 2 is associated with intellectual disability.

What Are the Genetic Causes of Berardinelli-Seip Congenital Lipodystrophy?

Berardinelli-Seip is an inherited condition and has an autosomal recessive pattern. That means that both parents are carriers of the gene, and both copies of the gene are in each cell. Parents do not have the condition. Changes in the *AGPAT2* and *BSCL2* genes cause the condition. Normally, these two genes produce proteins that play important roles in the storage of fat in fat cells called adipose cells. Mutations disrupt the normal function of storage of fat in the proper cells. People with type 2 have the mutation in *BSCL2* that affects the protein in the brain and testes. This might explain why intellectual disability is associated with type 2. *AGPAT2* is located on the long arm (q) of chromosome 9 at position 34.3. *BSCL2* is located on the long arm of chromosome 11 at position 13.

What Is the Treatment for Berardinelli-Seip Congenital Lipodystrophy?

There is no treatment for the disorder itself. However, the manifestations can be treated. Intake of fat is restricted to between 20% and 30% of total food to maintain normal triglyceride concentration. Diabetes mellitus is managed with insulin control.

Further Reading

Berardinelli, W. 1954. "An Undiagnosed Endocrinometabolic Syndrome: Report of Two Cases." *Journal of Clinical Research in Pediatric Endocrinology*. 14: 193–204. http://www.jcrpe.org/eng/makale/283/35/Full-Text. Accessed 5/7/12.

Van Maldergem, L. 2001. "Berardinelli-Seip Congenital Lipodystrophy." *Orphanet Encyclopedia*. November. http://www.orphan.net/data/patho/GB/uk-berard.pdf. Accessed 5/7/12.

Van Maldergem, L. 2010. "Berardinelli-Seip Congenital Lipodystrophy." *GeneReviews*. http://www.ncbi.nlm.nih.gov/books/NBK1212. Accessed 5/7/12.

Beta Thalassemia

Prevalence Fairly common worldwide; thousands are born with the condition each year, mostly in people from Mediterranean countries, North Africa, Middle East, India, Central Asia, and Southeast Asia

Other Names Cooley's anemia; erythroblastic anemia; Mediterranean anemia; microcytemia, beta type

Following is a case of a young, 22-year-old woman: She describes how she has felt tired all her life and has had heart palpitations, flatulence, difficulty concentrating, and fuzziness. She goes to sleep tired and wakes up tired. She is tired after 12 hours of sleep. She has been diagnosed with beta-thalassemia minor, one of the two conditions of this disorder.

What Is Beta Thalassemia?

The beta thalassemias are a group of syndromes that reduce the production of hemoglobin, the iron-containing substance in the red blood cells that carries oxygen. Alpha thalassemia is another type of disorder prevalent in people of western African descent and presently found in people living in Africa and in the Americas.

People with beta thalassemia usually appear pale, weak, and tired; other serious complications may occur such as blood clots. Two types exist:

- Beta thalassemia major, or Cooley's anemia: Signs appear the first two years of life, and the children develop severe anemia. They do not gain weight and may develop jaundice. The condition may also affect the spleen, liver, heart, and bones. Blood transfusions may be necessary, but these administrations can cause problems with iron buildup if administered for a long period of time.
- Beta thalassemia minor: The condition is much milder and symptoms appear in early childhood or later life. The person may be tired because of the anemic condition. Also, slow growth and bone abnormalities may develop.

The word "thalassemia" comes from two Greek words: *thalassa*, meaning "sea"; and *haema*, meaning "blood." The condition is one of those disorders named for its association with coastal areas around the Mediterranean Sea. Like sickle-cell disease, the trait may have evolved as a protection against malaria.

What Is the Genetic Basis for Beta Thalassemia?

Both beta thalassemias are inherited in an autosomal recessive pattern. In beta thalassemia major, both parents are carriers of the mutated gene, and the mutation is found in all cells. If the person has one mutated gene, he or she may develop beta thalassemia minor.

The *HBB* gene is responsible for thalassemia. Normally, *HBB* gives the instructions for making a protein called beta-globin, an important component in hemoglobin. Hemoglobin is made up of two subunits of beta-globin and two subunits of alpha-globin. Each of the subunits of hemoglobin carries an iron-containing molecule called *heme*, a substance essential for red blood cells to pick up oxygen in the lungs. When oxygen is attached to hemoglobin, it gives blood the bright red color. However, the changes in the *HBB* gene then stop the production of beta-globin. This genetic defect is a missense or nonsense mutation in *HBB* or may be due to deletion in the gene. When these mutations occur, hemoglobin in the blood does not function properly, the red blood cells do not develop normally, and the signs and symptoms of thalassemia occur. *HBB* is located on the short arm (p) of chromosome 11 at 15.5.

What Is the Treatment for Beta Thalassemia?

Iron supplements, iron chelation therapy, and blood transfusions to prevent fatal iron overload are considered. On July 13, 2010, Italian scientists announced a gene therapy treatment for beta thalassemia. The treatment cleared pre-clinical trials and was slated to go into initial trials with humans.

See also Alpha Thalassemia; Newborn Screening: A Special Topic

Further Reading

"Beta Thalassemia." 2010. Genetics Home Reference. National Library of Medicine (U.S.). http://ghr.nlm.nih.gov/condition/beta-thalassemia. Accessed 5/7/12.

"Gene Therapy Breakthrough Heralds Treatment for Beta-Thalassemia." 2010. Science Daily. July 13. http://www.sciencedaily.com/releases/2010/07/100713171559.htm. Accessed 5/7/12.

Takeshita, Kenichi. 2010. "Beta Thalassemia." Medscape. September 27. http://emedicine.medscape.com/article/206490-overview. Accessed 5/7/12.

Bloom Syndrome

Prevalence Very rare disorder in most populations; more common in people of Central and Eastern Europe (Ashkenazi) Jewish background, among whom 1 in 48,000 are affected

Other Names Bloom's syndrome; Bloom-Torre-Machacek syndrome

A number of genetic disorders appear in specific areas or the world or in certain ethnic groups. The Ashkenaz were a group of Jewish people that settled along the Rhine River in Germany during the early Middle Ages. The group sought to maintain their identity as a sect and lived in close-knit conclaves. Through the years, the groups spread throughout Europe and became respected merchants and bankers. However, they still lived in groups, did not marry out of their faith, and kept their Jewish traditions. Presently, the center of Ashkenazi Jews is in the United States.

Because of the intermarriage and close contact with only Jews of like faith, descendants of the Ashkenazi Jews of eastern Europe have a number of genetic disorders that are present in that population. A number of these diseases are very serious and may result in the death of the child. It is estimated that one in four individuals of Ashkenazi descent carries at least one of several really bad genetic conditions. Bloom syndrome is one of those conditions.

What Is Bloom Syndrome?

Bloom syndrome (BSyn) was first described by Dr. David Bloom in 1954. A registry of known diagnosed cases has been kept since that time and now reflects five decades of following the clinical courses of people with the condition. The main features of BSyn are the following:

- Exceptionally small size: Little adipose or fat tissue making the person look wasted; face is striking because of small narrow skull, underdeveloped jaws, but prominent nose and ears.
- Facial rash: Appears as a butterfly pattern after first exposure to sun; color of rash may vary from faint to bright red; rash may be on other parts of the body.
- High-pitched voice.
- Infections: Repeated bouts of ear infections and pneumonia.
- Fertility issues: Men with BSyn are infertile; women may bear children but enter menopause prematurely.
- Intelligence varies: The majority of children with BSyn appear not interested in school and learning requiring abstract thought.
- Many medical complications such as diabetes, chronic obstructive pulmonary disease (COPD), and lower urinary infections in men.

Affected individuals may have a predisposition to cancer. Although persons with BSyn may develop cancer at any age, the average age is about 25 years old.

What Is the Genetic Basis for Bloom Syndrome?

Like so many of the genetic disorders in people of Ashkenazi descent, Bloom syndrome is inherited in an autosomal recessive pattern. Both the father and mother carry a copy of the mutated gene. The parents do not display any of the symptoms of the disease. The carrier frequency is about 1 in 100.

Bloom syndrome is caused by mutations in the *BLM* gene. Normally, this gene gives the instructions for producing the BLM protein, which is a member of the DNA helicase family. DNA helicase, an enzyme, instructs for the two strands of the DNA molecule to unwind so that the DNA may perform the processes of synthesis, transcription, and repair. When the cell prepares to divide, chromosomes are duplicated, a process known as DNA replication. However, if errors are made during the replication process, a mutation may occur in the *BLM* gene, and the BLM protein's DNA helicase activity is not active or is nullified. This leads to chromosome instability and possible breakage. The increase in chromosome breakage and rearrangement leads to the symptoms of BSyn. *BLM* is located on the long arm (q) of chromosome 15 at position 26.1.

What Is the Treatment for Bloom Syndrome?

It is important for people of Ashkenazi Jewish background to have genetic testing not only for Bloom syndrome, but also for other genetic diseases that are prevalent in the population. There is no cure for Bloom syndrome. However, the manifestations of BSyn can be treated. For example, ear infections and pneumonia respond to antibiotics. Diabetes and cancer can be treated, but because some of the strong medicine damages DNA, the dosage and duration of the treatment may be reduced.

Further Reading

Bloom, D. 1954. "Congenital Telangiectatic Erythema Resembling Lupus Erythematosus in Dwarfs; Probably a Syndrome Entry." *AMA Am J Dis Child*. 88: 754–758.

"Bloom Syndrome." 2010. Genetics Home Reference. National Library of Medicine (U.S.). http://ghr.nlm.nih.gov/condition/bloom-syndrome. Accessed 5/7/12.

Sanz, M., and J. German. 2010. "Bloom's Syndrome." http://www.ncbi.nlm.nih.gov/books/NBK1398. Accessed 5/7/12.

Boomerang Dysplasia

Prevalence Exceedingly rare; only 10 persons have been identified with the disorder

Other Names Piepkorn dysplasia

A boomerang is a common tool that is used by the native people of Australia for hunting. The instrument has a special shape that returns to the thrower. It is this shape that inspired physicians to give this unusual name. Looking at the X-rays of persons with the disease, the bones appear in the shape of a boomerang. This led them to call it boomerang dysplasia.

What Is Boomerang Dysplasia?

The term "dysplasia" comes from two Greek words: *dys*, meaning "with difficulty"; and *plas*, meaning "form." A dysplasia is a malformation, and in this condition the bones throughout the body are misshaped and malformed. Pictures of the legs are bowed like the shape of a boomerang. Following are the symptoms of the disorder:

- Severe bowing of the femurs or upper leg bones, giving them a boomerang appearance
- Inward and upward turning feet, called clubfeet
- Dislocated bones in hips, knees, and elbows
- Underdeveloped or missing bones in spine, ribs, and pelvis; an underdeveloped rib cage affects breathing
- Limbs underdeveloped or missing
- Distinct facial appearance; individuals have a broad nose with very small nostrils and an underdeveloped area between the nostrils.
- Sac-like protrusions that may develop in the brain; called an encephalocele, a sac-like protrusion may occur in the abdominal wall that causes organs to protrude through the navel, called an omphalocele

Because of the many physical disorders, many of these children are stillborn or die shortly after birth from respiratory failure.

What Is the Genetic Cause of Boomerang Dysplasia?

Mutations in the *FLNB* gene, officially known as the "filamin B, beta" gene, cause boomerang dysplasia. Normally, *FLNB* instructs for a protein called filamin B. This protein is part of a network of protein fibers that create the cytoskeleton of the cell and give them shape and the ability to move. Filamin B attaches to the protein actin to make the cytoskeleton and is extremely active in the development of the skeleton before birth. It is active in developing the cartilage-forming cells that later lead to the creation of bone.

Two mutations in the *FLNB* gene that cause dysplasia are the result of a change in a single protein building block and in a small deletion in the protein. This abnormal protein disrupts the action of the protein, interfering with the formation of chondrocytes and their change to bone. This disruption leads to the many symptoms of boomerang dysplasia. *FLNB* is inherited in an autosomal dominant pattern and is located on the short arm (p) of chromosome 3 at position 14.3.

What Is the Treatment for Boomerang Dysplasia?

This genetic condition is very serious and has many disorders that must be addressed. If there is a life-threatening condition, such as deformed ribs that compress the lungs, that condition must be managed first. Other conditions must be treated as the symptoms arise. The prognosis for this disorder is not positive.

Further Reading

"Boomerang Dysplasia." 2010. Genetics Home Reference. National Library of Medicine (U.S.). http://ghr.nlm.nih.gov/condition/boomerang-dysplasia. Accessed 1/1/12.

"Boomerang Dysplasia." 2012. Geneva Foundation for Medical Education and Research. http://www.gfmer.ch/genetic_diseases_v2/gendis_detail_list.php?cat3=1202. Accessed 1/1/12.

"FLNB-Related Disorders. 2008. *GeneReviews*. http://www.ncbi.nlm.nih.gov/books/NBK2534. Accessed 1/1/12.

Breast Cancer Genetics: A Special Topic

What Is Breast Cancer?

Breast cancer is a condition in which certain cells begin dividing wildly and form a tumor. Breast cancer is complex because it may be in several cells and include a variety of genetic causes. The cells lining the ducts that carry milk to the nipple are the most common places for cancer to appear. This form is called ductal cancer. Other forms begin in the glands that produce milk, called lobular cancer, or in other parts of the breast.

Who Gets Breast Cancer?

After skin cancer, breast cancer is the most commonly diagnosed cancer and is the second leading cause of death after lung cancer. According to the National Cancer Institute, an estimated 209,060 new cases were diagnosed in 2010, and 40,000 deaths will occur from this number. Breast cancer is not just for women; about 2,000 cases are diagnosed in men each year. Recently, a slight decrease of cases has been seen after early reports from the Women's Health Initiative that hormone replacement therapy (HRT) may be a risk.

In addition to genetics, the following risk factors may be at play:

- Age: Cumulative risk of sporadic cases of breast cancer increases with age, with most cancers occurring after age 50. Women who have the genetic susceptibility tend to have breast cancer at an earlier age.
- Reproductive and menstrual history: The risk of breast cancer increases with those beginning the menstrual cycle earlier and those experiencing late menopause.
- Oral contraceptives: Oral contraceptives may be a risk for breast cancer among long-term users.
- Hormone Replacement Therapy (HRT): Data exist from both observation and clinical tests that there is an association between breast cancer and HRT in postmenopausal women.

- Radiation exposure: There appears to be a connection between radiation exposure for treatments and breast cancer if there is a genetic susceptibility.
- Alcohol intake: Some studies have shown a connection between alcohol intake and breast cancer.
- Physical activity and weight gain: Obesity and weight gain are risks for breast cancer, especially in postmenopausal women.
- Smoking: Some studies show smoking is a factor for breast cancer in certain genetic groups.

It's All Done in the Cells

The Normal Cell. Before a cell divides, DNA is replicated. The cell cycle in the process of mitosis goes through five phases:

1. Interphase. This is the non-dividing or resting phase known as G_0 (G for gap), in which the cell carries out its normal metabolic functions. Signals tell the cell when to enter G_1 and make the molecules essential for the replication of DNA. It then enters the S phase to get ready to replicate the DNA. When all the materials are gathered, it enters G_2. Now mitosis M is ready to begin.
2. Prophase: In this phase, the chromosomes condense and appear as two identical chromatids or parts held together by a centromere. Each chromatid has a double-stranded molecule of DNA. When cell division occurs, the centrioles move to opposite poles of the nucleus, and spindles form that become a road or pathway for the chromatids or chromosomes to travel on. The chromatids then move along the spindles to the centrioles at the opposite ends.
3. Metaphase: The centromere of each double-stranded chromosome attaches to the spindle.
4. Anaphase: The centromere divides so that each one of the pair ends up at the spindle.
5. Telophase: The cell physically divides and becomes two daughter cells containing identical genetic material.

The cell enters interphase again, and the cell cycle starts over.

Very important: Progression through this cycle is highly regulated between G_1 and S and during G_2. Under normal conditions, cells do not enter S phase unless they have all the necessary molecules gathered for DNA synthesis. The order of movement is as follows: $G_0 \rightarrow G_1 \rightarrow S \rightarrow G_2 \rightarrow M$. If an error occurs, the cell will not go from G_2 to M.

Specific proteins known as proofreading and repair enzymes provide checkpoints along the cell cycle pathway. The process of apoptosis is also known as "programmed cell death" and is a natural process of cell self-destruction. Things must go right, or the cell dies.

What Can Go Wrong?

Cancers arise when the errors or mutations occur in replicating DNA. Some of the errors result in cell death. Often, however, the cell does not die and in time acquires many errors. The built-in controls have failed. The wayward cell begins to divide furiously and its daughter cells accumulate more quickly than surrounding cells, and a mass of cells group together within the tissue. The cells may form a benign tumor. The dividing cells become increasingly unstable and begin to invade normal tissue. The normal regulatory signals do not control them. Now they are malignant. What started out as a single cell now has daughter cells with great variability.

The malignant cells are survivors. They may have the ability to break away and go through the blood or lymph nodes to distant locations and survive. This condition is called "metastasis." Moving from the single cell to the metastatic condition is called "tumor progression." In order for the mass to live, it must acquire a blood supply; the tumor cells secrete substances called angiogenesis factors, which recruit new blood vessels into the tumor.

Anatomy of the Human Breast

Different cancers develop in the breast depending on the location. Special glands called lobules surround branches of six to nine independent ducts that run systems from the nipple back to the underlying pectoral muscle. Genetics determines the amount of breast tissue. Milk is made in the lobules and then carried through the duct system to the nipples. Most cancers begin in the lining of the duct called the epithelial cells. These cancers are called carcinomas.

The lobules and ducts are embedded in the stroma, the foundation supporting the tissues of the breast. The stroma is made of fat, fibrous tissues, blood vessels, and lymph vessels. The function of the lymph vessels is to drain fluid with waste matter from the breast in a series of filters called lymph nodes. The lymph nodes are also important in the immune system function, but when the cancerous cells invade, they tend to spread through the auxiliary nodes located under the armpits.

What Are the Symptoms of Breast Cancer?

Early detection may escape the individual because there is no pain or noticeable symptoms. As the cancer progresses, the following signs may occur:

- Lump or thickening in or near the breast
- Change in the size or shape of the breast
- Discharge in nipples
- Tenderness
- Retraction or turning inward of nipples
- Skin irritation

- Scaly-appearing skin on breast
- Dimpling of breast

The mammogram is a must procedure for determining cancer in early stages before it gets to later or metastatic stages.

What Are the Genetic Causes of Breast Cancer?

The genetics of breast cancer are complex. Studying the genetics of breast cancer is not looking at a condition like cystic fibrosis caused by a single gene. Some cases are sporadic with no apparent hereditary connection. These cancers are connected to mutations that happen over a person's lifetime. Many environmental conditions as listed in a previous section may be at play here. These cancers are caused by somatic mutations and are not inherited.

About 5% to 10% of breast cancer is traced to families. Much has been said about the mother-daughter connection and tracing breast cancer in the maternal family, but it is just as important to look at the father's family history. Of course, tracing family history may be difficult because of the accuracy and completeness. Small family size, early deaths for other reasons such as accidents, and not knowing relatives may make information difficult to obtain.

Hereditary conditions with particular mutations are common among people of certain geographic or ethnic groups. Among these groups are people of Ashkenazi Jewish descent. The Ashkenazi were a group of Jewish people that settled along the Rhine River in Germany during the early Middle Ages. The group sought to maintain their identity as a sect and lived in close-knit conclaves. Because of intermarriage and close contact with only Jews of like faith, descendants of the Ashkenazi Jews of Eastern European have a number of genetic disorders that are present in that population, and breast cancer appears to be one of these conditions. Presently, the center of Ashkenazi Jews is in the United States. Other regional and ethnic groups who may also have particular mutations associated with breast cancer include people of Norwegian, Icelandic, or Dutch heritage.

Mutations involving breast cancer are found in genes regulating the cell cycle. The two major regulatory classes are oncogenes and tumor suppressor genes. The first group of cancer genes, the oncogenes, is especially related to non-inherited cancer. In normal cells, genes called proto-oncogenes cause the cell to proceed thorough the cell cycle and divide. Many of the proto-oncogenes instruct for proteins such as the epidermal growth factor (EGF) and Her2/neu. Mutations in the proto-oncogenes cause uncontrolled growth. The second group of genes, the tumor suppressor genes, regulates cells and instructs them to stop dividing, or die (apoptosis). The tumor suppressor genes include *BRCA1*, *BRCA2*, and *p53*. Mutations in these genes are associated with hereditary cancer.

People with hereditary breast cancer have some things in common. Normally, it will be seen in families and will occur earlier in life around the age of 40. It will generally be bilateral, occurring in both breasts. The three important genes in this group are *BRCA1*, *BRCA2*, and *TP53*.

BRCA1

BRCA1 is also called "breast cancer 1, early onset gene." Normally, *BRCA1* instructs for a protein that is directly involved in repairing DNA. In the nuclei of normal cells, *BRCA1* interacts with proteins produced by the *RAD51* and *BARD1* genes to mend breaks in the DNA. The breaks can be natural, caused by radiation, or other environmental exposures. *BRCA1* acts to stabilize the cell's genetic information. The more than 1,000 mutations in *BRCA1* are associated with an increase of risk for breast cancer (and other cancers). There are large segments of DNA in the *BRCA1* gene, and these missing parts prevent the proper proteins from repairing DNA or fixing mutations in other genes. As the defects accumulate over time, the abnormal cells multiply rapidly. *BRCA1* is inherited in an autosomal dominant pattern and is located on the long arm (q) of chromosome 17 at position 21.

BRCA2

BRCA2 is also known as "breast cancer 2, early onset gene." Normally, this tumor suppressor gene *BRCA2* instructs for a protein involved directly in repairing damaged DNA. It interacts with other proteins produced by the genes *RAD51* and *PALB2* to mend breaks in the DNA, which may occur when chromosomes exchange material in cell division. In addition, BRCA2 protein may regulate several functions within the cells during cell division. The more than 800 mutations of the *BRCA2* gene are associated with a risk of breast cancer. The mutations may be caused by insertion or deletion or a small number of the amino acids. The genetic changes disrupt the process of repairing DNA, causing the defects to accumulate and the cells to divide rapidly. *BRCA2* is inherited in an autosomal dominant pattern and is located on the long arm (q) of chromosome 13 at position 12.3.

P53 or *TP53*

TP53 is also called "tumor protein p53 gene" Normally the gene *TP53* instructs for the tumor protein p53. This suppressor keeps the cells from growing wildly. Sometimes, scientists refer to this gene as "the guardian of the genome." The protein p53 is located in the nucleus of the cells and binds directly to DNA. When damage occurs, the suppressor works to repair damage and calls other cells to help in the repair. If the damage cannot be repaired, this protein keeps the cells from dividing, allowing them to self-destruct (apoptosis).

Mutations in this gene increase the risk of developing breast cancer. The inherited mutations are thought to account for less than 1% of cases. Changes in somatic mutations or non-inherited version of the *TP53* genes are more common. Just one change in a protein building block can disrupt the function of the p53 protein, causing it not to bind to DNA and regulating cell growth. People with recurring tumors are more likely to have mutations of this gene. *TP53* is inherited in an autosomal dominant pattern and is located on the short arm (p) of chromosome 17 at position 13.1.

Variations or mutations in other genes also increase the risk of developing breast cancer. Investigation in to the role of these genes is ongoing. They are the following:

- *CDH1*
- *PTEN*
- *STK11*
- *AR*
- *ATM*
- *BARD1*
- *BRIP1*
- *CHEK2*
- *DIRAS3*
- *ERBB2*
- *NBN*
- *PALB2*
- *RAD50*
- *RAD51*
- *ERB2*, also called Her2/neu

This long list of genes may interact with other genes to cause cancer. Many of the genes are also related to other cancer conditions and are usually connected with non-inherited conditions that happen over a period of the lifetime.

What Is the Treatment for Breast Cancer?

In this highly researched area, it is most difficult to keep up with new drugs and new treatments. For example, innovative drugs include targeted agents that focus specifically on the tumor, monoclonal antibodies, kinase inhibitors, cancer vaccines, hormone regulators, and novel versions of old chemotherapies. These areas of research are emerging so rapidly that even physicians may have difficulty keeping up. The trend of novel directions is expected to continue over the next decades.

Many of these new strategies use genetics as their basic targets. In the United States, women with metastatic breast cancer have a newly approved synthetic drug called Halaven (eribulin) that attacks mutant dividing cells. There is also new group of drugs called poly (ADPribose) polymerase (PARP) inhibitors that block repair of single-stranded DNA breaks, causing the cell not to replicate and self-destruct.

Knowledge of gene expression may lead to gene therapy experiments. The goal of gene therapy is to replace the defective gene with a normal one. Some experiments are underway to develop gene-based inhibitors of Her2/neu over expression or to block telomerase. However, mass trials of gene therapies that will lead for approval are in the far future.

Further Reading

"Breast Cancer." 2007. Genetics Home Reference. National Library of Medicine (U.S.). http://ghr.nlm.nih.gov/condition/breast-cancer/show/print. Accessed 5/9/12.

"Genetics of Breast and Ovarian Cancer (PDQ)." 2012. National Cancer Institute (U.S.). http://www.cancer.gov/cancertopics/pdq/genetics/breast-and-ovarian/HealthProfessional/page2/. Accessed 1/25/12.

Zimmerman, Barbara, PhD. 2004. *Understanding Breast Cancer Genetics*. Jackson: University of Mississippi Press.

Bruton Agammaglobulinemia

Prevalence About 1 in 200,000

Other Names agammaglobinemia

In 1952, Colonel Ogden Bruton, an army doctor at Walter Reed Hospital, noted that eight-year-old Joseph Hotoner had a history of pneumonia and other serious bacterial lung infections. When he treated the boy with intramuscular injections of immunoglobulins (IG), the boy survived and lived until his early 40s, when he died of chronic pulmonary disease. Bruton was the first to make the connection between severe infections and the absence of immunoglobulins. This condition was one of the first disorders of the immune system to be described. Researchers later found the gene that caused the condition located on the X chromosome. For his work in the field of immunology, he was honored with the name of the disease—Bruton's agammaglobulinemia—and the name of the gene Bruton agammaglobulinemia tyrosine kinase, or *BTK*.

What Is Bruton Agammaglobulinemia?

Bruton agammaglobulinemia is a condition that affects cells in the immune system. The word "agammaglobulinemia" explains the condition when broken into root words: *a* is a prefix meaning "without"; gammaglobulin is a protein essential for the immune system; "emia" comes from *heme*, meaning "blood." Thus agammaglobulinemia is a condition in which blood factors essential for the immune system are not present or are not working effectively.

The immune system controls how well a person reacts to outside invaders and is a complex system necessary for life. When something goes wrong, the person experiences serious difficulties and cannot fight infections. That is what happens with X-linked agammaglobinlinemia, or XLA. These people have few B cells, white blood cells that protect the body against infection and are instrumental in producing antibodies or immunoglobulins. These antibodies attach to foreign invaders such as bacteria or other foreign particles and destroy them. Because people with XLA do not have these disease-fighting mechanisms, they get sick often and are prone to infections.

Usually the male child is healthy the first few months of life because the immune factors acquired from his mother protect him. But children with XLA soon begin to experience serious ear infections or otitis, pink eye or conjunctivitis, sinus infections or sinusitis, chronic diarrhea, and bouts of pneumonia. They also can develop life-threatening bacterial infections, but with treatment, they can improve their quality of life.

What Is the Genetic Cause of Bruton Agammaglobulinemia?

Mutations in the Bruton tyrosine kinase gene, or *BTK* gene, cause XLA. The disorder is passed on by the mutated recessive *BTK* gene carried on the X chromosome. X-linked recessive means that the mother carries the gene, and then because one altered copy is in each cell, the male child will have the condition. In a normal individual, this gene gives the instructions for making bruton tyrosine kinase, an enzyme that is important for the development of B cells and for the proper functioning of the immune system to produce antibodies. Most of the 600 mutations in *BTK* reflect a change only a single base pair in this very large gene. Some people have the entire end of the gene removed and all of a neighboring gene called *TIMM8A*. The missing part causes additional symptoms of deafness-dytonia-optic neurnopathy syndrome (DDOBN). The gene *BTK* is located on the long arm (q) of the X chromosome at position 21.33-q22.

What Is the Treatment for Bruton Agammaglobulinemia?

Because this condition is genetic, there is no cure. However, it is important to treat the symptoms, such as ear infections, sinus infections, or pneumonia, as soon as possible.

Further Reading

Bruton, O. C. 1952. "Agammaglobulinemia." *Pediatrics*. June. (6): 722–728. http://www.pediatrics.aappublications.org/content/102/Supplement_1/213.full.pdf. Accessed 5/8/12.

"Pediatric Bruton Agammaglobulinemia." 2010. Medscape. http://www.emedicine.medscape.com/article/885625-overview. Accessed 5/9/12.

"X-Linked Agammaglobulinemia." 2010. Genetics Home Reference. National Library of Medicine (U.S.). http://ghr.nlm.nih.gov/condition/x-linked-agammaglobulinemia. Accessed 5/9/12.

Burger-Grutz Syndrome

Prevalence Affects about 1 per million people worldwide; more common in some areas of Quebec in Canada

Other Names familial fat-induced hypertriglyceridemia; familial lipoprotein lipase deficiency; familial LPL deficiency; hyperchylomicronemia Type I; Lipase D deficiency; LIPD deficiency; lipoprotein lipase deficiency, familial

Early in the twentieth century, Max Burger, a German physician, noted a family whose infants failed to thrive and had yellow fatty-like deposits under the skin. He speculated this condition was an inherited metabolic disorder. Later, the condition called Burger-Grutz syndrome became more commonly known as familial lipoprotein lipase deficiency.

What Is Burger-Grutz Syndrome?

This metabolic disorder is a condition in which the person is missing an enzyme, lipoprotein lipase, which is an essential protein needed to break down fat molecules in digested food. Therefore, fat particles called chylomicrons builds up in the bloodstream.

The following symptoms and signs characterize this disorder:

- Failure to thrive in infancy—lots of abdominal pain, which may be interpreted as colic
- Yellow, fatty deposits in the skin called xanthomas
- No appetite; nausea and vomiting
- Pain in muscles and bones
- Episodes of pancreatitis
- Enlarged liver and spleen
- Very high triglycerides in the blood
- Jaundice in eyes and skin
- Pale retinas with white-colored blood vessels in retina

The condition appears to run in families, indicating a genetic cause.

What Is the Genetic Cause of Burger-Grutz Syndrome?

Mutations in the "lipoprotein lipase" gene, or *LPL* gene, cause Burger-Grutz syndrome. Normally *LPL* instructs for making the enzyme lipoprotein lipase. Made primarily in the fat or adipose tissue and in muscle, the protein is critical in carrying fats and breaking down the molecules that carry fat called lipoproteins. This very important enzyme removes fate from two types of fat protein: chylomicrons and very low-density lipoproteins, or VLDLs. When one eats, the chylomicrons carry fat from the intestine into the bloodstream, and VLDLs carry fat and cholesterol from the liver to tissues in the body. The enzyme lipoprotein lipase works with other enzymes to break down fat for energy use or for storage as fat.

However, more than 20 mutations in the *LPL* gene keep the enzyme from working properly. The enzyme lipoprotein lipase does not work properly. Lipoproteins are not broken down, and these fatty lipoproteins build up in the bloodstream. The person may then exhibit the signs and symptoms of familial lipoprotein lipase deficiency. The genetic condition is inherited in an autosomal recessive pattern. Each parent must carry a copy of the mutated gene, although neither has signs and symptoms of the condition. *LPL* gene is located on the short arm (p) of chromosome 8 at position 22.

What Is the Treatment for Burger-Grutz Syndrome?

Because this condition is a genetic disorder, there is no prevention or cure. Treatment involves controlling the triglyceride levels and a very low-fat diet. The person should eat no more than 20 grams of fat in the diet. Supplement of fat-soluble vitamins—A, D, E, K—are recommended. If the person will stick to the low-fat diet and other recommended regimens, the prognosis for a positive quality of life is very good.

Further Reading

"Familial Lipoprotein Lipase Deficiency." 2010. Genetics Home Reference. National Library of Medicine (U.S.). http://ghr.nlm.nih.gov/condition/familial-lipoprotein-lipase-deficiency. Accessed 5/8/12.

"Familial Lipoprotein Lipase Deficiency." 2010. Medline Plus. National Library of Medicine (U.S.). http://www.nlm.nih.gov/medlineplus/ency/article/000408.htm. Accessed 5/8/12.

Semenkovich, C. F. 2007. "Disorders of Lipid Metabolism." In L. Goldman and D. Ausiello, Eds., *Cecil Medicine*, 23rd ed. Chap. 217. Philadelphia, PA: Saunders Elsevier. http://www.umm.edu/ency/article/000408.htm. Accessed 5/8/12.

C

CADASIL

Prevalence The prevalence of CADASIL is unknown; estimated prevalence of CADASIL in Scotland is 1 in 50,000; however, this estimate cannot be generalized worldwide

Other Names cerebral arteriopathy with subcortical infracts and leukoencephalopathy; familial vascular leukoencephalopathy; lacuna dementias; multi-infarct dementia

In a 2004 Spanish film, *The Sea Inside*, the main character Julia, an attorney, had an unusual disability involving dementia, psychiatric disturbances, and migraines. The condition that she had was called cerebral autosomal dominant arteriopathy with subcortical infarcts and leukoencephalopathy. The syndrome was described 30 years ago in a Swedish family but the acronym was not given until the 1990s. Because the name is so complicated, it is referred to as CADASIL, an acronym created from the first letters of the name.

What Is CADASIL?

CADASIL is a condition that involves the degeneration of the smooth muscle cells in blood vessels. Smooth muscles line involuntary muscles such as those of the stomach, intestine, and all other organs that are not voluntary muscle or heart (cardiac) muscle. The symptoms start with migraines, and then the person has transient ischemic strokes or TIAs, which begins usually between 40 and 50 years. The prognosis of the disease is a progressive deterioration leading to dementia and depression.

A breakdown of each of the components of CADASIL follows:

- *C—Cerebral*: The word cerebral comes from the Latin term meaning "brain." An unusual accumulation of a substance called Notch 3 is in the cytoplasm of the smooth blood vessels in the brain and in other blood vessels outside the brain.

- *AD—autosomal dominant*: The syndrome is inherited in an autosomal dominant pattern, meaning that the likelihood of inheriting the gene from one parent with the condition is 50%.
- *A—Arteriopathy*: The Greek root word for "disease" is *patho*; thus the syndrome is characterized by a disease of the arteries.
- *SI—Subcortical infarcts*: The cerebral cortex is the thin convoluted area of gray matter that covers the cerebrum; it consists of the cell bodies of neurons arranged in five layers. The subcortical region lies just below this area of the cerebral cortex and is composed of white matter. White matter consists of bundles of myelinated nerve cells, or the axon parts of the nerves. White matter forms the bulk of the deep parts of the brain. The infarct is a failure of blood supply to the brain, generally called a stroke. These recurrent ministrokes damage the myelin or fatty sheaths that cover the neuron.
- *L—Leukoencephalopathies*: This disease belongs to a family known are the leukoencephalopathies, which involve the white matter of the brain. The word "leukoencephalopathy" comes from three Greek root words: *leuko*, meaning "white"; *encephalo*, meaning "brain"; and *path*, meaning "disease." Thus leukoencephalopathy is a disease of the white matter of the brain.

The first clinical symptom of CADASIL usually begins between ages 35 and 55, with attacks of migraine with aura or subcortical transient ischemic attacks or ministrokes. Cognitive and physical impairments are progressive and may include variable degrees of weakness, problems with walking, palsy or shaking, urinary incontinence, depression, and psychosis. The reduced blood supply caused by the abnormality of the smooth muscles gradually destroys the brain cells. CADASIL is not associated with risk factors normally associated with stroke and heart disease, such as high cholesterol or high blood pressure; however, some of the people with the condition may have these risk factors.

What Is the Genetic Cause of CADASIL?

CADASIL is traced to disruptions in the *NOTCH3* gene, officially known as the "Notch homolog 3 (Drosophila)" gene. Normally, this gene gives instructions for producing the Notch3 receptor protein that is important for the function of vascular smooth muscle. Special molecules bind to Notch3 receptors to send signals to the nucleus of the cells. These signals control the action of other genes within the smooth muscle, telling them to turn on. *NOTCH3* was discovered in 1996.

Mutations or changes in the *NOTCH3* gene can disrupt the protein receptor and lead to cell death or apoptosis. Most cases are inherited in an autosomal dominant pattern when one parent has the condition. There are a few instances of new mutations in the gene in families with no history of the condition. *NOTCH3* is located on the short arm (p) of chromosomes 19 at position 13.2-p13.1.

What Is the Treatment for CADASIL?

Although there is no treatment for the condition, certain anti-platelet strategies such as aspirin might help prevent the strokes. Any reasonable prevention may help delay dementia and stroke. Some of the medications used for patients with Alzheimer disease may help delay memory loss. An MRI is able to detect signs of the disease prior to clinical symptoms and may be helpful in developing preventive strategies.

Further Reading

Bousser, M.-G. 1996. "Notch 3 Mutation in CADASIL, a Hereditary Adult-Onset Condition Causing Stroke and Dementia." *Nature*. 383: 707–710. http://www.nature.com/nature/journal/v383/n6602/abs/383707a0.html. Accessed 5/9/12.

"CADASIL (Cerebral Autosomal Dominant Arteriopathy with Subcortical Infarcts and Leukoencephalopathy)." 2010. Medscape. http://emedicine.medscape.com/article/1423170-overview. Accessed 5/9/12.

"Cerebral Autosomal Dominant Arteriopathy with Subcortical Infarcts and Leukoencephalopathy." 2010. Genetics Home Reference. National Library of Medicine (U.S.). http://ghr.nlm.nih.gov/condition/cerebral-autosomal-dominant-arteriopaty-with-subcortical. Accessed 5/9/12.

Campomelic Dysplasia

Prevalence Ranges from 1 in 40,000 to 200,000 births

Other Names campomelic dwarfism; campomelic syndrome; camptomelic dysplasia; CMD1; CMPD1

Campomelic dysplasia was first described in the 1950s but not recognized as a separate disorder until the 1970s. The term "campomelia" comes from the Greek roots *campos*, meaning "curved," and *melia*, meaning "limb." Both camptomelia and campomelia are used interchangeably. The condition is also known as CMD1 and CMPD1.

What Is Campomelic Dysplasia?

Campomelic dysplasia is a rare disorder that affects the skeletal and reproductive systems. As a rare form of osteochondrodysplasia (bone/cartilage disorder) and congenital dwarfism, the disorder is characterized specifically by the bowing of the legs, which causes characteristic skin dimples to form over the curved bone. Other symptoms may be present: short legs, dislocated hips, severe angles of the bones that look like fractures, bell-shaped narrow chest, and 11 pairs of ribs. The child may have a chronic cough since birth and occasional bouts with pneumonia.

He or she may have a distinct look. The face is broad and flat with a high forehead, small jaw, long philtrum (the groove between the nose and upper lip), cleft palate, and low set ears.

The reproductive system may be affected. They may have ambiguous genitalia, meaning they possess external genitalia that appear as neither male nor female. The internal organs may not be the same as the appearance of the external genitalia. Occasionally, there is a combination of both ovaries and testes.

Children with campomelic dysplasia usually die in infancy. However, those who do survive may develop severe curvature of the spine that presses on the spinal cord. The individual will always be very short and have possible hearing loss.

What Is the Genetic Cause of Campomelic Dysplasia?

Mutations in the *SOX9* gene, officially known as the "SRY (sex determining region Y)-box 9" gene, causes campomelic dysplasia. The condition is inherited in an autosomal dominant pattern, meaning that only one gene has to be in the cells to cause the disorder. The disorder is usually caused by a new mutation in the gene, and neither of the parents has the disorder. Normally, this gene gives the instructions for a critical protein that plays a role in developing many different tissues and organs during the embryonic period. *SOX9* regulates many other genes and is responsible for development of both the skeletal and reproductive systems. However, a mutation in the *SOX9* gene can disrupt the activity of the protein and cause the characteristic symptoms of campomelic dysplasia. About 5% of cases occur from disruptions in the chromosome around the *SOX9* gene. *SOX9* is located on the long arm (q) of chromosome 17 at position 23.

What Is the Treatment for Campomelic Dysplasia?

A sonogram can reveal this severe disorder and the option of termination may be offered. After birth, special attention to respiratory problems that may develop must be taken. Caring for symptoms as they appear is important for managing this disorder.

Further Reading

"Campomelic Dysplasia." 2010. MedicineNet.com. http://www.medicinenet.com/campomelic_dysplasia/article.htm. Accessed 5/9/12.

"Campomelic Dysplasia." 2010. Nemours Children's Hospital. http://www.nemours.org. Accessed 5/9/12.

Unger, S.; G. Scherer; and A. Superti-Furga. 2008. "Campomelic Dysplasia." *GeneReviews*. July. http://www.ncbi.nlm.nih.gov/books/NBK1760. Accessed 5/9/12.

Camurati-Englemann Disease

Prevalence About 200 cases worldwide; actual prevalence unknown

Other Names Camurati-Englemann syndrome; CED; diaphyseal dysplasia, progressive; diaphyseal hyperstosis; Englemann's disease; PDD; Ribbing disease

Many genetic disorders affect only a small number of people. However, because of mass communications and the people's interest in investigating their medical own conditions and disorders, awareness of rare disorders has certainly expanded. The public hears information about diseases that was once confined to doctors, researchers, or a small number of affected individuals. If the person is a celebrity, the press and users of the Internet may write about the condition, and the public probably hears about it for the first time. For example, John Belluso, writer for the CBS television show *Ghost Whisperer*, had Camurati-Englemann syndrome and had been confined to a wheelchair since the age of 13. He died in February 2006 at the age of 36.

Camurati-Englemann disease was named for two doctors who were interested in this condition. M. Camurati, an Italian physician, wrote about the condition in 1922, and German doctor G. Englemann published a paper describing a condition in which the bones are dense and heavy in 1929. However, it was not until 2006 that researchers found the molecular data to link it to a genetic condition.

What Is Camurati-Englemann Disease?

Camurati-Englemann disease, or CED, is a rare bone disease. The underlying cause of the disease is heavy and thickened bones, especially in the long bones of the arms and legs. The thickness is along the shafts of the bones, a condition known as diasphyseal dysplasia. Bones of the skull and hip may also thicken narrowing the nerves and blood vessels and leading to many serious symptoms.

Thickening of the shafts of the various bones may cause the following symptoms:

- Severe pain in the arms and legs
- A waddling walk
- Muscle weakness and extreme tiredness
- Headaches
- Vision and hearing loss
- Dizziness or vertigo
- Ringing of the ears or tinnitus
- Curvature of the spine, a condition known as scoliosis
- Joint deformities

The person may appear to have very long limbs in proportion to height, a decrease in muscle mass and body fat, and delayed development at puberty. The condition may start at different ages. However, most people with CED develop pain during adolescence when the bones really begin to change.

The pain is really severe during a "flare-up," which can last from just a few hours to several weeks. Patients may require a wheelchair for mobility and life functions. CED may cause an enlarged spleen and liver and affect, causing a host of new symptoms.

What Is the Genetic Cause of Camurati-Englemann Disease?

The gene "transforming growth factor beta 1" or *TGFB1* causes CED. Normally, *TGFB1* gives the instructions for producing a protein called "transforming growth factor beta-1" (TGFβ-1). This protein is responsible for growth and cell division and differentiation. TGFβ-1 also aids in cell movement and signals the cells to stop growing or to self-destruct. Found throughout the body, the TGFβ-1 protein is active before birth, forming blood vessels and regulating muscle and body fat. It is also important for wound healing and immune system function. But its major role is in the skeletal system, where it regulates bone growth and forms the lattice, working in the spaces between the bone cells called the extracellular matrix. Within the cells, the protein is turned off when its work is done and turned on when it is needed.

The gene is inherited in an autosomal dominant pattern, meaning that only one copy of the mutated gene in each cell can cause the disease. In some cases, new mutations may result from parents with no history of the condition. About 10 mutations in the *TGFB-1* gene are responsible for CED. All the mutations turn on the TGFβ-1 protein, making it overactive and causing the abnormal thickening of the bones and the abnormal symptoms of CED. The *TGFB1* gene is located on the long arm (q) of chromosome 19 at position 13.2.

What Is the Treatment for Camurati-Englemann Disease?

CED is somewhat treatable. Anti-inflammatories, such as glucocorticosteroids, and immunosuppressives are used for treatment. However, these medications, which assist in building bone strength, may have serious side effects. Other treatments, including massage, relaxation techniques, spa baths, gentle stretching, and other alternative treatments, have been effective in treating symptoms.

Further Reading

"Camurati-Englemann Disease." 2005. Orphanet. http://www.orpha.net/consor/cgi-bin/OC_Exp.php?Expert=1328. Accessed 5/9/12.

"Camurati-Englemann Disease." 2010. Office of Rare Disease. National Institutes of Health (U.S.). http://rarediseases.info.nih.gov/GARD/Disease.aspx?PageID=4&DiseaseID=1072. Accessed 5/9/12.

"*TGFB1*." 2010. Genetics Home Reference. National Library of Medicine (U.S.). http://ghr.nlm.nih.gov/gene/TGFB1. Accessed 10/15/10.

Canavan Disease

Prevalence 1 in 6,400 to 13,000 people of Ashkenazi Jewish population; the incidence in other populations in unknown

Other Names ACY2 deficiency; Aminoacylase 2 deficiency; Aspa deficiency; aspartoacylase deficiency; Asp deficiency; Canavan–Van Bogaert–Bertrand disease; leukodystrophy, spongiform; spongy degeneration of the brain; spongy degeneration of the central nervous system; spongy degeneration of white matter in infancy; Van Bogaert–Bertrand disease; Van Bogaert–Bertrand syndrome

Myrtelle May Canavan, a graduate of Women's Medical College of Pennsylvania, became interested in neuropathology and diseases and disorders of the nervous system. In 1931, Dr. Canavan published an article in the *Archives of Neurology and Psychiatry* in which she described a 16-and-one-half-month-old child with a progressive form of encephalitis. The disorder, Canavan disease, is now named after this pioneer of children's nervous system disorders.

The discovery of the gene for Canavan disease became the subject of a major lawsuit and ethics controversy in the 1990s. In 1987, Daniel Greenberg and his wife, who had two children with Canavan disease, gave money and tissue samples from a number of donors to Dr. Reuben Matalon, a researcher at the Miami Children's Hospital. In 1993, he successfully identified the gene and developed a test for genetic counseling of prospective couples, who had the disorder in their families. The Canavan Foundation then offered free testing to those who had a possible family history of the disorder. However, the Miami Children's Hospital had secretly patented the gene and claimed royalties on the test. The Greenbergs and Canavan Foundation brought a lawsuit against the hospital, and it was resolved with an out-of-court settlement in September, 2003. The confidential settlement provided for continued royalty-based testing by licensed laboratories but royalty-free research for scientists and institutions doing basic research for a cure. They also agreed that the plaintiffs would not challenge Miami Children's Hospital's ownership and licensing of the Canavan gene patent. This case established the arguments about the appropriateness of patenting genes.

What Is Canavan Disease?

Canavan disease (CD) is a rare, but always fatal, degenerative disorder that causes progressive damage to the nerve cells of the brain. This disease is one of a group of leukodystrophies, characterized by the deterioration of the myelin sheath, the fatty covering that covers the nerve fibers. The disorder primarily affects Ashkenazi Jews of eastern European descent, which includes about 90% of the Jewish population in the United States. Estimates are that 1 in 40 Ashkenazi Jews carry the gene for Canavan disease.

The life expectancy of the people with Canavan varies, but most with the disorder die in childhood. As an infant, the baby appears normal; but within three to nine months, the symptoms begin to appear. The parent may first notice that the child cannot turn over as other children, and he or she has delayed development.

The following symptoms may develop:

- Rapidly increasing head development, a condition called macrocephaly
- Lack of head control
- Reduced visual responsiveness
- Abnormal muscle tone, such as stiffness or floppiness
- Cannot crawl, walk, sit, or talk
- Over time, may suffer seizures and become paralyzed
- Feeding and swallowing difficulties
- Sleep disturbances
- Mental retardation

What Is the Genetic Cause of Canavan Disease?

The genetic cause of Canavan is related to the aspartoacylase or *ASPA* gene. Normally, *ASPA* gives the instructions for an enzyme aspartoacylase, which breaks down N-acetyl-1-L-aspartic acid (NAA) into two compounds: aspartic acid (an amino acid that is the building block of many proteins) and acetic acid. The breakdown of NAA is critical for maintaining the brain's white matter, which consists of the myelin sheaths that cover the nerve fibers of the brain. The myelin sheath has several functions: insulating and protecting the nerve cells, making fats or lipids to product the myelin, and transporting molecules of water out of the nerve cells.

More than 55 mutations in *ASPA* are known to cause Canavan. One of these most common mutations involves only one base unit change; glutamic acid is replaced by the amino acid alanine at position 285 of the enzyme. These mutations in the *ASPA* gene reduce or completely stop the activity of aspartoacylase, thereby preventing the normal breakdown of NAA in the brain. This change disrupts the protective function of myelin, and without this protection, the nerve fibers die; the brain is damaged and the serious signs of Canavan appear. This condition is considered one of a group of leukodystrophies, which affect the myelin sheath. Canavan disease is inherited in an autosomal recessive pattern. That means both parents must carry the gene and pass it on to the offspring. *ASPA* is located on the short arm (p) of chromosome 17 at position 13.3.

What Is the Treatment for Canavan Disease?

The treatment is to alleviate the symptoms of the disorder; however, several experimental treatments are being investigated. One is the use of lithium citrate, a substance that has proven effective in a rat genetic model of Canavan. Jacob's Cure is a foundation dedicated to finding a cure for Canavan; it has been active in investigating gene therapy for this condition.

Further Reading

"Canavan Disease." Genetics Home Reference. National Library of Medicine (U.S.). http://ghr.nlm.nih.gov/condition/canavan-disease. Accessed 5/9/12.

Canavan Research Foundation. 2012. http://www.canavan.org. Accessed 1/26/12.

Jacob's Cure. 2012. http://jacobscure.org/about-jacobscure.php. Accessed 1/26/12.

Carney Complex

Prevalence Very rare; fewer than 750 individuals worldwide

Other Names Carney syndrome; LAMB—Lentigines, atrial myxoma, mucocutaneous myoma, blue nevus syndrome; NAME—Nevi, atrial myxoma, skin myxoma, ephelides syndrome

In 1985, Dr. J. Aidan Carney and a team studied a disorder that was characterized by an increased risk of many types of tumors, dark blotches on the skin, and endocrine overactivity. The disorder has had several names, such as the acronyms LAMB and NAME. The name that is recognized by the medical professionals is Carney complex, after Dr. Carney.

What Is Carney Complex?

Three elements make up the disorder called Carney complex: several types of tumors, changes in skin coloring or pigment, and overactivity in the endocrine glands. The symptoms may begin during the teen years or during early adulthood.

Several Types Of Tumors Or Myxomas

- Benign heart tumors: These myxomas may be found in any of the four chambers of the heart and may block the flow of blood as it passes through the heart. These blockages can be life-threatening.
- Skin tumors: These small bumps can occur any place on the skin. They tend to come back when they are removed.

- Endocrine tumors: These tumors especially develop on the hormone-producing glands, such as the adrenal glands.
- Other glands: Other endocrine tumors may develop in the thyroid, testes, and ovary.
- Rare nerve tumor: A tumor called a psammomatous melanotic schwannoma may wrap around certain cells called a schwannoma that insulate nerves. The tumor is benign but can become cancerous.

Skin Coloring or Pigmentation

Several types of pigmentation may occur.

- Brown spots called lentigines: These spots may appear anywhere on the body, especially the lips, eyes, or genitalia.
- A blue-black mole called a blue nevus: At least one of these is present.

Endocrine Disorders

- Adrenal disorders: When the tumors are present on the adrenal gland, a disease called primary pigments nodular adrenocortical disease, or PPNAD, may develop. This special pigment causes the glands to produce an excess amount of the hormone cortisol, leading to the development of Cushing syndrome. This syndrome causes many health problems such as slow growth, fragile skin, and weight gain in the upper part of the body.
- Pituitary gland: A growth called an adenoma may lead to excess growth hormone and cause acromegaly, where feet and hands grow. The person may also develop arthritis and distinct facial features.

What Is the Genetic Cause of Carney Complex?

Mutations in the *PRKAR1A* gene, officially known as the "protein kinase, cAMP-dependent, regulatory, type I, alpha (tissue specific extinguisher 1)" gene, causes Carney complex. Normally, *PRKAR1A* instructs for making type 1 of four parts of the enzyme protein kinase A. When the regulatory subunits are attached to two of the four parts, protein kinase A is turned off. When it breaks away from the other parts, it is turned on.

Over 117 mutations in *PRKAR1A* cause Carney complex. The mutations make an abnormal type 1 that is more quickly broken down by the cell. Because protein kinase A acts as a regulatory enzyme, the missing action of the subunit 1 causes the enzyme to turn on more often resulting in the uncontrolled growth of cells forming tumors. *PRKAR1A* is inherited in an autosomal dominant pattern and is located on the long arm (q) of chromosome 17 at position 23-q24.

What Is the Treatment for Carney Complex?

Treating of the heart condition is critical. Open heart surgery may remove the myxomas. Some of the skin tumors may be removed but have a tendency to recur. Surgery may be essential for other adrenal tumors. Medication may be given for management of Cushing syndrome.

Further Reading

"Carney Complex." 2010. *GeneReviews*. http://www.ncbi.nlm.nih.gov/books/NBK1286. Accessed 12/23/11.

"Carney Complex." 2010. Genetics Home Reference. National Library of Medicine (U.S.). http://ghr.nlm.nih.gov/condition/carney-complex. Accessed 12/23/11.

"Carney Complex." 2011. Medscape. http://emedicine.medscape.com/article/160000-overview#showall. Accessed 12/23/11.

Celiac Disease

Prevalence Common worldwide disorder; affects about 1 in 100 people worldwide

Other Names celiac sprue; gluten enteropathy; gluten intolerance; non-tropical sprue; sprue

Aretaeus of Cappadocia in Turkey first described a condition in which people had a reaction to wheat products and experienced major intestinal problems. He used the term *koilakos* that is translated "abdominal." A translation in the nineteenth century called it celiac from the description of Aretaeus. The name has stuck.

What Is Celiac Disease?

Celiac disease is a disorder of the intestinal system. It is both an autoimmune disease and a disease of malabsorption, in which people are very sensitive to gluten, a protein found in wheat, rye, and barley. In addition, gluten may be found also in everyday products such as medicines, vitamins, and lip balm. The malabsorption leads to inflammation, which damages the villi, the small fingerlike projections on the inside of the small intestine.

The person will experience the following symptoms:

- Chronic diarrhea: The villi have become flattened and do not absorb nutrients. The result may be pale, foul-smelling, or fatty stools.
- Weight loss: Poor absorption of nutrients lead to loss of weight

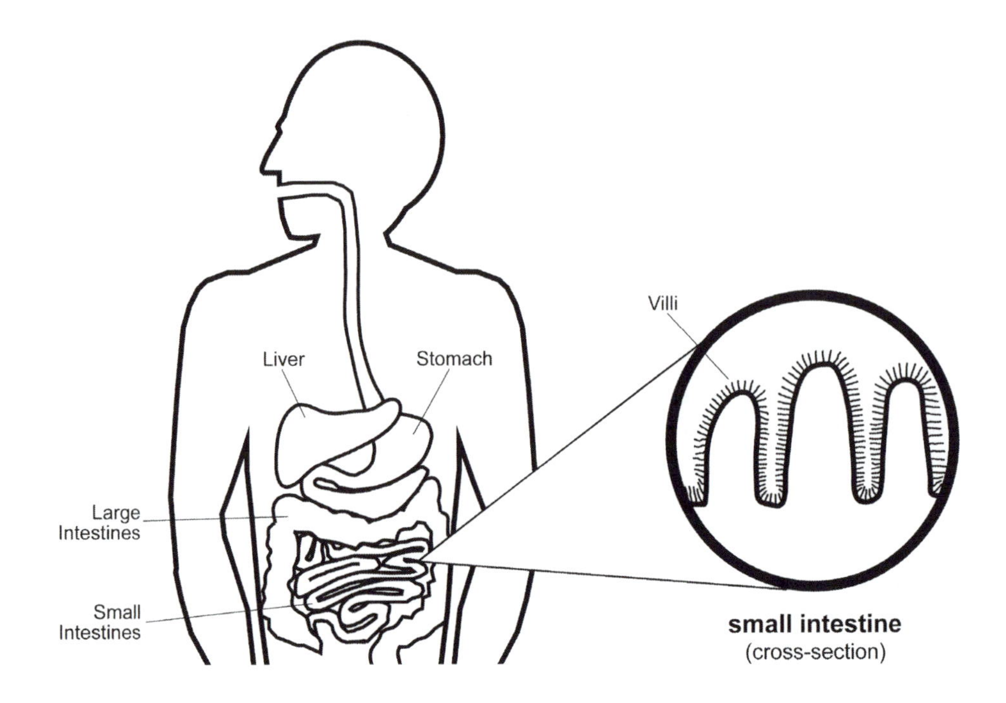

Villi are part of the lining on the intestine wall that aids in digestion. Villi become damaged in people with celiac disease. (ABC-CLIO)

- Abdominal pain
- Swelling of abdominal area
- Vomiting
- Food intolerances
- Unexplained iron-deficiency anemia
- Arthritis
- Cancer: Inflammation may lead to increased risk of developing gastrointestinal cancers, especially of the small intestine or the esophagus
- Other health problems: The person may develop type 1 diabetes; autoimmune thyroid disease; autoimmune liver disease; rheumatoid arthritis; Addison disease; and Sjogren's syndrome, in which the person loses the ability to produce tears and saliva

Celiac disease can develop any time the person starts eating foods containing gluten. The disease also may go undiagnosed because the symptoms are nonspecific.

What Is the Genetic Cause of Celiac Disease?

Mutations in two genes cause Celiac disease. Other genes, as well as factors from the environment, may also be involved.

HLA-DQA1

The first gene is the *HLA-DQA1* gene, officially called the "major histocompatibility complex, class II, DQ alpha 1" gene. Normally, *HLA-DQA1* provides instructions for making a critical protein necessary for the function of the immune system. This gene is part of a large family of genes called the human leukocyte antigen (HLA) complex, which helps the body tell the difference between its own cells and invading bacteria or viruses. The HLA complex is the human version of the major histocompatibility complex (MHC), which is made up of three basic classes of genes. The class to which *HLA-DQA1* belongs helps the body trigger an attack on invading viruses and bacteria.

Variations of the gene are associated with celiac disease and cause an inappropriate immune response to part of the gluten protein called gliadin. The response causes the inflammation and leads to the signs of the disease. *HLA-DQA1* is inherited in an autosomal dominant pattern and is located on the short arm (p) of chromosome 6 at position 21.3.

HLA-DQB1

The *HLA-DQB1* gene, officially known as the "major histocompatibility complex, class II, DQ beta 1" gene, is also associated with celiac disease. Normally, this gene is also important in the immune system. As a member of the HLA complex, it works with the *HLA-DQA1* gene to aid in attacking invading viruses and bacteria. Mutations or specific versions of the genetic products appear to cause the immune system to malfunction and lead to the symptoms of celiac disease. *HLA-DQB1* is inherited in an autosomal dominant pattern and is located on the short arm (p) of chromosome 6 at position 21.3.

What Is the Treatment for Celiac Disease?

A strict, lifelong adherence to a gluten-free diet is essential. The person must avoid food or any product with wheat, rye, or barley. The person must be constantly aware of nutrition and how to make up for any deficits caused by the strict diet. All of the other conditions can be treated according to the symptoms.

Further Reading

"Celiac Disease." 2010. *GeneReviews*. http://www.ncbi.nlm.nih.gov/books/NBK1727. Accessed 1/3/12.

"Celiac Disease." 2010. Genetics Home Reference. National Library of Medicine (U.S.). http://ghr.nlm.nih.gov/condition/celiac-disease. Accessed 1/3/12.

"Celiac Disease." 2010. National Institute for Diabetes and Digestive and Kidney Disease. National Institutes of Health (U.S.). http://digestive.niddk.nih.gov/ddiseases/pubs/celiac. Accessed 1/3/12.

Cerebrotendinous Xanthomatosis

Prevalence About 3 to 5 per 100,000 worldwide; common among Jewish population of Morocco with incidence of 1 in 108

Other Names CTX; Van Bogaert–Scherer–Epstein disease; xanthomatosis, cerebrotendinous

In 1936, L. van Bogaert and two other French physicians, published an article in a French medical journal that described a condition in which yellow growths called xanthomata developed, along with other serious conditions that included seizures and dementia. He determined that it was an inborn error of metabolism. Later scientists described the clinical symptoms, the biochemistry, and the genetics of the condition. In 1968, Menkes found that accumulations of cholestanol were found in the brain and nervous system. It was not until 1980 that the connection was made with the mitochondrial action.

What Is Cerebrotendinous Xanthomatosis?

Cerebrotendinous xanthomatosis, or CTX, is a rare metabolic disorder of cholesterol and bile acid metabolism that affects many body parts. The term, cerebrotendinous xanthomatosis, itself describes the clinical conditions. The word "cerebrotendinous" comes from two Greek root words: *cerebro*, meaning "brain," and *tendinous*, referring to the "connective tendons," which attach muscle to bone. The word "xanthomatosis" comes from three Greek roots: *xantho*, meaning "yellow"; *oma*, meaning "tumor"; and *osis*, meaning "condition of."

The problem with this disorder appears to be that people cannot break down lipids. This failure to break down allows the fats to accumulate, especially certain forms of cholesterol. When the fatty materials accumulate, yellow nodules develop. When these nodules locate in the brain and connective tissue, they cause the condition cerebrotendinous xanthomatosis.

Following are the symptoms of cerebrotendinous xanthomatosis:

- Diarrhea: The first clue to the disorder is chronic diarrhea, which begins in infancy.
- Cataracts: These eye growths become evident in childhood or adolescence.
- Progressively brittle bones that break easily.

- Xanthomas: Beginning in the second or third decades of life, the person develops the yellow nodules over the body.
- Xanthomas: These occur in the tendons, especially the Achilles tendon in the heel.
- Neurological disorders: In about the third decade of life, the person begins to have seizures, hallucinations, depression, dementia, and difficulty with coordination. This condition becomes progressively worse until the sixth decade of life. The person may die if he does not receive treatment. The neurological disorders are thought to be the result of the accumulation of the xanthomas in the fatty myelin sheath that cover parts of the neurons. In addition, the deposits of fats contribute to these problems.
- Increased risk for heart disorders.

The symptoms of the condition can vary widely.

What Is the Genetic Cause of Cerebrotendinous Xanthomatosis?

Mutations in the *CYP27A1* gene, officially known as the "cytochrome P450, family 27, subfamily A, polypeptide 1" gene, cause cerebrotendinous xanthomatosis. Normally, *CYP27A1* provides instructions for the enzyme sterol 27-hydroxylase, which is located in the mitochondria. The mitochrondria are the power houses of the cell. The enzyme helps to form bile acids that digest fats. Specifically, the enzyme breaks down cholesterol to make chenodeoxycholic acid and, in addition, plays a key role in helping the body keep cholesterol levels normal.

About 50 mutations in *CYP27A1* cause cerebrotendinous xanthomatosis. Mutations disrupt the sterol 27-hydroxylase enzyme, keeping it from breaking down cholesterol to a specific bile acid called chenodeoxycholic acid. As a result, a molecule called cholestanol, which is similar to cholesterol, is overproduced and forms xanthomas in the blood, nerve cells, and the brain. Cholesterol levels are not increased in the blood, but they are elevated in various tissues throughout the body. The accumulation of cholesterol and cholestanol in the brain, tendons, and other tissues causes the signs and symptoms of cerebrotendinous xanthomatosis

Most of the mutations involve changes in only one amino acid building block in the enzyme. The most common occurs when cysteine replaces arginine. These changes in the enzyme disrupt the function of breaking down cholesterol. Other mutations reduce the amount of the enzyme, allowing cholestanol to accumulate throughout the body, especially in the brain and tendons. *CYP27A1* is inherited in an autosomal recessive pattern and is located on the long arm (q) of chromosome 2 at position 33-qter.

What Is the Treatment for Cerebrotendinous Xanthomatosis?

Patients with this disease overproduce cholestanol and bile acid precursors because of the block in synthesis. Replacement with chenodeoxycholic acid shuts down abnormal pathways, reduces elevated level of cholestanol, and improves the clinical syndrome. Since this disorder relates to metabolism, the neurological problems may require long-term treatment with chenodeoxycholic acid (CDCA), which normalizes bile acid synthesis and equalizes plasma. Dealing with the symptoms, such as cataracts and heart disorders is with conventional treatment.

Further Reading

"Cerebrotendinous Xanthomatosis." 2008. Genetics Health Reference. National Library of Medicine. (U.S.). http://ghr.nlm.nih.gov/condition/cerebrotendinous-xanthomatosis. Accessed 1/8/12.

"Cerebrotendinous Xanthomatosis." 2010. Medscape. http://emedicine.medscape.com/article/1418820-overview. Accessed 1/8/12.

Centronuclear Myopathy. *See* Myopathy, Centronuclear

Charcot-Marie-Tooth Disease

Prevalence One of the must common neurological disorders; affects an estimated 350,000 in the United States; worldwide, affects 1 in 2,500 people of all races and ethnic groups

Other Names Charcot-Marie-Tooth neuropathy; hereditary motor and sensory neuropathy (HMSN); hereditary sensorimotor neuropathy (HSMN); Morbus Charcot-Marie-Tooth; peroneal muscular atrophy

Several researchers have given their names to Charcot-Marie-Tooth disorder. The first description of the condition was in 1884 by the German doctor Friedrich Schultze, who was convinced he was dealing with a muscle disorder. Following soon in 1886 were the classic descriptions published in a French journal by Jean-Martin Charcot and his pupil Pierre Marie. In this article they described five cases of progressive muscular atrophy and assumed it was related to the nervous system. That same year, Howard H. Tooth in London wrote about the peroneal type of progressive muscular atrophy related to neuropathy. Thus, the disorder then was introduced as a neuropathy rather than just a muscle problem and called Charcot-Marie-Tooth disease.

What Is Charcot-Marie-Tooth Disease?

Charcot-Marie-Tooth disease, or CMT, is characterized by slowly progressive wasting and weakness of peripheral muscles of the arms and feet. The peripheral nerves connect the brain and spinal cord to these muscles and also to the sensory cells that detect heat, cold, and touch. The condition usually begins in late childhood or early adolescence, but onset can begin at any time even into late adulthood. The following symptoms of the disease are caused by degeneration of the peripheral nerves, nerve roots, and even the spinal cord:

- Earliest symptoms involve muscle weakness in the foot, causing extremely high arches or curled toes, called hammer toes.
- Abnormal feet make it difficult to flex the foot and may cause the person to develop a high step or gait and increase the risk for falls.
- Lower legs may take on an "inverted champagne bottle" appearance because of the loss of muscle bulk.
- As the disease progresses, weakness in lower limbs may require the person to use a wheelchair.
- Muscle loss in the hands may make it impossible to write, fasten buttons, or even open doors.
- Sensory cells are disturbed, and the person may experience burning or numbing sensation in the feet and hands.
- Sensitivity to touch, heat, and cold is decreased.
- In rare cases, there is a loss vision and hearing.

The symptoms of CMT may vary even within a family. Several types of CMT exist, including CMT1, CMT2, CMT3, CMT4, and CMTX. There are subtypes of several of these types.

What Is the Genetic Cause of Charcot-Marie-Tooth Disease?

CMT has a rather complicated system of inheritance depending upon the type and subtype. Following is an outline of the types of CMT and the pattern of inheritance:

- CMT-1 Type 1: caused by mutations in *PMP22* (subtypes 1A and 1E, *MZB* (subtype 1B), *LITF* (subtype 1 C), *ERG2* (subtype 1D), and *NEFL* (subtype 1F). *CMT1A* and *CMT1B* are inherited in an autosomal dominant pattern. *CMT1A* results from a duplication on chromosome 17. The gene provides instructions for producing peripheral myelin protein-22, a critical component of the myelin sheath.
- CMT2: There are many subtypes designated from A to L. The striking feature is abnormalities of the axon of the peripheral nerve rather than the myelin sheath. The gene here is unknown. There has been some connection between

this condition and one that codes for kinesins, which are proteins that aid in transporting materials to the cell. The following genes may be involved: *MFN2*, *KIFB*, *RAB7A*, *LMNA*, *BSCL2*, *GARS*, *GDAP1*, *NEFL*, *HSPB1*, *MPZ*, *GDAP1*, *HSPB8*, and *DNM2*.

- CMT3: Also known as Dejerine-Sottas disease, this is a severe neuropathy that attacks the myelin sheath beginning in infancy; the newborn will have severe muscle atrophy, weakness, and sensory problems. Mutations in the *PMP-22* gene cause this rare disorder.
- CMT4: This has several subtypes of autosomal recessive genes that cause problems with the myelin and also sensory disorders. Each subtype is related to a different genetic mutation, none of which have been identified. However, the following genes are under investigation: *GDAP1*, *MTMR2*, *SBF2*, *SH3TC2*, *NDRG1*, *EGR2*, *PRX*, *FGD4*, and *FIG4*.
- CMTX: This is the only one of the CMT types connected to the X chromosome. This condition is inherited in a dominant pattern and related to a point mutation in the *connein-32* gene on the X chromosome. Males who inherit this gene develop moderate to severe symptoms.

What Is the Treatment for Charcot-Marie-Tooth Disease?

Although no cure exists for the condition, occupational therapy, physical therapy, braces, and orthopedic devices may help. Pain-killing drugs can be prescribed for those in severe pain. The National Institute of Neurological Disorders and Stroke (NINDS) supports research on CMT and other peripheral neuropathies and is working to treat, prevent, and ultimately cure the disorder. Ongoing research is being conducted to identify gene mutations that cause the various subtypes.

Further Reading

"Charcot-Marie-Tooth Disease." 2011. Charcot-Marie-Tooth Association (CMTA). http://www.charcot-marie-tooth.org. Accessed 5/10/12.

"Charcot-Marie Tooth Disease." 2010. Mayo Clinic. http://www.mayoclinic.com/health/charcot-marie-tooth-disease/DS00557. Accessed 5/10/12.

"Charcot-Marie-Tooth Fact Sheet." 2011. National Institute of Neurological Disorders and Stroke. http://www.ninds.nih.gov/disorders/charcot_marie_tooth/detail_charcot_-marie_tooth.htm. Accessed 5/10/12.

CHARGE Syndrome

Prevalence 1 in 8,500 to 10,000

Other Names CHARGE association; Hall-Hittner syndrome

In 1979, Dr. B. D. Hall wrote an article in the *Journal of Pediatrics* about a group of children who had abnormal closure of the choanae, which are passages from the back of the nose to the throat. The medical term for this condition is choanal atresia. Also, Dr. H. M. Hittner observed 10 children who had not only choanal atresia, but also coloboma (an eye defect), heart defect, and hearing loss.

Two years later, in 1981, Dr. R. A. Pagon coined the acronym CHARGE to refer to a cluster of features seen in a number of children. At first, the condition was considered just an association, but later genetics studies traced the condition to a specific gene, and thus it became known as CHARGE syndrome.

What Is CHARGE Syndrome?

When Dr. Pagon wrote the article in the *Journal of Pediatrics*, he put together a set of somewhat different symptoms under one umbrella. The acronym somewhat describes the features of the disorder and at one time were used to officially diagnose the syndrome. Now several other symptoms may be added, but according to the CHARGE Foundation, the name will not be changed. The following symptoms make up the acronym:

- *C—Coloboma* of the eye with central nervous system anomalies: A coloboma looks like a keyhole in the side of the iris or colored part of the eye. This cleft or fissure may extend to the back of the eye to the retina. Vision is affected.
- *H—Heart defects*: These defects can be of any type and many are complex, such as the tetralogy of Fallot.
- *A—Atresia* of the choannae: These passages from the back of the nose to the throat may be narrow (stenosis) or blocked (atresia). The blockage can be on both sides and can be bony or covered with a membrane.
- *R—Retardation*: Growth deficiency may be from disruption of growth hormone; about 70% are short in stature.
- *G—Genital and or urinary defects*: Males and females may have small external genitalia; males may have undescended testicles and females no uterus.
- *E—Ear* anomalies and/or deafness: Ears may have an unusual shape.

The patterns of malformations vary among individuals. A diagnosis is made on a combination of major symptoms, which may be listed in the acronym, and a collection of minor characteristics such as cleft palate, esophageal blockage, kidney abnormalities, and prominent CHARGE face. These facial characteristics include a square face, prominent forehead, arched eyebrows, droopy eyelids, thick nostrils, flat midface, and a small chin that gets larger with age. Younger individuals may also have obsessive compulsive behavior (OCD).

What Is the Genetic Cause of CHARGE Syndrome?

Changes in the *CHD7* gene, officially known as the "chromodomain helicase DNA binding protein 7" gene, cause CHARGE syndrome. Normally, the *CHD7* protein plays a role in the organization of chromatin, the complex of DNA and protein that makes up chromosomes. The *CHD7* protein regulates genes express using a process called chromatin remodeling. In this process, chromatin is changed or remodeled to determine how tightly it is packaged. When tightly packaged, gene expression is lower than when DNA is loosely packaged.

Mutations in the *CHD7* gene cause an abnormally short and nonworking protein. The chromatin remodeling system is disrupted during the embryonic stage of development, causing the cluster of varying symptoms. In August 2004, a group of scientists in the Netherlands located the first *CHD7* gene. CHARGE syndrome is inherited in an autosomal dominant pattern, which means that only one copy of the altered *CHD7* gene will cause the condition. Most of the cases occur as new mutations. *CHD7* is located on the long arm (q) of chromosome 8 at position 12.2. Even though the gene has been discovered, the gene test is very expensive and is not perfect as only about two-thirds of the people with CHARGE have a positive test.

What Is the Treatment for CHARGE Syndrome?

Infants with serious medical problems such as a heart defect have a serious life-threatening condition; however, a number of surgical procedures are available. Children with CHARGE will need special intervention of speech, physical, and occupational therapy.

Further Reading

Charge Syndrome Foundation. 2005. "About CHARGE." http://www.chargesyndrome.org/about-charge.asp. Accessed 5/10/12.

"*CHD7*." 2010. Genetics Home Reference. National Library of Medicine (U.S.). http://ghr.nlm.nih.gov/gene/CHD7. Accessed 5/10/12.

Kugler, Mary. 2007. "CHARGE Syndrome."About.com: Rare Diseases. http://rarediseases.about.com/od/rarediseasesc/a/chargesyndrome.htm. Accessed 5/10/12.

Char Syndrome. *See* Patent Ductus Arteriosus (Char Syndrome)

Chediak-Higashi Syndrome (CHS)

Prevalence Fewer than 200 cases have been reported worldwide

Other Names Griscelli syndrome

It is unusual to see a syndrome that occurs on humans, white tigers, cattle, blue Persian cats, and a captive albino orca. But Chediak-Higashi syndrome is such a disorder. Named for the Cuban physician Alexander Chediak and Japanese pediatrician Otokata Higashi, the syndrome affects many body parts and is characterized by oculocutaneous albinism, which causes light eyes and hair, light pigment in the skin, and vision problems.

What Is Chediak-Higashi Syndrome (CHS)?

Chediak-Higashi (CHS) is a disorder that targets the immune and other body systems. Damage to the immune system cells keep them from performing their functions, which is to fight off bacteria and viruses. At a cellular level, CHS targets the lysosomes, the large bodies in the cells that act as a recycling and digestive system within the individual cell. Few people with CHS live to adulthood.

The following symptoms are common to most individuals with Chediak-Higashi syndrome:

- Persistent infections
- Oculocutaneous albinism with light skin and silvery hair, vision problems, involuntary eye movements, and light sensitivity
- Problems with blood clotting, leading to bruising and abnormal bleeding
- Neuropathy beginning in teenage years
- Seizures
- Susceptibility to bacteria, especially *Staphylococcus aureus* and fungi

A milder form may occur in adulthood, affecting the nervous system and causing difficulty in walking

Eventually, the condition develops into an end-stage accelerated phase thought to be triggered by the Epstein Barr virus (EBV). In this last life-threatening stage, the white blood cells divide rapidly and invade other body organs. The person develops high fever and organ failure, and then dies.

What Is the Genetic Cause of Chediak-Higashi Syndrome (CHS)?

The disorder is caused by a mutation in the "lysosomal trafficking regulator" gene, or *LYST*. Normally, the *LYST* gene controls the movement of materials in and out of the globular structures called lysosomes. Lysosomes are recycling centers of the cells acting to break down poisonous substances, digest bacteria that enter the cells, and recycle materials from worn-out or destroyed parts. The protein may also act as a regulator of the size of the lysosome and controller of what goes into the lysosome.

Mutations in *LYST* disturb the functioning of the lysosome trafficking regulator. The lysosomes become enlarged with undestroyed bacteria and other debris and then do not function normally. Closely related to lysosomes are melanosomes, the structures that produce melanin or pigment in the skin. In CHS, melanin becomes trapped within the giant cells, and the pigment that is normally in skin and hair is not available. The fact that the lysosome structures within red blood cells are not functioning may cause the person to bleed abnormally or bruise easily.

About 30 mutations exist. One mutation that affects early childhood appears to be an abnormally shortened version of the gene. The mutation that causes a milder form in older people is from the change of a single protein base pair in the gene. The *LYST* gene is inherited in an autosomal recessive pattern, meaning that the parents carried the gene but did not show the symptoms of the disorder. The *LYST* gene is located on the long arm (q) of chromosome 1 at position 42-42.2.

What Is the Treatment for Chediak-Higashi Syndrome (CHS)?

There is no specific treatment for CHS. Bone marrow transplants have been successful in some patients. Treating infection with various drugs helps to prolong life. Vitamin C therapy has improved the clotting and immune function in some patients.

See also Albinism

Further Reading

"Albinism Database." 2001. International Albinism Center, University of Minnesota. http://albinismdb.med.umn.edu/chs1mut.html. Accessed 5/10/12.

"Chediak-Higashi Syndrome." 2008. The Merck Manual for Health Care Professionals.

http://www.merck.com/mmpe/sec13/ch164/ch164d.html. Accessed 5/10/12.

Nowicki, Roman. 2010. "Chediak-Higashi Syndrome." Medscape. http://emedicine.medscape.com/article/1114607-overview. Accessed 5/10/12.

Cockayne Syndrome

Prevalence About 2 per million newborns in the United States and Canada

Other Names CS; dwarfism-retinal atrophy-deafness syndrome; Neill-Dingwall syndrome; progeria-like syndrome; progeroid nanism; Weber-Cockayne syndrome

Edward Cockayne (1880–1956), a London physician, found his niche in studying children with rare inherited diseases. In 1933 he published a paper, "Inherited Abnormalities of the Skin and its Appendages," to call attention to the fruitful research in this untouched field. In 1936, he described a condition in which abnormalities of the skin appear with a host of other symptoms, including dwarfism and premature aging. In addition to his work in genetics, Cockayne was an enthusiastic entomologist and built up a massive collection of butterflies and moths, which are now displayed in a zoological museum near London. Not only is the condition named after Dr. Cockayne, the butterfly is the symbol for the Share and Care Cockayne Syndrome Network.

What Is Cockayne Syndrome?

Cockayne syndrome (CS) is a complex disorder with a long list of clinical features. Prominent among the symptoms are sensitivity to sunlight, short stature, and premature aging. The following are four basic types of the condition:

- *CS Type I*: In this classic form, the child is born with normal fetal growth but within the first two years of life, abnormalities begin to appear. The child is small, and parents note a failure to gain weight and grow at a normal rate. This lack of growth is called "failure to thrive." Parents begin to note that even minor exposure to the sun can cause severe sunburn. Vision and hearing are noticeably impaired, and the central and peripheral nervous systems being to degenerate. The child begins to look like an older person, a condition of premature aging called progeria. The lifespan is between 10 and 20 years.
- *CS Type II*: In this most serious form, the symptoms appear at birth. The child has the abnormalities at birth and develops little after birth. This type is sometimes called cerebro-oculo-facio-skeletal (COFS) syndrome or Pena-Shokeir syndrome type II. The child seldom lives past seven years of age.
- *CS Type III*: This type is the least serious of the three and may have later onset, fewer symptoms, and slower progression. The life span may extend to 40 or 50 years of age.
- *Xeroderma-pigmentosum-Cockayne syndrome (XP-CS)*: This skin condition, another DNA repair disease, may also occur in conjunction with Cockayne. The skin has a wide range of changes from freckling to skin cancer on areas exposed to sunlight.

The symptoms of CS are many and can vary within the individual. However, all have the short stature, premature aging, and sensitivity to light. The following characteristics may vary:

- A pleasant and sociable personality
- Small head, called microcephaly
- Problems with joints called contractures
- Spasticity and unsteady gait
- Rounded back
- Deep-set eyes that give the appearance of aging
- Tooth decay
- Problems with vision; cataracts

- Poor circulation; hand and feet are cold
- Small penis in males
- Feeding problems
- Sleeping with eyes open
- Tremors
- Liver abnormalities
- High blood pressure
- Severe itchiness

What Are the Genetic Causes of Cockayne Syndrome?

Changes in the *ERCC6* (also known as the CSB gene) and the *ERCC8* (the CSA gene) cause Cockayne syndrome. Normally, *ERCC6* and *ERCC8* give instructions for making two proteins—CSB and CSA—that repair damaged DNA. DNA can be damaged in several ways: ultraviolet (UV) rays from the sun, toxic chemicals, radiation, and molecules that may circulate in the blood stream called free radicals. When encountering these damaging conditions, working cell mechanisms can usually repair the DNA damage before problems are caused.

Mutations in the *ERCC6* and *ERCC8* genes disrupt the function of DNA repair. More and more abnormalities in the DNA build up, and the cells then do not work properly and eventually die. For example, after exposure to sunlight, people with CS cannot perform a type of DNA repair, known as "Transcription-coupled repair" This repair occurs when the DNA that codes for proteins is being replicated. Both the CSA and CSB genes interfere with the transcription process and with repairing DNA and lead to premature aging and other the symptoms of CS.

The official name of *ERCC8* is "exclusion repair cross-complimenting rodent repair deficiency, complementation group 8." Many of these mutations are a short version of the gene, and others have just one change in the base building blocks. ERCC8 is located on the long arm (q) of chromosome 5 at position 12.1. The official name of *ERCC6* is "exclusion repair cross-complimenting rodent repair deficiency, complementation group 6." *ERCC6* is located on the long arm (q) of chromosome 10 at position 11.23. A problem with either one of the genes may cause the symptoms of CS.

CS is inherited in an autosomal recessive pattern, meaning that both parents must carry a copy of the gene. The parents will not have signs and symptoms of the disease.

What Is the Treatment for Cockayne Syndrome?

No specific treatment is currently available for CS. However, the symptoms can be treated. Physical, occupational, speech, vision, and hearing therapy are most beneficial.

Further Reading

"Cockayne Syndrome." 2010. Genetics Home Reference. National Library of Medicine (U.S.). http://www.ghr.nlm.nih.gov/condition/cockayne-syndrome. Accessed 5/10/12.

"Cockayne Syndrome." 2010. National Center for Biotechnology Information (U.S.).

http://www.ncbi.nlm.nih.gov/bookshelf/br.fcgi?book=gnd&part=cockaynesyndrome. Accessed 5/10/12.

Share and Care Cockayne Syndrome Network. 2007. http://www.cockayne-syndrome.org. Accessed 5/10/12.

Coffin-Lowry Syndrome (CLS)

Prevalence Uncertain; possibly 1 in 40,000 to 50,000 people

Other Names CLS; mental retardation with osteocartilaginous abnormalities

In 1966, Dr. C. S. Coffin described a condition that was characterized by dwarfism, severe mental retardation, and muscle weaknesses. Writing about the condition in an article "Mental retardation with osteocartilaginous anomalies" in the *American Journal of Diseases of Children*, he noted how these children have specific looks that make them more like each other than their parents. Several years later, in 1971, Dr. R. B. Lowry wrote about a new dominant genetic mental retardation syndrome associated with some unusual body features. The condition was named after the two researchers: Coffin-Lowry syndrome.

What Is Coffin-Lowry Syndrome (CLS)?

Coffin-Lowry syndrome (CLS) is a disorder with a set of specific medical symptoms that affect many parts of the body. Although females with the disorder may show some mild-to-severe symptoms, males with CLS always show quite serious disorders. Following are the symptoms related to CLS:

- Profound intellectual disability
- Head and facial abnormalities including a small head (microcephaly), an underdeveloped upper jaw, broad and short nose with a wide tip, protruding nostrils, very prominent brows, widely spaced and downward slanting eyes, large low-set ears, and thick eyebrows
- Skeletal abnormalities including front-to-back and side-to-side curvature of the spine, prominent breastbone (pigeon chest), dental problems, and short, hyperextensible, and tapered fingers
- Feeding and respiratory problems
- Hearing impairment

- Awkward gait
- Heart and kidney disorder

Beginning in adolescence, the person with CLS may experience brief episodes of collapse when excited or startled by a loud noise. Such attacks are called stimulus-induced drop episodes (SIDES).

What Is the Genetic Cause of Coffin-Lowry Syndrome (CLS)?

Mutations in the *RPS6KA3* gene cause CLS. The official name of *RPS6KA3* is "ribosomal protein S6 kinase, 90kDa, polypeptide 3." Normally, the *RPS6KA3* gene gives instructions for making a protein that is part of a family called ribosomal S6 kinases (RSKs), which help regulate the activity of certain genes and are involved in signaling within cells. The RSK proteins play a role in important cellular structures called ribosomes. The RSK protein is required for learning and for nerve development.

Coffin-Lowry syndrome is caused by a mutation in a *RPS6K3* gene that is carried on the X chromosome and is an X-linked dominant disorder. The female carrier may exhibit some of the symptoms of the disorder; but the males, who have only one X chromosome, will have the severe symptoms. According to the Coffin-Lowry Syndrome Foundation:

- In males with the CLS-causing mutation, 100% will be affected
- Females with CLS-causing mutation are carriers and are at high risk for developmental delay and mild physical symptoms of CLS.

About 70% to 80% of cases have no family history of CLS, whereas 20% to 30% have one or more affected members.

The more than 125 mutations in the *RPS6KA3* gene have been identified in people with CLS. The mutations reduce or eliminate the activity of the RPS6KA3 protein and lead to the signs and symptoms of Coffin-Lowry syndrome. The gene is located on the short arm (p) of chromosome X at position 22.2-p22.1

What Is the Treatment for Coffin-Lowry Syndrome (CLS)?

There is no cure and no standard course of treatment for the syndrome. However, the symptoms may be treated with supportive interventions, which may include physical and speech therapy and educational services. The prognosis for individuals with CS depends on the severity of symptoms and early intervention. Life span is in reduced in some individuals.

Further Reading

"Coffin-Lowry Syndrome." 2010. RightDiagnosis.com. http://www.rightdiagnosis.com/c/coffin_lowry_syndrome/intro.htm. Accessed 5/10/12.

Coffin-Lowry Syndrome Foundation. http://clsf.info. Accessed 5/10/12.

"NINDS Coffin Lowry Syndrome Information Page." 2008. National Institute of Neurological Disorders and Stroke. http://www.ninds.nih.gov/disorders/coffin_lowry/coffin_lowry.htm. Accessed 5/10/12.

Collagen: A Special Topic

Collagen is the fibrous protein that is the most abundant protein in nature. In mammals, about one-quarter of the protein in the body is collagen. The word "collagen" comes from the Greek roots *kolla*, meaning "glue," and *gen*, meaning "giving birth to." Thus, the idea of glue or a substance of strong tensile strength holding matter together is implied. Without collagen, the body would literally fall apart.

Collagen is fibrous in nature and supports the following body tissues:

- Ligaments
- Tendons
- Connective tissue
- Skin elasticity; works with keratin to provide strength, resilience, and flexibility to the skin
- Cornea of the eye where it is present in crystalline form
- Teeth, inside the dentin or inner part of the tooth

Collagen is present both inside and outside the cells. It works with another protein, elastin, to support body tissues including the lungs and bones. The breakdown of collagen as one ages is responsible for the wrinkles. Collagen makes up about 25% to 35% of the protein in the human body.

What Is the Structure of Collagen?

Scientists have studied collagen since the mid-1930s. Several Nobel laureates have been honored for their study of this structure. They have found that collagen rolls three polypeptide chains into a triple helix. The three chains have a regular arrangement of amino acids, but with an unusual structure. The arrangement of collagen is as follows:

- Glycine (Gly), the smallest amino acid molecule with no side chain, is found at every third position.
- Proline (Pro) is found in about 17% of collagen.
- Two other amino acids, which are not directly inserted by ribosomes during translation, are Hydroxyproline (Hyp), derived from proline, and Hydroxylysine (Hyl), derived from Lysine (Lys). Both of these products require vitamin C for their translation.

- Sequence of collagen often follows the pattern Gly-Pro-X-Hyp (X is any amino acid residues).
- The white collagen of most connective tissue is made of interwoven fibers of the protein collagen. The fibers consist of globular units of a collagen subunit called tropocollagen. Tropocollagen twists to the left, but the other strands twist to the right to form a triple helix.
- Tropocollagen subunits self-assemble and have regularly staggered ends. The collagen fibrils are arranged in different combinations and concentrations.
- The triple helix coils around very tightly and resists stretching. Collagen is therefore valuable for structure and support.

There are 29 different kinds of collagen. In this encyclopedia, many disorders of collagen are caused by mutations in the genes that then disrupt the ability of the protein to form collagen.

The following are the types of collagen, the genes that control them, and the disorders related to the collagen type:

- *Type I*: This most abundant form is found in tendons, skin, artery wall, the endomysium of myofibrils (inner part of muscle fibers), fibrocartilage, and inner parts of the bones and teeth. Related genes are "collagen I, alpha 1," or *COL1A1*, and "collagen I, alpha 2," or *COL1A2*. Disorders include osteogenesis imperfecta, Ehlers-Danlos syndrome, and hyperstosis or Caffey's disease.
- *Type II*: Hyaline cartilage makes up about 50% of cartilage; found also in the vitreous humor of the eye. Related gene is "collagen 2alpha1" or *COL2A1*. Disorders connected to this gene are the collagenopathies, types II and XI, including Kniest syndrome.
- *Type III*: This collagen is related to granulation tissue, the connective tissue that replaces a fibrin clot in healing wounds. It is produced quickly by young fibroblasts and is also found in artery walls, skin, intestines, and the uterus. Related gene is "collagen 3, alpha 1" or *COL3A1*. Mutations in this gene lead to certain types of Ehlers-Danlos syndrome.
- *Type IV*: This collagen makes up the basal lamina (a layer on which the epidermis of the skin lies), eye lens, filtration system in the capillaries, and in the cells of the kidney. Several genes are related here: *COL4A1*, *COL4A2*, *COL4A3*, *COL4A4*, *COL4A5*, and *COL4A6*. Disorders connected to this type are Alport syndrome, a disease of the kidneys, and Goodpasture syndrome, a kidney disorder.
- *Type V*: This collagen makes up most interstitial tissue, which lies between other tissues or organs and is associated with type I and with the placenta. Genes are *COL5A1*, *COL5A2*, and *COL5A3*. Classical Ehlers-Danlos syndrome is a result of mutations of these genes.
- *Type VI*: This collagen makes up interstitial tissue and is also associated with type I. Genes are *COL6A1*, *OL6A2*, and *COL6A3*. Related conditions are two

forms of congenital muscular dystrophy called Ulrich myopathy and Bethlem myopathy.

- *Type VII*: This type of collagen anchors fibrils in the dermis of the skin. Related gene is *COL7A1*. Mutations in the gene cause epidermolysis bullosa dystrophica.
- *Type VIII*: This type of collagen is found in some endothelial cells, which line the blood vessels and is associated with type I. Genes related to this type are *COL8A1* and *COL8A2*. Condition related is posterior polymorphous corneal dystrophy 2, an eye disease.
- *Type IX*: This type affects FACIT collagen and cartilage, and is associated with type II and XI fibrils. Related genes are *COL9A1*, *COL9A2*, and *COL9A3*. The disease correlated with the genes is epiphyseal dysplasia.
- *Type X*: This type is found in mineralizing cartilage. The related gene is *COL10A1*. A type of dysplasia called Schmid dysplasia is the result of mutations in this gene.
- *Type XI*: This type is found in cartilage. Genes are *COL12A1* and *COL12A2*. Certain types of collagenopathy, such as types II and XI, are the result of mutations in this gene.
- *Type XII*: The term Fibril Associated Collagens with Interrupted Triple helices, or FACIT, refers to a type of collagen that does not following the typical pattern of the triple helix. This type interacts with the abundant Type I. It is also another substance called a proteoglycan. Gene is *COL12A1*. There are no mutations related to genetic disorders.
- *Type XIII*: This is a transmembrane collagen that interacts with several components to form basement membranes of many organs. Gene is *COL13A1*. No mutations are related to genetic disorders.
- *Type XIV*: This type relates to FACIT collagen. Gene is *COL14A1*. No genetic conditions are indicated here.
- *Type XV*: This type of collagen is being studied. Gene is *COL15A1*. No genetic conditions are known.
- *Type XVI*: This collagen is being studied. *COL16A1* is the gene. No genetic conditions are known.
- *Type XVII*: This collagen is a transmembrane collagen. Gene is *COL17A1*. Mutations in this gene cause bullous pemphigoid and some kinds of epidermolysis bullosa.
- *Type XVIII*: This collagen is a source of endostatin. The gene is *COL18A1*. No genetic defects are known.
- *Type XIX*: This collage is a FACIT collagen. The gene is *COL19A1*. No known disorders are related to mutations in this gene.
- *Type XX*: Gene is *COL20A1*. No known mutations are related to this gene.

- *Type XXI*: This collagen is a FACIT collagen. Gene is *COL21A1*
- *Type XXII*: Gene is *COL21A1*.
- *Type XXIII*: Multiple triple-helix domains and interruptions or MACIT collagen are formed by *COL23A1*. No genetic disorders are related here.
- *Types XXIV, XXV, XXVI, XXVII, and XXVIII*: These are related to the genes bearing their numbers. No genetic disorders have been connected to these genes.
- *Type XXIX*: This collage is epidermal collagen. The gene is *COL29A1*. Atopic dermatitis is caused by mutations in this gene.

Collagen is used in many products and processes such as cosmetics, photography, and pharmaceuticals. A disease that was common in sailors in the nineteenth century was scurvy, a condition in which the men developed severe weakness, bleeding gums, loss of teeth, and serious skin issues. Scurvy was caused by a lack of vitamin C, the substance essential for the making of collagen.

Further Reading

"Collagen." 2005. 3DChem.com. http://www.3dchem.com/molecules.asp?ID=195. Accessed 2/10/12.

"Collagen: The Fibrous Protein of the Matrix." 2000. In H. Lodish et al., *Molecular Cell Biology*. 4th ed. New York: W.H. Freeman and Company. http://www.ncbi.nlm.nih.gov/books/NBK21582. Accessed 2/14/12.

"What Is Collagen?" 2012. http://www.news-medical.net/health/Collagen-What-is-Collagen.aspx. Accessed 2/14/12.

Congenital Hypertrichosis

Prevalence Very rare; known in a rare collection of families throughout the world

Other Names Ambras syndrome; an informal name not accepted by the medical profession is Werewolf syndrome

In 1648, Aldrovandus recorded the first case of hypertrichosis. Petrus Gonzales, who lived on the Canary Islands, was a member of a family who had two daughters, a son, and a grandchild with hair growing on the face, the body, and other areas where hair does not grow. Portraits of the family were found in the Ambras castle, near Innsbruck, Austria. During the next 300 years, over 50 cases were seen. In 1873 the German scientist Rudolf Virchow related hypertrichosis to another symptom called gingival hyperplasia, a malformation of the gums.

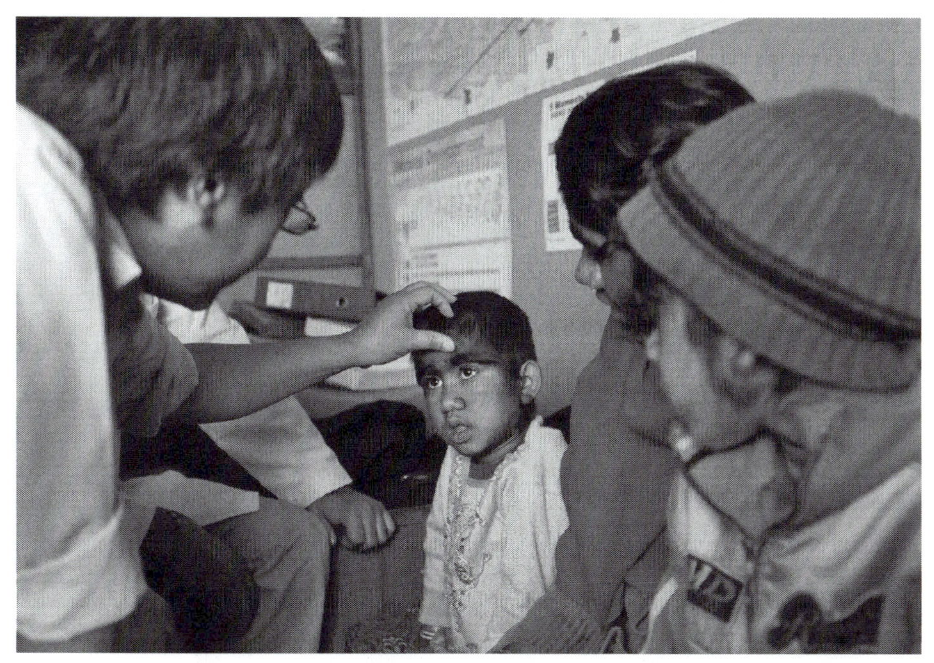

A Nepalese doctor inspects a child who suffers from hypertrichosis, at a hospital in Kathmandu in 2012. A family suffering from this rare genetic condition in which hair grows all over the face arrived in the Nepalese capital of Kathmandu for treatment of their "werewolf-like" appearance. The mother, her daughters, and son say they have endured constant humiliation because of the disease. (Getty Images)

Although the condition is rare, people with congenital hypertrichosis do not fare well in society. At one time, some people referred to the disorder as the "werewolf syndrome." Some individuals found word in circus sideshows of the nineteenth and twentieth centuries. Julia Pastrana called herself "the Bearded Lady." Recently, families with hypertrichosis have been found in Mexico, Burma, and Thailand.

What Is Congenital Hypertrichosis?

Congenital hypertrichosis is a condition in which hair covering the body is present at birth. The word "hypertrichosis" comes from three Greek words: "hyper," meaning "above" or "beyond"; *trich*, meaning "hair"; and *osis*, meaning "condition of." The growth of hair is present in areas of the skin that are not dependent upon male hormones. These areas include the pubic area, face, and the axillary regions or underarm regions,

Hypertrichosis must not be confused with hirsutism. Women and children with hirsutism have adult male hair growth, often on the chest and back. It is linked to male hormones in women and may also show symptoms of acne, deep voice, and irregular menstrual periods. This condition may be both congenital and acquired. This disorder may be treated with medications that reduce male hormones.

This rare condition is always present at birth. There are several different kinds of the disorder:

- Hypertrichosis lanuginosa: Lanuga hair is extremely fine hair that normally covers the embryo, perhaps to protect it from the strong amniotic fluid. It is replaced just before birth by vellus hair. The person with congenital hypertrichosis keeps the lanuga hair after birth, except for the palms, feet, and inside of the mouth. The hair is usually silver or blond in color and grows away from the body midline. As the child ages, the lanugo hair may remain only in limited areas.
- Generalized hypertrichosis: Males have lots of facial and upper body hair; women have less hair, which may appear in asymmetrical patterns.
- Terminal hypertrichosis: The person has thick, dark hair all over the body, as well as severe gum disease. This form is the one connected with the werewolf syndrome and other anomalies.
- Circumscribed hypertrichosis: This type is related to thick vellus hair on the upper part of the body or on certain body parts. One type, called the Hairy Elbow syndrome, has excessive growth of hair on the elbows.
- Localized hypertrichosis: Hair is thick and dense only on certain body parts.
- Acquired hypertrichosis: This type appears after birth and can be from side effects of drugs, cancers, or eating disorders. This type affects the cheeks, upper lip, and chin.

What Are the Genetic Causes of Congenital Hypertrichosis?

Three genetic causes have been identified: mutations in an X-linked gene and mutations on chromosomes 8 and 17.

X-linked dominant gene: Males with this disorder appear when the mother is a carrier and has a 50-50 chance of passing the trait onto her offspring. A male carrier will pass to his daughter but never to the sons. The generalized form is inherited in a dominant pattern and is located on the long arm (q) of the X chromosome at position 24-q27.1.

Chromosome 8: The type, congenital hypertrichosis lanuginosa, may be related to a mutation on this chromosome; however, it may be the result of a spontaneous mutations rather than inheritance.

Chromosome 17: The terminal type of hypertrichosis is related to mutations in chromosome 17, which result in the removal of many amino acid building blocks. The gene *MAP2K6* may be a factor that relates to transcription of genes on this chromosome.

What Is the Treatment for Congenital Hypertrichosis?

No cure exists for the congenital forms of hypertrichosis. For acquired conditions, it is important to determine the underlying cause. Following are some of the ways to treat the condition:

- Temporary removal using shaving and chemical depilatories.
- Epilation methods such as plucking, electrology, waxing, sugaring, and threading may remove hair from the roots and last several weeks.
- Laser hair removal, devices for which the Food and Drug Administration (FDA) has approved for permanent hair reduction.
- Certain medications are being tested. One option suppresses testosterone by increasing the sex-hormone-binding globulin.

Further Reading

"Congenital Hypertrichosis Lanuginosa." 2011. Medscape.

http://emedicine.medscape.com/article/1072987-overview. Accessed 12/31/11.

"Hypertrichosis." 2003. DermNet NZ. http://dermnetnz.org/hair-nails-sweat/hypertrichosis.html. Accessed 12/31/11.

"Hypertrichosis Information." 2011. http://www.hypertrichosis.com. Accessed 12/31/11.

Cornelia de Lange Syndrome (CdLS)

Prevalence Approximately 1 in 10,000 to 30,000 newborns

Other Names BDLS; Brachmann-DeLange syndrome; CDLS; De Lange syndrome

In 1916, Dr. W. Brachmann described a patient with specific facial symptoms and problems with the upper arms, but he stopped his investigations into a possible new syndrome. It was not until 1933 that Dr. Cornelia de Lange, a Dutch pediatrician, had two patients, one 17 months of age and the other 6 months. Surprisingly, these children looked more like each other than members of their families. The first was admitted to Emma Children's Hospital with pneumonia and a history of feeding difficulties. Dr. de Lange noted that in addition to common medical problems, the two had very small heads and distinct facial features. Later, the condition was named for this Dutch physician, Cornelia de Lange syndrome.

What Is Cornelia de Lange Syndrome (CdLS)?

Cornelia de Lange syndrome (CdLS) is a congenital condition affecting many parts of the body. Although the features may vary widely, certain common signs and symptoms are noted at birth or soon after. Several of the following features characterize CdLS:

- Low birth weight, often under 5 pounds/2.5 kilograms
- Developmental delay with small stature and slow growth
- Distinctive facial appearances including: thick eyebrows that meet at midline (synophrys); long, curly eyelashes; short, turned-up nose; long philtrum; small, widely spaced teeth; low-set ears; and cleft palate with feeding problems
- Hearing and vision abnormalities
- Missing limbs or portions of limbs; small hands and feet; partial joining of the second and third toes; incurved fifth finger
- Gastroesophageal reflux
- Excessive body hair (hirsutism)
- Heart defects and seizures

Children with CdLS may exhibit a number of behavior problems, including autistic-like behaviors. They may be aggressive, injure themselves, or develop a strong preference to certain routines.

What Is the Genetic Cause of Cornelia de Lange Syndrome (CdLS)?

Three genes appear to be involved in CdLS: *NIPBL*, *SMC1A*, and *SMC3*. Normally, these genes produce proteins that are essential for directing development before birth. Within the cells, the proteins regulate and organize chromosomes that repair DNA. These genes also control development of the limbs, face, and other body parts.

Mutations in these three genes disrupt the normal function of gene regulation during early development. Following are the multiple genes involved in this condition:

- *NIPBL*, officially known as the "Nipped-B homolog (Drosophila)" gene. This gene accounts for about 50% of cases and is inherited in an autosomal dominant pattern. Almost all are from new mutations of the gene occurring in children whose family has no history of the condition. This gene is found on the short arm (p) of chromosome 5 at position 13.2. The gene was discovered jointly in 2004 by researchers in Philadelphia and in the United Kingdom.
- *SMC3*, officially known as the "structural maintenance of chromosomes 3" gene. This gene is responsible for a milder form of the syndrome and is found in about 1% of cases. This inheritance pattern is also autosomal dominant. It was discovered in 2007 by a Philadelphia research team. The gene is located on the long arm (q) of chromosome10 at position 25.

- *SMC1A*, officially known as the "structural maintenance of chromosomes 1A" gene. This gene is different from the first two in that it is found on the X chromosome. However, unlike most X-linked conditions, CdLS appears to affect males and females in a similar manner. These generally appear as new mutations with no history of the condition in the family. The gene is located on the short arm (p) of the X chromosome at position 11.22-p11.21.
- In about 35% of cases, the cause is unknown. A possible link may be to chromosome 3q26.3.

What Is the Treatment for Cornelia de Lange Syndrome (CdLS)?

There is no cure for this disorder. However, attention to the medical issues that arise can alleviate the symptoms. The Cornelia de Lange Syndrome Foundation is a support group that provides public information and assistance.

Further Reading

"Cornelia de Lange Syndrome." 2012. Genetics Home Reference. National Library of Medicine (U.S.) http://ghr.nlm.nih.gov/condition/cornelia-de-lange-syndrome. Accessed 2/1/12.

"Cornelia de Lange Syndrome." 2011. Medscape. http://emedicine.medscape.com/article/942792-overview. Accessed 5/10/12.

The Cornelia de Lange Syndrome Foundation. 2011. http://www.cdlsusa.org. Accessed 5/10/12.

Costello Syndrome (CS)

Prevalence Rare; an estimated 200 to 300 people worldwide

Other Names faciocutaneoskeletal syndrome; FCS syndrome

Some conditions are very similar to others in their symptoms making identifying the condition difficult, especially in infancy. Grouping disorders is also difficult because there are so many mutations. Two other conditions—cardiofaciocutaneous (CFC) and Noonan syndrome—have overlapping signs and symptoms with Costello syndrome.

The symptoms of Costello syndrome were first described in 1977, but it was not until 2005 that the genetic mutation of the gene *HRAS* was pinpointed as the cause. Spanish researchers developed a Costello mouse in 2008, and Italian and Japanese researchers developed a Costello zebra fish in late 2008. The creation of these models may assist in developing treatments for the disorder.

What Is Costello Syndrome (CS)?

Like so many other syndromes, Costello syndrome (CS) affects many parts of the body. The mother may note the first sign of CS when an excess of amniotic fluid, which is a condition known as polyhydramniosis, develops. Many of these babies begin life with a more than normal weight and large heads, but soon after birth, problems develop. They develop feeding problems and do not grow and thrive like most babies. They begin to lose weight and become ill.

The following symptoms may appear:

- Facial features: Coarse with full lips; epicanthal folds; wide nasal bridge with a short full nose
- Loose, soft skin, deep creases on the palm and foot, curly or sparse, fine hair
- Heart abnormalities: These include a very fast heart beat (tachycardia), structural heart defects, and overgrowth of the heart muscle, a condition called hypertrophic cardiomyopathy
- Cancerous and noncancerous lesions: Small wart-like growths called papillomas develop around the nose and mouth or near the anus. A cancerous tumor of muscle tissue around the nose and mouth called a rhabdomyosarcoma is frequent. Children may have a bladder cancer called transitional cell carcinoma, which is usually seen in older adults.

The children are usually developmentally delayed but are friendly and sociable.

What Are the Genetic Causes of Costello Syndrome (CS)?

The gene *HRAS* is implicated in 80% to 90% of cases of CS. The official name of *HRAS* is "v-Ha-ras Harvey rat sarcoma viral oncogene homolog." Normally, the gene instructs for making a protein called H-Ras that regulates cell division. A process called signal transduction tells the cell what to do, when to grow, and when to divide.

HRAS is a member of a class of oncogenes, which are genes that cause normal cells to become cancerous. Mutations in *HRAS* may cause the H-Ras protein to be permanently active. Instead of sending the proper signals, the overactive proteins tell cells to grow and keep replicating. The unchecked multiplication may cause the individuals to develop benign and malignant tumors, as well as other signs and symptoms of CS. CS is inherited in an autosomal dominant pattern. *HRAS* is located on the short arm (p) of chromosome 11 at position 15.5.

What Is the Treatment for Costello Syndrome (CS)?

At present, there is no cure and caring for the symptoms is important. Research into several medications is ongoing.

Further Reading

Costello Kids. 2008. http://costellokids.com. Accessed 5/10/12.

Gripp, K., and A. Lin. 2006. "Costello Syndrome." *GeneReviews*. http://www.ncbi.nlm.nih.gov/bookshelf/br.fcgi?book=gene&part=costello. Accessed 5/10/12.

Sweet, Kevin. 2002. "Costello Syndrome." http://www.healthline.com/galecontent/costello-syndrome. Accessed 5/10/12.

Cowden Syndrome (CS)

Prevalence About 1 in 200,000 people; exact prevalence unknown because of the difficulty of diagnosis

Other Names Cowden's disease; CS; MHAM; multiple hamartoma syndrome

Naming diseases and disorders varies with the history of the discovery of the condition. Many diseases are named after the people who first write about them, such as Alzheimer disease, or after a famous person, such as Lou Gehrig disease. Cowden syndrome is named after the surname of the patient who had the disease. In 1963, Drs. K. Lloyd and M. Dennis wrote about one of their patients who had a possible new complex of symptoms that involved many systems. The article published in the *Annals of Internal Medicine* referred to the patient, a Mr. Cowden, who had the disease. The name Cowden syndrome has stuck.

What Is Cowden Syndrome (CS)?

Cowden syndrome (CS) is a disorder characterized by small, noncancerous growths on the skin called a hamartoma. The word "hamartoma" comes from two Greek words: *hamartia*, meaning "defect," and *oma*, meaning tumor; thus the hamartoma is a defective tumor. In CS, hamartomas are found not only on the external skin but on the lining of the mouth and nose, intestinal tract, or other body parts. The cells grow spontaneously, reach maturity, and then stop growing. Hence, they are contained and considered benign. The growths usually appear in individuals with CS in their 20s. However, people with these hamartomas have an increased risk for cancer.

Many signs and symptoms may be associated with CS, which may vary with individuals. Symptoms include:

- Learning disabilities, autism, and mental retardation.
- A large head, including a rare form of noncancerous brain tumor called Lhermitte-Duclos disease.

- Many lesions or papules (bumps) on the skin. Types of lesion, which a dermatologist can recognize, are: trichelemmomas (benign tumors on the external root of the hair follicle; papillomatous lesions that can have a cobblestone appearance; and keratoses or hard growths on the palms of hands or soles of feet.
- High risk of developing other cancerous and noncancerous lesions.

Cancer risks are high in people with CS. According to the M. D. Anderson Cancer Center, Houston, Texas, men and women with CS have a 10% lifetime risk to develop thyroid cancer, compared to 1% among the general population. Women with CS have a 25% to 50% lifetime risk to develop breast cancer and at a younger age, compared to 12% among the general population. Benign tumor risks include goiters; polyps in the stomach, small intestine, or colon; uterine fibroids; fibrocystic breast changes; and lipomas (benign fatty tumors) and fibromas. Because an individual with CS is at an increased risk to develop cancer, it is important to get consistent screening for cancer often.

What Is the Genetic Cause of Cowden Syndrome (CS)?

Mutations in the *PTEN* gene cause Cowden syndrome. The official name of this gene is "phosphatase and tensin homolog." Normally, *PTEN* gives instructions for making a protein that acts as a tumor suppressor and is found in almost all tissues of the body. A tumor suppressor regulates cell division, keeping them from growing and dividing too rapidly or in a controlled way. *PTEN* protein acts as a regulator and controller of cell death.

Over 100 mutations in *PTEN* cause Cowden syndrome. These changes probably involve a small number of DNA building blocks or base pairs and lead to the production of an enzyme that does not function properly. Because the activity of the enzyme is disrupted, cells multiply abnormally and contribute to the development of noncancerous growth and tumors.

The condition is inherited in an autosomal dominant pattern. It may be inherited from a parent or may be the result of a new mutation. *PTEN* is located on the long arm (q) of chromosome 10 at position 23.3.

What Is the Treatment for Cowden Syndrome (CS)?

There is no cure for CS. Cancer prevention is a main management strategy. The following screenings should be performed in people with CS: breast cancer screening, every six months beginning at age 25; thyroid screening in both men and women beginning at age 18; and other screenings including yearly skin exams and uterine cancer risk. Because CS is a genetic disorder, other members of the family should also be carefully screened.

Further Reading

Adkisson, Kendall. 2010. "Cowden Syndrome (Multiple Hamartoma Syndrome)." Medscape. http://emedicine.medscape.com/article/1093383-overview. Accessed 5/10/12.

"Cowden Syndrome." 2010. Genetics Home Reference. National Library of Medicine (U.S.). http://ghr.nlm.nih.gov/condition/cowden-syndrome. Accessed 5/10/12.

"Cowden Syndrome." 2010. M. D. Anderson Cancer Center. http://www2.mdanderson.org/app/pe/index.cfm?pageName=opendoc&docid=2190. Accessed 5/10/12.

Creutzfeldt-Jakob Disease. *See* Prion Disease

Cri-du-Chat Syndrome

Prevalence One in 20,000 to 50,000

Other Names cat cry syndrome; chromosome 5p syndrome; 5p deletion syndrome; monosomy 5p; 5p-syndrome; LeJeune's syndrome

What does the mewing of a little kitten and a serious genetic disorder have in common? Answer: The cry of a child with cri-du-chat syndrome. First described by French physician Jerome Lejeune in 1963, the sounds of the infant reminded him of the cry of a little cat. Cri-du-chat is the French term for cat-cry or cry of the cat. The syndrome gets its name from this characteristic cry, which is due to problems with the larynx and the nervous system. About a third of the children lose the cry by age 2.

What Is Cri-du-Chat Syndrome?

Cri-du-chat syndrome is a condition in which a piece of chromosome 5 is missing. The high-pitched cry that sounds like that of a cat is the first major sign of the condition. Intellectual disability is almost always present. This syndrome may account for up to 1% of those with mental retardation. Several other symptoms may also be present:

- Very small head (microcephaly)
- Low birth weight and slow growth
- Weak muscle tone
- Distinct facial features that may include downward slant to the eyes, small jaw, wide-set eyes, rounded face, skin tags in front of the ears, and extra skin folds over the inner corner of the eye
- Body abnormalities may include partial webbing or fused fingers and toes, heart defects, inguinal hernia, and separated abdominal muscles
- Slow or incomplete development of motor skills

Skull X-ray may reveal an abnormal angle to the skull. Genetic tests show that part of chromosome 5 is missing.

What Is the Genetic Cause of Cri-du-Chat Syndrome?

Cri-du-chat is caused by a chromosomal aberration. The aberration can happen in two ways. First, a piece at the end of chromosome 5 called TERT (telomerase reverse transcriptase) is missing and is a random happening during the formation of eggs or sperm or possibly in early fetal development. Second, an unaffected parent may carry a chromosome rearrangement called a balanced translocation. However, the translocation can become unbalanced when passed to an offspring. Regardless of the means, children with cri-du-chat are missing genetic material that is involved in the control of cell growth and that plays a role in the development of the symptoms of the syndrome.

The area that is missing is a small region on the short arm (p) of chromosome 5 at position 12.3. Two genes, Semaphorine F (*SEMA5A*) and delta catelin (*CTNND2*), are in the region and possibly related to brain development. Deletion of TERT may also contribute to the characteristic conditions of the syndrome.

What Is the Treatment for Cri-du-Chat Syndrome?

No specific treatment is available for the condition itself, but symptoms may be treated. Parents should have genetic counseling and a karyotype test to determine if one parent has a rearrangement of chromosome 5. There is a support group called 5p-Society.

Further Reading

"Cri-du-Chat Syndrome." 2010. Learn.Genetics, University of Utah. http://learn.genetics.utah.edu/content/disorders/whataregd/cdc. Accessed 5/10/12.

"Cri-du-Chat Syndrome." 2010. MedlinePlus. National Library of Medicine (U.S.). http://www.nlm.nih.gov/medlineplus/ency/article/001593.htm. Accessed 5/10/12.

Cri du Chat Syndrome Support Group (UK). 2009. http://www.criduchat.org.uk. Accessed 5/10/12.

Crohn Disease

Prevalence Common in Western Europe and North America; affects 100 to 150 in 100,000 people; about one million Americans have disorder; more common among Caucasian and people of Ashkenazi Jewish descent than among other ethnic backgrounds

Other Names colitis, granulomatous; Crohn's disease; Crohn's enteritis; enteritis granulomatous; enteritis regional; ileitis; ileocolitis; inflammatory bowel disease (IBD)

The first writings in medical history talk of intestinal and bowel troubles; however, it was not until the eighteenth century that the Italian Giovanni Morgagni (1682–1771) suspected that inflammation was the cause. A Polish surgeon, Antoni Lesniowski, first described it as ileitis terminalis in 1904. When Dr. Burrill Crohn, a gastroenterologist, described a group of patients who had inflammation of the terminal ileum in 1932, the disorder became known as Lesniowski-Crohn's disease. However, because of the cumbersome name and the fact the Crohn's name was first in the alphabet, medical literature began to call the disorder simply Crohn's disease. Only in Poland does it still have the Polish name. Crohn disease and a related disease ulcerative colitis belong to a larger group of illnesses called inflammatory bowel disease (IBD).

What Is Crohn Disease?

Crohn disease is a chronic disorder that involves inflammation of the digestive tract. Although it may affect any part of the digestive tract from the mouth to the anus, the disorder primarily involves inflammation of the intestinal walls, in which swelling may cause thick and abnormal tissues. Crohn disease is sometimes called regional ileitis or just ileitis. The ileum is the last part of the small intestine before it joins to the large intestine. It is similar to a disease called ulcerative colitis that affects only the colon. However, Crohn disease can affect any part of the digestive system.

Crohn disease may occur at any age, but the usual time of onset is between 15 and 30 years of age. The individual may have some symptoms prior to diagnosis, and initial symptoms may be vague and sporadic. The inflamed mucous lining can cause the inner surface of the digestive tract to develop open sores or ulcers. One of the first symptoms is abdominal pain and cramping, but then the person may experience diarrhea, weight loss, and fever. Some people may experience bleeding from the tissues and over time can lose red blood cells, causing anemia. The signs and symptoms of Crohn disease tend to flare up and then subside.

Over time, the chronic disease takes a toll over the digestive system. Looking at the intestinal lining has been described as viewing a cobblestone street with areas of ulceration separated by healthy tissue. Intestinal blockages from the buildup of scar tissue may cause the individual to develop fistulae, which are abnormal connections between intestine and other tissues. Ulcers in the intestine break through and form passages to nearby organs such as the bladder, vagina, or skin. Also, inflammation damages the lining so that nutrients, water, and fats from food cannot be absorbed. The malabsorption causes malnutrition, dehydration, vitamin and mineral deficiencies, gallstones, and kidney stones.

Burrill B. Crohn (1884–1983), a gastroenterologist, was one of the people who discovered the disease that became known as Crohn disease. (National Institutes of Health)

Crohn disease can affect many other organs and systems. Following are some of the systems that are associated with Crohn:

- Uveitis, causing eye pain in the interior portion of the eye
- Inflammation of the white part of the eye or sclera
- Seronegative spondylarthropathy, a type of arthritis characterized by inflammation of one or more joints
- Ankylosing spondylitis
- Skin conditions such as erythema nodosum (red nodules) and a gangrene condition
- Risk for blood clots
- Autoimmune conditions that attack red blood cells
- Fatigue and tiredness, associated with anemia
- Clubbing, a deformity at the end of the fingers
- Osteoporosis
- Neurological complications

What Are the Genetic Causes of Crohn Disease?

A combination of factors is involved in Crohn disease. Some of the factors may be environmental. For example, smokers have a higher risk of developing the disease. Other causes relate to the immune system and certain bacteria that are present in the intestines. However, although the inheritance pattern is not known, genetic factors are also involved as evidenced by its cluster in families. Having one family member with the disorder is a significant risk for developing the disease.

Recent studies have implicated a number of mutations in several genes that may increase the development of Crohn disease. These genes are *ATG16LI*, *IRGM*, *NOD2*, and *IL23R*, as well as chromosomes 5 and 10. Normally, these genes give the instructions for making proteins that are involved in the function of the immune system. Following is the breakdown of genetic involvement in the disorder:

- Chromosome 5: A combination of genes on the long arm (q) at position 31 and variation in the short arm at position 31 has been shown to increase risk for developing Crohn disease. Some of the areas of chromosome 5 have been referred to as a "gene desert" because it contains no known genes.
- Chromosome 10: This chromosome is another "gene desert" with no known genes, but it possibly contains some DNA that regulates nearby genes such as *ERG2*. Position 21.1 at 10q has been associated with risk of Crohn disease.

- *IL23R*: *IL23R*'s official name is "interleukin 23 receptor," and it is located on the short arm of chromosome 1 at position 31.3. The association of risk of this gene with Crohn disease is found primarily in Caucasians. The gene appears to have a protective factor against developing the disorder; mutations in the gene disrupt this protection.
- *ATG16L*: The name of this gene is "ATG16 autophagy related 16-like 1 (S.cerevisiae)." It is located on the long arm (q) of chromosome 2 at position 37.1. This gene has at least one variation where alanine replaces threonine and makes the difference in the action of the protein. These changes affect a process called autophagy in the cell, which allows worn-out parts and persistent bacteria to exist, leading to an improper immune system response and the problems of the disorder.
- *IRGM*: The name of this gene is "immunity-related GTPase family, M." It is located on the long arm (q) of the infamous chromosome 5 at position 33.1. Normally, this gene gives the instructions for a protein that is important in the immune system. The protein is also important in the autophagy process. Several variations of this gene have been associated with the risk of Crohn disease.
- *NOD2*: The official name of this gene is "nucleotide-binding oligomerization domain containing 2." It is located on the long arm (q) of chromosome 16 at position 21. Normally, *NOD2* gives instructions for the protein nucleotide-binding oligomerization domain containing 2, which is active in the immune system to protect against foreign invaders. More than 30 mutations of the *NOD2* genes have been associated with the chromic inflammation and digestive problems of Crohn disease.

What Is the Treatment for Crohn Disease?

Crohn disease is a very complex disorder with still a lot of unanswered questions about the function. However, treatment includes medicines, nutrition supplements, surgery, or a combination of the three. Some people have long periods of remission when they are free of symptoms.

Further Reading

"Crohn Disease." 2010. Genetics Home Reference. National Library of Medicine (U.S.).

http://ghr.nlm.nih.gov/condition/crohn-disease. Accessed 5/10/12.

Crohn's and Colitis Foundation of America. 2009. http://www.ccfa.org/info/about/crohns. Accessed 5/10/12.

"Crohn's Disease." 2010. MedlinePlus. National Library of Medicine (U.S.). http://www.nlm.nih.gov/medlineplus/crohnsdisease.html. Accessed 5/10/12.

Crouzon Syndrome

Prevalence About 16 per million newborn

Other Names craniofacial dysarthrosis; craniofacial dysostosis; craniofacial dysostosis, type 1 (CFD1); craniofacial dysostosis syndrome; Crouzon dysostosis; Crouzon's disease; Crouzons disease

In 1912, Octave Crouzon, a French physician, noted that a mother and her daughter had similar abnormal head and facial features. The obvious similarity of the bulging eyes and downward-slanting eyelids made the doctor suspect this was a genetic disorder of some kind. Writing in a French medical journal, Crouzon called the condition a hereditary craniofacial disorder or dysostosis. He had just described the most common craniosynostosis or premature growing together of the sutures in the skull. The disorder now bears his name: Crouzon syndrome.

What Is Crouzon Syndrome?

Crouzon syndrome is a condition in which the bones of the skull fuse together prematurely. It is one of a group of conditions known as craniosynostosis. The word "craniosynostosis" comes from the Greek root words: *cranio*, meaning "skull" or "head"; *syn*, meaning "together"; *ost*, meaning "bone"; and *osis*, meaning "condition of." Thus the syndrome is characterized by the abnormal growth that causes the skull bones to close together prematurely, causing the symptoms and signs of the disorder. Following are the characteristics of Crouzon syndrome:

- Premature fusion of the skull, causing the head to appear abnormal
- Wide-set, bulging eyes due to shallow eye sockets, leading to vision problems
- Eyes that do not point in same direction (strabismus or crossed-eyes)
- Small, underdeveloped jaw, causing serious dental disorders
- A beaked nose, with a parrot-like appearance
- High, narrow arched palate; may have a cleft palate
- Darkened, rough patches of skin called *Acanthosis Nigricans*, which appears between the ages of 2 and 4
- Downward-slanting eyelids

The symptoms may vary with the individual. People with Crouzon usually have normal intelligence.

What Is the Genetic Cause of Crouzon Syndrome?

Changes in the gene *FGFR2*, officially known as the "fibroblast growth factor receptor 2" gene, are associated with Crouzon syndrome. The gene programs for the protein-fibroblast growth factor receptor 2, one of the major proteins involved in important growth processes. These processes include: cell division, cell growth regulation, maturation, blood vessel formation, wound healing, and embryonic development. The FGFR2 protein extends through the cell membrane, with one end remaining inside the cell and the other protruding from the cell's surface. The projecting part interacts with growth factors outside the cells and helps the cell undergo certain changes to take on specialized functions. The protein signals the developing embryo when to become bone cells in the head, hands, and feet.

Mutations in the gene disrupt the function, causing the premature fusion and the characteristic symptoms of the disorder. The condition is inherited in an autosomal dominant pattern, which means one copy of the altered gene is in each cells. Because it is dominant, there is a 50% chance that an affected parent will pass the condition for the offspring. However, there may be new mutations when neither parent has the disorder. Scientists have identified at least 35 mutations in the *FGFR2* gene. *FGFR2* is located on the long arm (q) of chromosome 10 at position 26.

What Is the Treatment for Crouzon Syndrome?

The symptoms of Crouzon may be treated with surgery and medical attention to specific disorders. Usually the condition is determined at birth, and surgery may repair the area that has closed prematurely. Depending on the severity of the condition, the child may need frontal orbital advancement to allow the skull to grow properly, jaw surgery, and surgery on the mid-face area.

FACES is the National Craniofacial Association, a support group that provides information, networking, and some financial support for nonmedical expenses.

Further Reading

"*FGFR2*." 2010. Genetics Home Reference. National Library of Medicine (U.S.). http://ghr.nlm.nih.gov/gene/FGFR2. Accessed 5/10/12.

"Genetics of Crouzon Syndrome." 2010. MedScape. http://emedicine.medscape.com/article/942989-overview. Accessed 5/10/12.

The National Craniofacial Association (FACES). "Crouzon Syndrome." http://www.faces-cranio.org/Disord/Crouzon.htm. Accessed 5/10/12.

Cyclic Vomiting Syndrome (CVS)

Prevalence About 4 to 2,000 per 100,000 children; less common in adults, although recent studies show it may be more prevalent than previously thought

Other Names CVS; cyclical vomiting; cyclical vomiting syndrome; periodic vomiting

Cyclic vomiting syndrome (CVS) is one condition whose name describes the symptoms. Severe vomiting comes in cycles or intervals with no symptoms. It was originally thought to be a pediatric disease but now appears to occur in all age groups. Researchers believe CVS and migraines headaches are related.

What Is Cyclic Vomiting Syndrome (CVS)?

Cyclic vomiting syndrome is a condition characterized by episodes of vomiting that can last for hours or even days, and then there are days with no symptoms. It is different from other conditions in which one may vomit for a short period of time. In this condition, the individual may retch 6 to 12 times an hour; the median episode duration is 41 hours. The majority of individuals can identify certain triggers that lead to the attack. Most common triggers are certain smells, sounds, infections such as colds, menstrual period, lack of sleep, and stress.

Each episode of CVS is similar to the previous one. The episodes tend to start at the same time each day, last the same amount of time, and occur with the same symptoms and level of intensity. According to the National Institute of Diabetes and Digestive and Kidney Diseases (NIDDK), CVS has the following four phases:

- Symptom-free phase: In this period there are no symptoms.
- Prodrome phase: Something signals that the vomiting is about to begin. This phase may be marked with nausea or with abdominal pain. Taking medicine during this phase may help.
- Vomiting phase: The individual experiences nausea and vomiting; they may not be able to eat, drink, or take medicines.
- Recovery phase: Nausea and vomiting stop; normal living resumes.

The condition usually appears between the ages of three and seven, although the exact number of people with the condition is not known. The current belief is that it can occur in all age groups, although some children with CVS tend to develop migraine as they age. The severe vomiting is a risk for many complications such as dehydration, electrolyte or salts imbalance, injury to the esophagus, and tooth decay. Because vomiting occurs in many conditions, it is often difficult to diagnose until several cycles are experienced.

What Is the Genetic Cause of Cyclic Vomiting Syndrome (CVS)?

Researchers believe that CVS is a migraine-like condition related to changes in signaling between nerve cells in certain areas of the brain. The inheritance of this disorder is different from others is that it is likely related to mutations in the mitochondrial DNA. The mitochondria are bean-shaped structures that are the powerhouses of the cells. They convert the energy from food into a form that cells can use. Most DNA is within the cell, but the mitochondria have their own bits of DNA known as mitochondrial DNA, or mDNA.

Like the genes on chromosomes that instruct for various cell functions, the DNA on the mitochondria instruct for normal mitochondrial function. mDNA has 37 genes with 13 giving instructions for making enzymes and adenosine triphosphate, the cell's main energy source. The remaining genes instruct for making transfer RNA (tRNA) and ribosomal RNA (rRNA).

Most cases of CVS are related to genetic changes in mDNA. Some of the changes are in only one of the base nucleotide pair. The mutations disrupt the ability of the mitochondria to produce energy and lead to periods when the body requires more energy. An example of this is when the immune system is fighting infection. However, exactly how changes in the mDNA are related to episodes of nausea and vomiting is unclear.

What Is the Treatment for Cyclic Vomiting Syndrome (CVS)?

Treatment may vary, but the person generally improves after learning to control the symptoms. Each phase has certain treatments. For example, during the prodrome phase, certain medications may help.

Further Reading

"Cyclic Vomiting Syndrome." 2008. National Institute of Diabetes and Digestive and Kidney Disorders. http://digestive.niddk.nih.gov/ddiseases/pubs/cvs. Accessed 5/10/12.

"Cyclic Vomiting Syndrome." 2010. Genetics Home Reference. National Library of Medicine (U.S.). http://www.ghr.nlm.nih.gov/condition/cyclic-vomiting-syndrome. Accessed 5/10/12.

Cyclic Vomiting Syndrome Association. 2010. http://www.cvsaonline.org. Accessed 5/10/12.

Cystic Fibrosis (CF)

Prevalence Affects about 1 in 2,500 to 3,500 Caucasian newborns; about 1 in 17,000 African Americans; and 1 in 31,000 Asian Americans

Other Names CF; cystic fibrosis of pancreas; fibrocystic disease of pancreas; mucoviscidosis

The symptoms and signs of cystic fibrosis (CF) were known before a name was ascribed to the disease. Some observers in eighteenth-century Germany wrote, "Woe is the child who tastes salty from a kiss on the brow, for he is cursed, and soon must die." In the nineteenth century, several others described some of the

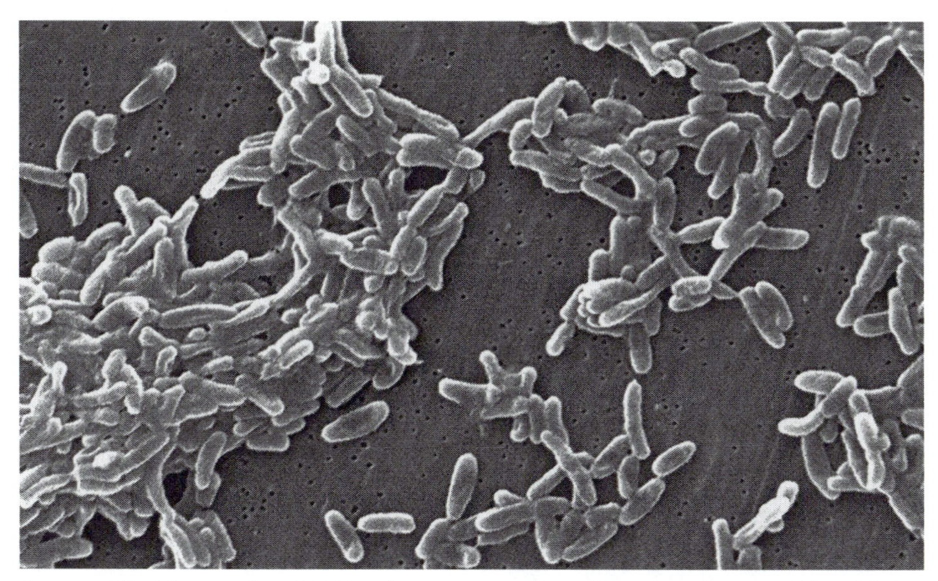

A magnified image (by 6,740x, from a scanning electron micrograph), which depicts a grouping of Ralstonia mannitolilytica bacteria. Ralstonia spp. are gram-negative bacteria found in the environment, primarily in water, soil, and on plants, but have been found in clinical samples of respiratory secretions of cystic fibrosis patients. (CDC/ Judith Noble-Wang, Ph.D.)

unusual symptoms. However, in 1938, Dr. Dorothy Hansine Andersen published an article called "Cystic Fibrosis of the Pancreas and Its Relation to Celiac Disease." She was the first to describe cystic fibrosis of the pancreas and connect it with the lung and intestinal issues. She also surmised that CF was a recessive disease and used a pancreatic enzyme to treat children with the disorder. In 1952, Paul di Sant Agnese found abnormalities in the sweat production and develop a test for CF using this in the 1960s. In the 1950s, children with CF could expect to live only into elementary school; today, many people with CF may live into their 40s.

The mutation to the CF gene is thought to be about 52,000 years old. Several ideas have developed as to why such a lethal mutation has persisted and spread. The concept is called heterozygote advantage, which means that carriers are protected from other diseases. Theories speculate that CFTR arose as possible protection from cholera, typhoid, diarrhea, and tuberculosis. Francis Collins, an important player in the Human Genome Project, found the first mutation on chromosome 7; researchers have now found over 1,000 different mutations.

What Is Cystic Fibrosis (CF)?

Cystic fibrosis is a childhood disease, which affects the entire body and often leads to an early death. The name "cystic fibrosis" refers to fibrous tissue or scarring of the pancreas. The condition is characterized by the production of thick, sticky mucus. Normally, mucus is a slippery, fluid substance made by the lining of body tissues that keeps the lining of certain organs from drying out and getting infections. When the mucus becomes thick and heavy, it clogs the lungs, leading to serious infections, and obstructs the pancreas, stopping the normal production of enzymes that are essential for digestion of food. CF causes sweat to become very salty and the body tissues consequently need large amounts of salt. This sweating upsets the electrolyte balance of minerals in the blood and causes numerous health problems.

Following are the symptoms of CF:

- Poor growth and weight gain, although the food intake is normal
- Infections that block airways and cause frequent coughing
- Frequent sinus infection, bronchial infections, and pneumonia
- Diarrhea
- Oily, foul-smelling feces
- Severe constipation
- Pancreatitis
- Liver disease
- Diabetes
- Gallstones
- Dehydration
- Reproductive infertility

The symptoms often appear in infancy or childhood. A bowel obstruction called meconium ileus may be in newborns. Meconium is the first feces of a newborn infant and is normally passed within 24 hours. In infants with CF, the meconium may be impacted in the intestines. As the child grows, the thick mucus begins to obstruct the airways.

What Is the Genetic Cause of Cystic Fibrosis (CF)?

Cystic fibrosis is caused by a mutation in the *CFTR* gene, officially known as the "cystic fibrosis transmembrane conductance regulator (ATP-binding cassette subfamily C, member 7) gene. Normally, *CFTR* instructs for a protein called the cystic fibrosis transmembrane conductase regulator, which functions to produce mucus, sweat, saliva, tears, and digestive enzymes. Chloride ions, which are negatively charged atoms, are ushered in and out of the cells through a chloride channel that controls the movement of water in tissues. Water is essential for the normal production of mucus and other substances.

More than 1,000 mutations in *CFTR* have been found in people with CF. The most common mutation is DF508, which is the result of a deletion of three base nucleotides resulting in the loss of the amino acid phenylananine; 90% of the cases in the United States are related to this mutation. These mutations disrupt the structure and function of the chloride channel and the movement of water in and out of the cells. The result is the thick and sticky mucus that causes the many problems of CF.

Everyone inherits two *CFTR* genes, one from each parent. The gene is inherited in an autosomal recessive pattern. The child with CF has two mutated genes, one from each parent. If a person has one of the genes, he or she will be a carrier and not have the disease; however, it can be passed on to their offspring. *CFTR* is found on the long arm (q) of chromosome 7 at position 31.2.

What Is the Treatment for Cystic Fibrosis (CF)?

Because CF is a genetic disorder, there is no cure; so good management and proactive treatment, especially of the airways, is important. Mechanical methods and medications may help the flow of mucus. Good nutrition and an active lifestyle aids in maximizing organ function.

Further Reading

"Cystic Fibrosis." 2010. MedicineNet. http://www.medicinenet.com/cystic_fibrosis/article.htm. Accessed 5/10/12.

Cystic Fibrosis Foundation. 2010. http://www.cff.org. Accessed 5/10/12.

"What Is Cystic Fibrosis?" 2010. National Heart Lung and Blood Institute. National Institutes of Health. http://www.nhlbi.nih.gov/health/dci/Diseases/cf/cf_all.html. Accessed 5/10/12.

D

Dandy-Walker Syndrome (DWS)

Prevalence One in 25,000 births, mostly females

Other Names Dandy-Walker complex; DWS

In 1914, Walter Edward Dandy (1886–1946), an American neurosurgeon, and a colleague noted a 13-month-old child with several neurological problems. The child had hydrocephalus, which is a collection of fluid on the brain, cysts in the back of the skull, and notable underdevelopment in the area of the cerebellum. The cerebellum, the lower portion of the brain, is about the size of an orange and coordinates voluntary movements. The two lobes of the cerebellum are joined by a structure called the vermis and surrounded by fluid. Later in 1921, Dandy described the clinical features and symptoms of the condition. In 1942, Arthur Walker (1907–1995) presented case findings involving six cases and referred to the Dandy study at a conference in Chicago. The name Dandy-Walker was given to the syndrome in 1954.

What Is Dandy-Walker Syndrome (DWS)?

Dandy-Walker syndrome (DWS) is a congenital brain malformation involving the cerebellum and the fluid areas that surround the back part of the brain. In the brain are series of ventricles or spaces that create channels for fluid to flow freely between the areas of the central nervous system. In DWS, a complete or partial absence of the vermis disrupts the flow of fluid. This leads to the increasing in size of another area called the fourth ventricle so that more fluid collects, increasing pressure on the brain. In addition, cysts may form at the base of the brain.

Symptoms of the syndrome usually occur in early infancy and can appear suddenly or developed unnoticed. Following are the signs of the disorder:

- Vomiting, extreme irritability, convulsion
- Slow motor development

- Progressive enlargement of the skull
- Lack of muscle coordination
- Jerky movements of the eyes
- Increased head circumference
- Bulging at the back of the head

DWS is often associated with other disorders of the nervous system including malformation of the corpus callosum, the area of nerve fibers that connect the hemispheres of the brain. Malformations of the heart, face, limbs, fingers, and toes may also be present. The condition develops *in utero* before birth.

What Is the Cause of Dandy-Walker Syndrome (DWS)?

The exact cause of DWS is unknown, although genetic factors may be involved. Several incidences of occurrences in families have been reported. Individuals with one child with DWS should seek genetic counseling. Multiple factors are probably involved and certain predisposing factors include the following:

- Exposure to rubella, the virus that causes German measles
- Cytomegalovirus (CMV), a sexually transmitted virus
- Toxoplasmosis, the condition sometimes called "cat-scratch fever"
- Alcohol
- Use of isotretinoin or Accutane during the first trimester
- Use of warfarin or coumidin

What Is the Treatment for Dandy-Walker Syndrome (DWS)?

Treatment for DWS consists of focusing on associated problems. For example, a shunt may be required to drain off excess fluid. Longevity depends of the severity of the syndromes. Some children may have normal learning ability, but others can also have severe disability and short life.

Further Reading

Dandy-Walker Alliance. 2010. http://www.dandy-walker.org. Accessed 5/10/12.

Elquist, Marty, and MaryAnn Demchak. 2010. "Dandy-Walker Syndrome." http://www.cde.state.co.us/cdesped/download/pdf/blv-DandyWalkerSyndrome.pdf. Accessed 5/10/12.

NINDS Dandy-Walker Syndrome Information Page. 2010. National Institute of Neurological Disorders and Stroke. http://www.ninds.nih.gov/disorders/dandywalker/dandywalker.htm. Accessed 5/10/12.

Danon Disease

Prevalence Rare; exact prevalence unknown

Other Names glycogen storage disease type 2B; glycogen storage disease type IIb; lysosomal glycogen storage disease with normal acid maltase

In 1981, Dr. Danon described two boys with heart and skeletal muscle disease, as well as mental retardation. When looking under the microscope, he determined that the muscles were similar to those of another rare genetic condition called Pompe disease. However, the fibers were much smaller, and tests for the specific conditions of Pompe disease were not present in these patients. He had found the rare disorder that carries his name—Danon disease.

What Is Danon Disease?

Danon disease has three major symptoms: a weak heart muscle, weak skeletal muscles, and mental retardation. Males usually develop the symptoms earlier and are more severely affected than females. Following are the different symptoms of both males and females.

Males

- Early onset of heart problems: Weakness in the heart muscle, a condition called cardiomyopathy, appears in childhood or adolescence. When the heart muscle is weak, it cannot pump blood efficiently. Some of the cardiac problems are related to the electrical conduction of the heart. The heart disease can be so severe as to require medication or even a heart transplant.
- Early onset of muscle problems: Usually, the weakness begins in the head and neck. However, muscle weakness can be so severe that the person is not able to walk.
- Intellectual disabilities: The child has difficulty learning.
- Vision: Some boys have vision difficulties, stemming to a problem with the pigment in the retina.

The average life expectancy for males with this disorder is to the age of 19.

Females

Symptoms are less severe than in males.

- Onset of symptoms: The onset occurs at a later age, perhaps in late adolescence or adulthood.
- Muscle weakness absent or subtle: The person may not have muscle weakness but may tire easily with exertion.
- No learning problems: Intellectual disabilities are usually not present.

- Heart disease: Heart issues do not appear early in girls but may occur later in adulthood. The problems with electrical conduction may still happen.
- Eye issues: Some females will have problems with vision and pigment in the back of the eyes on the retina.

The average life expectancy for females is the age of 34.

Danon disease is so rare that it is unfamiliar even to most physicians. It is often misdiagnosed and mistaken for other kinds of heart disease or muscular dystrophies.

What Is the Genetic Cause of Danon Disease?

Mutations in the *LAMP2* gene, officially known as the "lysosomal-associated membrane protein 2" gene, cause Danon disease. Normally, *LAMP2* provides instructions for a protein called the lysosomal associated membrane protein-2 (LAMP-2). This protein is found in the lysosomes, the structures in the cells that recycles old cell material and digests other materials.

Although the role of *LAMP2* is unclear, some researchers think that the protein carries cellular material or digestive enzymes into the lysosome. When the material to be recycled or destroyed gets into the cell, a vacuole surrounds the material. This structure is called an autophagic vacuole. Then this vacuole joins to a lysosome to transfer the material to be broken down. The LAMP-2 protein is involved in that fusion.

Many mutations in the *LAMP2* cause Danon disease. The mutations in the gene lead to very little or no LAMP-2 protein. Without the LAMP-2 protein, fusion between the autophagic vacuoles and lysosomes is very slow, leading to an accumulation of the vacuoles. People with Danon disease have a large number of these vacuoles in the heart and skeletal muscle cells, causing the weaknesses seen in the disorder. *LAMP-2* is inherited in an X-linked dominant pattern and is found on the long arm (q) of the X chromosome at position 24.

What Is the Treatment for Danon Disease?

This disorder is often mistaken for other conditions. Treatment for this rare disorder is symptomatic. The most pressing problems involve the heart, which can be treated with conventional drugs and treatment.

Further Reading

"Danon Disease." 2010. Genetic and Rare Disease Information Center. National Institutes of Health. http://rarediseases.info.nih.gov/GARD/Disease.aspx?PageID=4&DiseaseID=9730 Accessed 1/4/12.

"Danon Disease." 2010. Medscape. http://emedicine.medscape.com/article/952782. Accessed 1/4/12.

"Danon Disease." 2011. Genetics Home Reference. National Library of Medicine (U.S.). http://ghr.nlm.nih.gov/condition/danon-disease. Accessed 1/4/12.

Darier Disease (DAR)

Prevalence Unknown worldwide; 1 in 30,000 in Scotland; 1 in 36,000 in northern England; 1 in 100,000 in Denmark

Other Names Darier's disease; Darier-White disease; keratosis follicularis

Each area of medicine has historical figures that develop the clinical specialty. Ferdinand-Jean Darier is known as the "Father of Dermatology," the branch of medicine that deals with diseases and disorders of the skin. Born in Budapest, Hungary, to French parents, Darier studied and worked in France, where he discovered several diseases. In 1909, he wrote a textbook of dermatology, which was translated into German, Spanish, and English. Several diseases were named after him, the most notable of which is Darier disease.

What Is Darier Disease (DAR)?

Darier disease (DAR) is characterized by yellowish, crusty patches of wart-like appearance on various parts of the body. The patches are hard but greasy to the feel. The lesions are especially common in seborrheic regions such as the forehead, scalp, upper arms, knees, chest, elbows, back, and back of the ears. About 80% of patients have growths in the groin, underarms, and in women under the breast. These lesions are quite bothersome because they can emit a very strong odor.

Most people with Darier disease have a family history of the disorder. Some of the clinical features follow:

- Condition can occur from late childhood to early adulthood, but usually are typical during teenage years.
- Triggers for the disorder include heat, perspiration, sunlight, UVB exposure, medication such as lithium or corticosteroids; some females report flare-ups during menstruation.
- This chronic, reoccurring condition may improve with age.
- About 95% of people with DAR have lesions on the hands or lesions on the palms.
- Nails may display longitudinal ridges, red and white lines, and V-shaped nicks.
- One form of DAR is characterized by linear or segmented blemishes on localized areas of the skin that are not as widespread as in the classic disease.
- Some neurological conditions such as epilepsy, mood disorders, and mental impairment have been associated with the condition, although there is no genetic evidence to support the connection.

DAR affects both men and women and is not contagious.

What Is the Genetic Cause of Darier Disease (DAR)?

The official name of the gene involved in Darier disease is "ATPase, Ca++ transporting, cardiac muscle, slow twitch 2" or *ATP2A2*. This gene is a member of a superfamily of genes called *ATP*. Normally, *ATP2A2* instructs for making an enzyme called sarco (endo)plasmic reticulum calcium-ATPase 2 (SERCA2). This enzyme belongs to an important family of ATP enzymes that controls the level of calcium ions. At times these enzymes are referred to as calcium pumps. In each cell, structures called the endoplasmic reticulum and the sarcoplasmic reticulum are mazes within the cell that act as passageways for essential substances. The endoplasmic reticulum is involved in protein processing and transport; the sarcoplasmic reticulum aids in muscle concentration and relaxation by releasing and storing the calcium ions. It is the enzyme SERCA2 that controls the passage of calcium ions in and out of these passages in the cell.

Mutations in the *ATP2A2* disrupt the process. Most of the 130 mutations involve a change in a single protein base pair (amino acid) in the SERCA2 enzyme. The gene is thought to be inherited in an autosomal dominant pattern. Cells that have only one copy of the *ATP2A2* gene produce only half of the SERCA2 protein. Scientists have determined that these insufficient amounts of SERCA2 can then combine with environmental factors such as heat or minor injury to cause the symptoms of Darier disease. *ATP2A2* is located on the long arm (q) of chromosome 12 at position 24.11.

What is the Treatment for Darier Disease (DAR)?

Because this condition is genetic, there is no cure or treatment except for the symptoms. A preventive is to protect from heat or sunlight; in the summer sunscreens should be used. Accutane or retinoids may be used. During flare-ups, treatment may include topical or oral antibiotics.

Further Reading

"*ATP2A2*." 2010. Genetics Home Reference. National Library of Medicine (U.S.). http://ghr.nlm.nih.gov/gene/ATP2A2. Accessed 5/11/12.

"Darier Disease." 2010. Genetic and Rare Disease Information Center (GARD). National Institutes of Health (U.S.). 'http://rarediseases.info.nih.gov/GARD/Condition/6243/Darier's_disease.aspx. Accessed 5/11/12.

"Keratosis Follicularis (Darier Disease)." 2010. Medscape. http://emedicine.medscape.com/article/1107340-overview. Accessed 5/11/12.

Dentatorubral-Pallidoluysian Atrophy (DRPLA)

Prevalence Most common in the Japanese population, with an estimated 2 to 7 per million people; has been seen in North America and Europe

Other Names DRPLA; Haw River syndrome (found in a large African American family along the Haw River in North Carolina); Naito-Oyanagi disease

In North Carolina, members of a large African American family living along the Haw River displayed symptoms of a progressive condition characterized by jerky, involuntary movements, serious emotional problems, and progressive dementia. In 1958, a local doctor described a condition with the same symptoms in the Japanese population and in several Western countries. Later, it was decided that the two conditions were the same and given the cumbersome name dentatorubral-pallidoluysian atrophy from an area of the central nervous system that is affected.

What Is Dentatorubral-Pallidoluysian Atrophy (DRPLA)?

Dentatorubral-pallidoluysian atrophy (DRPLA) is a progressive brain disorder that causes degeneration of the brain and spine. Depending on the age of onset, the disease may display the following signs and symptoms:

- Juvenile-onset, occurring under 20 years of age: This condition is characterized by jerking and twitching of muscles, seizures similar to epilepsy, changes in behavior, intellectual decline, problems with balance and coordination (ataxia), sleep apnea, and autism.
- Early adult-onset, occurring between 20 and 40 years of age: The most frequent symptoms are loss of balance and coordination, extreme uncontrollable movements of the arms and limbs, seizures, and progressive deterioration of mental faculties.
- Late adult-onset: Uncontrollable movements, serious psychiatric symptoms such as delusions, deteriorating mental function, and dementia.

Serious reduction of central nervous system tissue occurs throughout the brain and spinal cord. The shrinkage of the brain and nervous system is attributed to the accumulation of the protein atrophin-1, which poisons the brain cells.

DRPLA is one of eight neurogenerative diseases that have CAG repeats. This means that in the gene nucleotide building blocks—cytosine (C), adenine (A), and guanine (G)—are repeated and appear as multiples in the row of protein base pairs. In a normal segment of CAG repeats on the gene, the trinucleotide repeat is about 6 to 35 times in a consecutive row. For example, a repeat of 6 would be: CAGCAGCAGCAGCAGCAG. DRPLA is similar to Huntington disease, the CAG repeat disorder that is famous in the history of genetics because it was one of the first genes located in the genome. Huntington disease is called the gold standard of genetics.

What Is the Genetic Cause of Dentatorubral-Pallidoluysian Atrophy (DRPLA)?

The cause of DRPLA is a mutation in the gene *ATN1*. Normally, *ATN1* instructs for the protein atrophin 1, which likely plays an important role in the neurons in the brain and spinal column. In people with DRPLA, the trinucleotide repeat—CAG—is more than 48 times. Because of the abnormal repeats, the function of the protein atrophin-1 is disrupted and collects in neurons leading to the death of the cell and causing the characteristic involuntary movements.

The gene is passed on in an autosomal dominant pattern, meaning only one copy from one parent may cause the disorder. Both males and females have the gene equally and have a 50/50 chance to pass the disorder to their children. However, the disorder appears more prominent if inherited from the father rather than from the mother. *ATN1* is located on the short arm (p) of chromosome 12 at position 13.31.

A transgenic mouse model has been developed, which has the human DRPLA phenotype. Mouse models permit extended research into CAG disorders.

What Is the Treatment for Dentatorubral-Pallidoluysian Atrophy (DRPLA)?

This condition is a serious genetic disorder that has no specific prescription for treatment. Treatment is symptomatic. Seizures are treated with anticonvulsants and psychiatric disorders with proper medications.

Further Reading

"*ATN1*." 2010. Genetics Home Reference. National Library of Medicine (U.S.). http://ghr.nlm.nih.gov/gene/ATN1. Accessed 5/11/12.

"Dentatorubral-Pallidoluysian Atrophy." 2012. Genetics Home Reference. National Library of Medicine (U.S.). http://www.ghr.nlm.nih.gov/condition/dentatorubral-pallidoluysian-atrophy. Accessed 5/11/12.

Diabetic Embryopathy

Prevalence Affects about 10% of diabetic mothers

Other Names Honeybee syndrome

Embryopathies are birth defects, problems that happen while the baby is developing during pregnancy. Certain metabolic conditions of the mother may be a source of problems with the fetus. Diabetes is one of these conditions.

What Is Diabetic Embryopathy?

Diabetic embryopathy is a diagnosis based on one of more of a complex of congenital anomalies or fetal/neonatal complications. The anomalies are attributed to the mother who has diabetes or who develops it during pregnancy. The following three types of diabetes mellitus pose risk not only for the offspring, but also for the mother:

- Type I: Type I is early onset, which at one time was called juvenile diabetes. The person is completely insulin dependent because the body does not produce any insulin. The person is exceedingly prone to ketosis, a condition in which ketones build up in the body and cause damage to cells and organs. There are many causes of Type I diabetes, but those with a family history are at greater risk. The person must take daily injections of insulin.
- Type II: Type II is adult onset and non-insulin dependent. Here, the body either does not produce enough insulin or the cells cannot use it properly. This type appears to be inherited as an incompletely penetrating recessive trait with many factors involved. Diet, exercise, and medications usually control this type.
- Type III (gestational diabetes): This type onsets during pregnancy and occurs in about 1% to 4% of all pregnancies. African American and Hispanic women have the highest incidence. Many factors may enter into the cause, with a family history indicating a higher risk. About 20% to 50% of women who develop diabetes during pregnancy will develop Type II during the next 5 to 10 years.

Diabetic embryopathy affects about 10% of children born to diabetic mothers. The high blood sugar or glucose levels and poisonous ketones pass through the placenta to the baby, increasing the chance of birth defects. The defects develop during that critical time of embryonic development between the fifth and eighth weeks of gestation. The extra sugar going into the baby also leads to a bigger baby that is hard to deliver. The jury is still out on whether administering insulin to the mother has a teratogenic effect; however, the outcome for the mother and possibly the child is much better when the diabetes is kept under control.

Both mother and child must be considered when a pregnant woman has diabetes. First, the mother herself may develop several serious conditions. She may have ketoacidosis, a condition in which the ketones and acids build up in the body and damage cells and organs. She may also have an abnormal amount of amniotic fluid, preeclampsia, or preterm labor, necessitating a Cesarean section.

Diabetic embryopathy can affect many organ systems of the child. Following are the complications that may happen to the child during pregnancy:

- Cardiac and major blood vessel anomalies.
- DiGeorge syndrome: This is a syndrome of disorders that results from abnormal neural crest migration during embryonic development. It affects development of the heart, thymus, and parathyroid glands.

- Neural tube defects (NTDs): During embryonic development, there is a period of folding. The most common neural tube defect is spina bifida, which occurs when an area of the spinal cord does not close. Another condition is anencephaly, in which the brain cavity is shut off and the front part of the brain is missing.
- Macrosomia: The word comes from the Greek roots *macro*, meaning "large," and *som*, meaning "body." About one-third of all children born to diabetic mothers have a lifelong struggle with obesity.
- Kidneys and gastrointestinal tract.
- Skeletal problems: Part of the lower spine may not develop, causing malformation of lower extremities. This response is thought to be due to improper embryonic folding.

Perinatal and Postnatal Problems

During birth, the child of the diabetic mother may have difficulty taking original breaths of air, which can cause cerebral palsy, heart problems, and gastrointestinal defects. In very large infants for whom the mother's diabetes was not controlled, the child may develop cardiomyopathy and/or an enlarged heart. Preterm births can lead to breathing problems; these occur in 30% of diabetic pregnancies even with great care. If the child is born with hypoglycemia, there is a potential for seizures and even a coma. Underproduction of calcium and magnesium can affect parathyroid function.

What Is the Treatment for and Prevention of Diabetic Embryopathy?

A woman with diabetes should be aware of the risks of pregnancy and institute rigid glycemic or sugar control and prepregnancy care. However, studies have shown that even with optimal care, the incidence of malformations is twice that of nondiabetic mothers. There may be a psychosocial implication: the mother may have guilt because she caused her baby such risks and complications.

See also Embryology: A Special Topic

Further Reading

Chang, T. I., and M. R. Loeken. 1999. "Genotoxicity and Diabetic Embryopathy: Impaired Expression of Developmental Control Genes as a Cause of Defective Morphogenesis." *Seminars in Reproductive Endocrinology*. 17(2): 153–165. http://www.ncbi.nlm.nih.gov/pubmed/10528366. Accessed 5/11/12.

Loeken, M. R. 2008. "Challenges in Understanding Diabetic Embryopathy." *Diabetes Journal*. December. 57(12): 3187–3188. http://diabetes.diabetesjournals.org/content/57/12/3187.full. Accessed 5/11/12.

Diamond-Blackfan Anemia (DBA)

Prevalence Approximately 5 million to 7 million newborns worldwide

Other Names Aase-Smith syndrome II; Aase syndrome; Anemia Diamond-Blackfan; BDA; BDS; congenital pure red cell aplasia; inherited erythroblastopenia

In 1938, L. K. Diamond and K. D. Blackfan described a congenital condition that they called hypoplastic anemia. In 1961, after following 30 of their original patients in a longitudinal study of a number of years, Diamond and his colleagues presented data that associated the anemic condition with skeletal disorders. The condition that bears the names of the two original researchers, Diamond-Blackfan anemia, has turned out to be a complex disorder with several genetic origins.

What Is Diamond-Blackfan Anemia (DBA)?

Diamond-Blackfan anemia (DBA) is a rare blood disorder caused by the failure of the bone marrow to produce red blood cells. The bone marrow is the area of production of the erythrocytes or red blood cells; in DBA, the cells that form the red cells are drastically reduced. Most children with the disorder are diagnosed during the first two years of life. The children have the general symptoms of anemia: paleness, weakness, sleepiness, and fatigue.

A serious side effect of DBA is that the individual may develop other abnormalities and disorders. Following are several of the complications:

- Head and face: About half of the children have a small head, a condition known as microcephaly. Their appearance may have a characteristically low frontal hairline, broad face, wide eyes, droopy eyelids, flat bridge of the nose, low-set ears, and a small lower jaw. They may also have a cleft palate.
- Body: They may have a short webbed neck and slowed growth leading to short stature.
- Disorders may include cataracts, glaucoma, kidney abnormalities, heart defects, and reproductive problems.
- Because of the malfunctioning bone marrow, the individual with DBA is at risk for myelodysplasic syndrome (MDS), which is a disorder affecting blood cell production; acute myeloid leukemia (AML); and a kind of bone cancer called osteosarcoma.

What Are the Genetic Causes of Diamond-Blackfan Anemia (DBA)?

Mutations in several genes cause DBA. These genes—*RPL5*, *RPL11*, *RPL35A*, *RPS17*, *RPS19*, and *RPS24*—are members of a group that instruct for making more than 75 ribosomal proteins. Ribosomes are parts of the cell that process information for making proteins. The ribosomal proteins are essential for the functioning of the ribosomes and may also act in regulating cell division and self-destruction or apoptosis of the cell.

Although most genetic conditions can be traced to one or two genes, this condition has six genes. Following are the suspect genes, their roles in DBA, and locations:

- *RLP11*: Instructs for making ribosomal protein L11. There are more than 25 identified gene mutations. *RLP11* is located on the short arm (p) of chromosome 1 at position 36 to p35.
- *RPL35A*: *RPL35A* instructs for the ribosomal protein L5A. There are more than five mutations. *RPL35A* is located on the long arm (q) of chromosome 3 at position 29 to *qter*.
- *RPL5*: The gene instructs for a ribosomal protein L5. There are about 30 mutations of this gene. The *RPL5* gene is located on the short arm (p) of chromosome 1 at position 22.1.
- *RPS17*: The specific function of ribosome protein S17 is unclear, but it has a similar function to other genes in the ribosomal protein (RPS) family. *RPS17* is located on the long arm (q) of chromosome 19.
- *RPS 19*: This gene also belongs to the RPS family. More than 80 mutations have been found in this gene relating to DBA. The gene is located on the long arm (q) of chromosome 19 at position 13.2.
- *RPS24*: Another of the family of RPS genes, *RPS24* has three mutations associated with DBA. It is located on the long arm (q) of chromosome 10.

There are many unanswered questions about the genes and how disruption of the proteins causes DBA. It does appear to be inherited in an autosomal dominant pattern, meaning one copy of the altered gene is in the cell. Many of the mutations occur in families with no history of the disorder.

What Is the Treatment for Diamond-Blackfan Anemia (DBA)?

Treating the symptoms of anemia may involve corticosteroid medications and blood transfusions. Bone marrow or stem cell transplants are also possibilities.

Further Reading

"About DBA." 2006. Diamond-Blackfan Anemia Foundation. http://www.dbafoundation.org/about.php. Accessed 5/11/12.

"Diamond Blackfan Anemia (DBA)." 2010. Centers for Disease Control and Prevention (U.S.). http://www.cdc.gov/ncbddd/dba/facts.html. Accessed 5/11/12.

"Diamond Blackfan Anemia and You." 2009. http://www.diamondblackfananemia.com/. Accessed 5/11/12.

Diastrophic Dysplasia

Prevalence Affects about 1 in 100,000 newborns; found in all populations, but very common in Finland

Other Names achondroplasia with club feet; arthrogyrposis multiplex congenita; diastrophic dwarfism; DTD

When most people see dwarfs, they think of them as just "little" people with short arms and limbs. However, there are over 200 recognized types of dwarfism. Lamy and Maroteaux first described this syndrome in 1960 and called it diastrophic dwarfism. They noted how the bones of the individual appeared bent out of shape and borrowed the term "diastrophic" from the geological concept of the earth's bending and reforming during its formative periods. The word "diastrophic" comes from the Greek root that means "distorted" or "bent," which is appropriate for this condition because in X-rays the skeleton appears bent or twisted. In 1977, the name was changed from diastrophic dwarfism to diastrophic dysplasia.

Matt Roloff, a businessman and motivational speaker, has diastrophic dysplasia. He promotes the acceptance of Little People as competent citizens capable of doing the things that the rest of the populace can do. Although he is interested in research on helping people with dwarfism, he is dedicated to improving their quality of life and integrating them into society.

Who Are the Little People of America (LPA)?

Little people are individuals with a medical or genetic condition that results in an adult height of 4′10″ or shorter. The condition is properly referred to as dwarfism. In 2009, the LPA held a press conference to protest the use of the word "midget" when referring to people with dwarfism. An actor named Jesse James had appeared on the television show *Celebrity Apprentice* in 2009 and had used the word many times. The complaint to the Federal Trade Commission stated how offensive the word is to people of short stature and should not be used. LPA said the unacceptable word "midget" dates back to 1865 and the "freak show" era when little people were used only for public entertainment.

The Little People of America started in 1957 when Billy Barty called upon individuals with dwarfism to get together in Reno, Nevada. As of 2010, the group had 6,900 members. The nonprofit group provides support, resources, and information to people living with dwarfism. For example, they may tell an individual how to adapt the counters in the kitchen, bathroom fixtures, and pedals for the car.

Source: Little People of America

What Is Diastrophic Dysplasia?

Diastrophic dysplasia is a disorder of the growth and remodeling of bone and cartilage that result in short stature. The idea of the bent skeleton is shown in many body parts. The condition is always characterized by a club or turned foot with the metatarsals or bones of the feet turning downward and inward. In addition, several other features can be present:

- Spine: About one-half of the children will have cervical (the vertebra in the neck) kyphosis that appears in early infancy. The individual can have scoliosis, a progressive curvature of the spine. These two conditions can cause so much skeletal imbalance that walking is awkward and difficult.
- Hands and feet: Hitchhiker thumb is a distorted abnormal position for the thumb. Foot deformities cause a stiff and artificial gait. Both the hands and feet display misalignment and bony malformations.
- Head and neck: Facial features can include prominent cheeks and narrow nasal bridge with or without flattening, and many teeth are visible when the person smiles. About 50% have a cleft palate. Ears can develop acute swelling at about three to six weeks; later in life, the cartilage in the pinnae or outside part of the ear hardens, giving the appearance of a "cauliflower ear." Vocal cords, which are also cartilage, may harden, causing a distorted, high voice.
- Major joints: Hip dysplasia is progressive and may result in a limited range of motion. Arthritic changes are possible in early-to-middle adulthood. Symptoms may involve hips, knees, ankles, shoulders, or elbows.
- Hernia: Cases of inguinal hernia have been reported with diastrophic dysplasia.

What Is the Genetic Cause of Diastrophic Dysplasia?

Several skeletal disorders are caused by mutations in the gene *SLC26A2*. Diastrophic dysplasia is one of these. Normally, the gene *SLC26A2* instructs for making a protein called the sodium-independent sulfate/chloride transporter that develops cartilage and directs its conversion to bone. Cartilage is that tough flexible tissue that makes up a large part of the skeleton during early development; in adulthood, cartilage is present at the ends of bones and in the joints, ears, and nose.

Mutations in the *SLC26A* gene disrupt this structure of cartilage, and bones are prevented from forming properly. The gene that is inherited is an autosomal recessive pattern, meaning that both parents must carry one copy of the defective gene; the parents are not affected. *SLC25A* was discovered in 1994 and called DTDST; however, the current name is now used. It is located on the long arm (q) of chromosome 5 at positions 31-q34.

What Is the Treatment for Diastrophic Dysplasia?

Because it is a genetic condition, the treatment must deal with the symptoms. For example, surgery may correct the cleft palate. Other surgical interventions may deal with club foot, scoliosis, and other orthopedic conditions.

Further Reading

"Frequently Asked Questions." 2010. Little People of America. http://www.lpaonline.org/mc/page.do?sitePageId=84634#Common. Accessed 5/11/12.

Donnai-Barrow Syndrome

Prevalence Extremely rare; only a few dozen cases have been reported worldwide; however, reports include northern and central European, Middle Eastern, African American, and Caucasian American

Other Names DBA; DBS/FOAR syndrome; diaphragmatic hernia-exomphalos-corpus callosum agenesis; diaphragmatic hernia-exomphalos-hypertelorism syndrome; facio-oculo-acoustico-renal syndrome; FOAR syndrome

Reports of children with a rare collection of facial abnormalities combined with rare abnormalities of the central trunk have appeared in the literature for many years. However, in 1993, D. Donnai and M. Barrow wrote an article on a suspected new autosomal recessive disorder in which a variety of symptoms were seen together. The disorder was called Donnai-Barrow syndrome after these two investigators.

What Is Donnai-Barrow Syndrome?

Donnai-Barrow syndrome is a genetic condition that affects many parts of the body. The individual has a distinct look that includes facial features with prominent wide-set, protruding eyes and outer corners of the eyes that point downward. The face is also flat with a short bulbous nose. Also, the ears are backwards, and the hairline ends with a widow's peak.

In addition to the facial appearance, several body abnormalities exist:

- Diaphragmatic hernia: A defect in the diaphragm allows the stomach and intestine to protrude into the chest, causing problems for the heart and lung. This condition is called exomphalos or umphocele.

- Abnormal brain development: The tissue that connects the hemispheres of the brain or right and left halves is called the corpus collosum. The tissue is absent, causing mild to moderate intellectual disability and developmental delay.
- Eye abnormalities: These include severe nearsightedness, retinal detachment, and iris coloboma, a condition in which the iris has a keyhole appearance.
- Abnormal heart, lungs, and other organs.

What Is the Genetic Cause of Donnai-Barrow Syndrome?

Mutations in the *LRP2* gene, officially known as the "low density lipoprotein receptor-related protein 2" gene, cause Donnai-Barrow syndrome. Normally, the *LRP2* gene instructs for making megalin, a very important protein that acts as a special receptor protein. The function of receptor proteins is to act like a ligand or connector that fits into other proteins. One might picture a ligand and its receptors like keys that fit into locks or interlocking pieces of puzzles. The ligands and receptors combination instructs for cell development and action. Megalin has many ligands that affect absorption of vitamins A and D, the immune system, stress responses, and transport of fat in the bloodstream. As megalin is embedded in the cell membrane, a little of it protrudes out and moves the ligands from the cell surface into the cell body for cell function. Megalin is active in may body parts including small tubes in the kidneys where urine is formed, the central nervous system, reproductive system, gastrointestinal systems, and sensory organs.

At least 12 mutations of the *LRP2* genes are related to Donnai-Barrow syndrome. The mutations disrupt the production of megalin and the necessary biochemical pathways for the ligand-receptor interaction. This disruption causes the characteristic symptoms of the condition. Another condition called facio-oculo-acoustico-renal (FOAR) disorder was once classified as a separate syndrome but is caused by the same *LRP2* gene and is now considered the same disorder.

Donnai-Barrow syndrome is inherited in an autosomal recessive pattern. The gene is located on the long arm (q) of chromosome 2 at position 24 through q31.

What Is the Treatment for Donnai-Barrow Syndrome?

Because Donnai-Barrow is a genetic disorder, there is no cure. However, each of the symptoms and disorders must be treated with surgery or other medical interventions.

Further Reading

"Donnai-Barrow Syndrome." 2010. Genetics Home Reference. National Library of Medicine (U.S.). http://ghr.nlm.nih.gov/condition/donnai-barrow-syndrome. Accessed 5/11/12.

"Donnai-Barrow Syndrome." 2010. Office of Rare Disease Research (GARD). National Institutes of Health (U.S.). http://rarediseases.info.nih.gov/GARD/Condition/1899/Donnai_Barrow_syndrome.aspx. Accessed 5/11/12.

"*LRP2*." 2010. Genetics Health Reference. National Library of Medicine (U.S.). http://ghr.nlm.nih.gov/gene/LRP2. Accessed 5/11/12.

Down Syndrome (DS)

Prevalence Occurs in about 1 in 740 newborns; chances of having a child with Down syndrome increase with age of mother; found among all ethnic groups; more than 400,000 people with DS living in the United States

Other Names Down's syndrome; trisomy 21; 47,X+21; 47,XY,+21

In the mid-1860s in England, people began to notice a group of children who looked more like each other than their parents. They had eyelids that turned inward, a condition called internal epicanthic folds and always some form of mental disability. English physician John Langdon Down first wrote about this condition in 1862. He used the term "mongoloid," a term that can be traced to the

A young man in Serbia with Down Syndrome attends a community center there. (Getty Images)

colonialism and thinking of the time, because he thought that children and people who had this condition looked like they were from the "Mongoloid" or Mongolian race. Because of the cognitive disability, the term "mongolian idiot" was later used, which was both racist and inaccurate.

Treatment of people with Down syndrome was certainly reprehensible. In the late nineteenth century, mental institutions became prominent, and Down syndrome was one of the most prevalent forms of mental disability found in the wretched asylums. Individuals were not treated or helped, and none of their associated medical problems were treated. Most of them died early in life. In the early 1900s, a movement called eugenics became popular. The aim of the eugenics movement was to improve the qualities of the populace by eliminating the undesirables. Out of the 48 states in the United States at the time, 33 developed laws to force sterilization of people with disabilities, including those with Down syndrome. However, the systematic murder of those with disorders rose to the ultimate evil in Nazi Germany with the "Action T4" program. Slowly after World War II, attitudes began to change, and the term "mongolian idiot" or "mongoloid" was changed officially to Down syndrome in 1972.

What Is Down Syndrome (DS)?

Down syndrome (DS) is a set of mental and physical symptoms that results from the presence of an extra chromosome 21 or extra material in that chromosome. Normally, a person has 23 pairs of chromosomes, but in these individuals, a third copy of the chromosome is present. Sometimes, the condition is referred to as trisomy 21. The word "trisomy" comes from two Greek root words—*tri*, meaning "three," and *some*, meaning "body." The presence of the extra chromosome causes the characteristic attributes of the disorder.

Children with Down syndrome look more like each other than members of the family. Following are some of the common features:

- A very round, flat face with an abnormally small chin
- Large or protruding tongue
- Eyes appear almond shaped, caused by an internal epicanthic fold of the eyelid
- Short limbs with poor muscle tone and loose ligaments
- Small hands and feet with a simian crease across the palms
- Short neck
- White spots on the iris of the eye known as Brushfield spots.

Intellectual disabilities vary. Generally, children have lower-than-average intellectual skills. Average IQs are around 50 compared to the normal average of 100.

Along with appearance and cognitive disabilities, people with DS may have higher risk for several physical complications:

- Congenital heart disease: About 50% of the children are born with a heart defect.
- Eye disorders and hearing problems: About half have crossed eyes or strabismus. Cataracts and glaucoma are common.
- Gastrointestinal problems: Gastroesophageal reflux disease, celiac disease, and blocked intestines and esophagus are common.
- Epilepsy and seizures.
- Alzheimer disease: Risk for individuals with DS is 10% to 25%, with signs appearing before age 50.
- Thyroid dysfunctions: Low thyroid or hypothyroidism is most common, occurring in almost one-third of those with DS.
- Skeletal problems.
- Infertility in both males and females: Males are usually unable to father children; females have difficulty with pregnancies ending in miscarriage, premature, difficult birth, and labor.

What Are the Genetic Causes of Down Syndrome (DS)?

Down syndrome is caused by an extra chromosome 21 or abnormal replication of parts of the chromosome. The following four different genetic patterns of chromosomes exist:

- Trisomy 21: Most cases of DS are caused by an extra copy of chromosome 21. There are three copies of the chromosome in all body cells instead of the normal two. When a sperm or egg is produced during meiosis, an extra copy of chromosome 21 is produced by an extra division. This is called a meiotic nondisconjunction happening, meaning that the division happens during meiosis in the creation of sperm or egg. The gamete—egg or sperm—has then 24 chromosomes. At fertilization, the embryo will have 47 chromosomes, instead of the normal 46, and an extra chromosome 21. Trisomy 21 is the cause of 95% of cases of DS, with 88% coming from the egg of the mother.
- Mosaicism: If some body cells are normal and others have the extra chromosome 21, the condition is called mosaic DS. Instead of the nondisjunction occurring in meiosis of the sperm or egg, the event occurs in the early cell division of the embryo or mitotic division. Some of the cells have the extra chromosome and others have a normal pattern. Because Mosaic DS is a random event occurring in early embryonic development, it cannot be inherited.
- Robertsonian or translocation DS: Translocation occurs when one part of a chromosome attaches to another chromosome. In this instance, the long arm (q) of chromosome 21 often attaches to chromosome 14. The individual looks normal and does not have the appearance of DS; however, the extra

chromosome may be packaged into an egg or sperm and produce a child with DS. Translocation DS is sometimes called familial DS and is the cause of about 2%–3% of cases. This event is not related to the age of the mother.
- Duplication: A very rare event may cause extra duplication of some of the genetic material on chromosome 21. If the duplicated parts are the ones relating to the mental and physical characteristics of DS, the person may have those symptoms.

A mouse model with the symptoms of DS has been created and is valuable in studying the condition.

What Is the Treatment for Down Syndrome (DS)?

Because it is a genetic disorder, no cure for DS exists, but because it is more common and well researched, several things are available. Pregnant women can be screened for the condition using amniocentesis, chorionic villus sampling (CVS), or percutaneous umbilical cord blood sampling (PUBS). The decision to terminate a pregnancy of a fetus with DS is a much-discussed topic in medical ethics considerations.

An October 17, 2011, *New York Times* article told of a less risky Down syndrome test that is being developed. This test can detect Down syndrome in a fetus by testing the mother's blood. Although most of the DNA in the blood is the mother's DNA, a small amount is from the fetus. The test uses high-speed DNA sequencers to determine the sequence of millions of fragments. The sequencer can determine which fragment the chromosome comes from. If the fetus has three copies of chromosome 21, it may be possible the child has Down syndrome. According to the researchers, the test picked up 98.6% of positive Down syndrome cases and only 0.2% of these were false positive. This test spares the mother from the invasive amniocentesis or CVS, both of which have been known to cause miscarriage.

Treatment of the child with DS depends on the severity of the case. Health issues such as heart defects or breathing can be addressed with medical intervention. Educators determine that high-quality care that begins soon after birth and early interventions in assisting the child in development can greatly improve cognitive skills. Today, the average life expectancy for people with DS is 55 years, but many live beyond that, and some into their 70s. Down syndrome is covered under the Americans with Disabilities Act and Individuals with Disabilities Education Act.

Further Reading

"Down Syndrome." 2012. Nemours Kids Health. http://kidshealth.org/parent/medical/genetic/down_syndrome.html. Accessed 3/7/12.

"Down Syndrome." 2012. National Institute of Child Health. National Institutes of Health (U.S.). http://www.nichd.nih.gov/health/topics/down_syndrome.cfm. Accessed 3/7/12.

National Down Syndrome Society. 2012. http://www.ndss.org. Accessed 3/7/12.

Duane-Radial Ray Syndrome

Prevalence Rare, prevalence unknown; only a few families reported worldwide

Other Names Okihiro syndrome

The late nineteenth century was a time of discovery in medicine, with physicians noting and reporting rare and unusual conditions. A disturbance of eye movement was first described and reported by Jakob Stilling, a German ophthalmologist, in 1887. However, Alexander Duane collected information about 14 cases and presented a study of 40 more of the similar cases in 1905. The medical community then added the name Duane to the syndrome. Later, Michael Okihiro found that these eye abnormalities were connected to bone condition of the hands. Sometimes, the condition is referred to as Okihiro's syndrome, but the name Duane-radial ray is most commonly used.

What is Duane-Radial Ray Syndrome?

Duane-radial ray syndrome is an unusual condition that affects the eyes and also the bones of the hands and arms. The chief characteristic is a problem with eye movement caused by improper development of the nerves that affect the muscles that move the eye. This problem, called Duane syndrome, limits the outward movement of the eye toward the ear and sometimes inward toward the nose. As eyes move inward, the opening becomes narrower, and the eyeball may slip back into the socket. A second set of abnormalities called radial ray malformations may include malformed or absent thumbs, an extra thumb that appears as a finger, or partial or complete absence of the bones of the forearm. The condition often appears in females.

Occasionally, people with this syndrome may have other signs and symptoms that include strangely shaped ears, hearing loss, heart and kidney defects, clubfoot, fused vertebrae, and a distinctive facial appearance. These signs and symptoms may overlap with other disorders, such as acro-renal-ocular syndrome and Holt-Oram syndrome. However, the disorders are traced to different genes.

What Is the Genetic Cause of Duane-Radial Ray Syndrome?

The genetic cause of Duane-radial ray syndrome is traced to mutations in the *SALL4* gene, officially known as the "sal-like 4 (Drosophila)" gene. Normally, *SALL4*, a member of the *SALL* family, instructs for making proteins involved in formation of certain organs and tissues during the embryonic period. The proteins appear to be transcription factors that bind to specific areas of DNA and control the activity of other genes. The protein appears to play a role in the development of limbs and for the nerves that control eye movement.

More than 25 mutations in *SALL4* stop the production of the protein and lead to the symptoms of Duane-radial ray syndrome. *SALL4* is inherited in an autosomal dominant pattern, meaning that one copy of the gene can cause the disorder.

It appears that the presence of only one altered copy can shut down the production of the necessary protein. Sometimes, the condition is present in the family and at other times the mutation occurs in people with no history of the disorder. *SALL4* is located on the long arm (q) of chromosome 20 at position 13.2.

What Is the Treatment for Duane-Radial Ray Syndrome?

Because this is a genetic disorder with no prevention or cure, treating and managing the disease depends on the specific symptoms that occur. Early diagnosis and routine medical care may prevent problems that occur later in life.

Further Reading

"Duane-Radial Ray Syndrome." 2010. Genetics Home Reference. National Library of Medicine (U.S.). http://ghr.nlm.nih.gov/condition/duane-radial-ray-syndrome. Accessed 5/11/12.

Okihiro, M. M., et al. 1977. "Duane Syndrome and Congenital Upper Limb Anomalies: A Familial Occurrence." *Archives of Neurology* (Chicago). 34: 174–179. http://www.ncbi.nlm.nih.gov/pubmed/843249. Accessed 5/11/12.

"*SALL4*." 2010. Genetics Home Reference. National Library of Medicine (U.S.). http://ghr.nlm.nih.gov/gene/SALL4. Accessed 5/11/12.

Duchenne/Becker Muscular Dystrophy

Prevalence Together, Duchenne and Becker dystrophies affect 1 in 3,500 to 5,000 newborn males; between 400 and 600 born each year in the United States; females are seldom affected

Other Names muscular dystrophy

Guillaume-Benjamin-Amand Duchenne (1806–1895) was one of the mysterious characters of medical history. As a French neurologist, he became especially interested in electricity and revived Galvani's famous research on electricity and muscle action. He spent an enormous amount of energy experimenting with the effect of electricity on the muscles of the face and the expression of emotions in people, and his book on the subject is a classic of medical history. However, his greatest contributions were in the field of myopathies and how they work, and in the muscular dystrophy that bears his name.

Peter Emil Becker (1908–2000), a German neurologist, lived much later but was fascinated with the work of Duchenne. He authored a genetics book that showed how muscular dystrophy is an X-linked gene. A form of muscular dystrophy now bears his name. The two dystrophies are different in some ways, but both involve the same gene.

What Are Duchenne and Becker Muscular Dystrophies?

Duchenne and Becker muscular dystrophies are both characterized by serious physical weakness. Both disorders are progressive, affecting the muscles nearest the trunk that are used for movement and the muscles of the heart. Both conditions affect boys but differ in their severity and age of onset.

In Duchenne muscular dystrophy, the first symptom is developmental delay, especially in learning to walk. These boys are not able to run, jump, or climb as normal toddlers do. At about two years of age, the gait becomes awkward, and the child looks like he is waddling. He also has problems getting up off the floor when he falls. As the child grows, shoulder muscles deteriorate and enlarge, but the muscles are flabby and weak. By the time the boy hits adolescence, he is usually confined to a wheelchair. The heart muscle also grows weak and large, a condition known as cardiomyopathy. About one-third of the boys also have learning disabilities. The prognosis for life is not good. Because of the weakness of the respiratory muscles, the boy is susceptible to pneumonia and other respiratory illness and usually does not live past the age of 20.

The symptoms of Becker muscular dystrophy are much less severe and disabling. Muscle weakness does not usually appear until adolescence and progresses much more slowly. Few young men are confined to wheelchairs and survive much longer—into their 30s or 40s. These boys develop cardiomyopathy, but it is later and not quite as severe.

What Are the Genetic Causes of Duchenne and Becker Muscular Dystrophies?

Both Duchenne and Becker muscular dystrophies are caused by mutations in the *DMD* gene, officially called the "dystorphin" gene. Normally, *DMD* instructs for a protein named dystophin, which stabilizes and protects muscle fibers. The protein also may play a role in the chemical signaling within cells. Mutations in the gene disrupt the production of dystorphin. In boys with Duchenne, the protein is almost completely lacking, and the muscles become weak and eventually damaged with repeated use. Boys with Becker produce dystorphin, but it does not function properly because the protein is altered. Both skeletal and heart muscles are affected.

This mutated gene is passed by the mother in an X-linked recessive pattern. Only one copy passed to males will cause the condition. Hundreds of mutations in the *DMD* gene have been identified. Most of the mutations involve a deleted part of the DNA and others involve abnormal duplication. *DMD* is located on the short arm (p) of chromosome X at position 21.2.

What Is the Treatment for Duchenne and Becker Muscular Dystrophies?

Because they are genetic conditions, neither can be cured. As in all genetic disorders with no cure and shortened lives, parents and caretakers should receive psychological counseling for the shortened life of their child. Individuals are encouraged to take physical therapy, exercise, and use orthopedic devices. Prednisone may be prescribed, but long-term use of the drug can produce undesirable side effects. Surgery may be used to avoid tight and painful muscles. Gene therapy to enable muscles to produce dystorphin is in an experimental stage in animal models.

See also Emery-Dreifuss Muscular Dystrophy (EDMD); Fascioscapulohumeral Muscular Dystrophy (FSHMD); Ullrich Congenital Muscular Dystrophy (UCMD)

Further Reading

"Distrophinopathies: Duchenne Muscular Dystrophy." 2011. Neuromuscular Disease Center. Washington University. http://neuromuscular.wustl.edu/musdist/dmd.html. Accessed 5/11/12.

"*DMD*." Genetics Home Reference. National Library of Medicine (U.S.). http://ghr.nlm.nih.gov/gene/DMD. Accessed 5/11/12.

"Duchenne and Becker Muscular Dystrophies." 2008. http://merckmanuals.com/home/sec05/ch073/ch073b.html. Accessed 5/11/12.

"Muscular Dystrophy." 2011. Centers for Disease Control (U.S.). http://www.cdc.gov/ncbddd/duchenne. Accessed 1/26/12.

VanderWeele, Misty. 2008. *In Your Face: Duchenne Muscular Dystrophy . . . All Pain . . . All Glory!* Raleigh, NC: Lulu Press.

Dystrophic Epidermolysis Bullosa (EB)

Prevalence Incidence 1 per 50,000; about one per million newborns worldwide; 6.5 per million newborns in the United States

Other Names EB; epidermolysis bullosa, dystrophic; epidermolysis bullosa dystrophica

A condition so serious that the skin falls off is difficult to understand, but with the blisters that form in this condition, that is literally what happens. It is like living consistently with third-degree burns. Children with this are often called "Butterfly Children" because their skin is pictured as being as fragile as a butterfly's wing. In fact, the Dystrophic Epidermolysis Bullosa Research Association of America (DEBRA) uses the butterfly in its logo. Other names used to describe the condition include "Cotton Wool Babies" and "Crystal Skin Children" in South America.

What Is Dystrophic Epidermolysis Bullosa (EB)?

Dystrophic epidermolysis bullosa (EB) is a condition in which blisters that form in the connective tissue causes sores in the skin and mucous membrane. The term "dystrophic epidermolysis bullosa" can be broken down as follows: "Dystrophic"

Jonny Kennedy: "The Boy Whose Skin Fell Off"

Some people have experienced painful sunburns when blisters form on the skin and sheets of the skin can be peeled away. Some might have experienced drinking a very hot coffee and burning the roof of the mouth. Fortunately, these inconveniences heal in a short period of time. However, a person with a classic case of EB has this suffering 24 hours a day and seven days a week. There is no possibility of healing.

Jonny Kennedy, who resided in Great Britain, lived for 36 years with the classic form of EB. Jonny's case of EB was so severe that the touch of a finger, contact with light clothing, a small bump or bruise, or slight scratch would cause his skin to actually fall off. He had painful sores and scars all over his body.

For 35 years, Jonny and mother Edna dealt daily with the condition, but when he was diagnosed with the fatal squamous cell carcinoma, he decided to do something different. He wanted the final months of his life to be documented on film to raise awareness of the condition and raise money for the charity DEBRA. The documentary, *The Boy Whose Skin Fell Off,* was first broadcast in the United Kingdom in March 2004. He was extremely candid about his life and upcoming death. He saw death "as a freedom and an escape." While the documentary was being filmed, Jonny died in his wheelchair on a train after visiting with Cherie Blair, wife of Tony Blair, prime minister of the UK.

refers to defects in body metabolism; "epidermolysis" comes from the Greek roots *epi*, meaning "upon," *dermis*, meaning "skin," *lysis*, meaning "breaking down," and *bullosa*, meaning "blister." Thus, this genetic disorder is a metabolic condition resulting from the breaking down of layers of skin giving rise to blisters.

Skin has two layers: the outer layer or epidermis, and an inner layer called the dermis. In people with EB, blisters form and make the outer layer of the epidermis peel away. Thus, the skin is extremely fragile, and lesions may form with a small amount of friction or minor injury. Symptoms may range from mild to life-threatening. Researchers classify ED into three groups:

- Recessive dystrophic epidermolysis bullosa, Hallopeau-Siemens (RDEB-HS): This classic form, inherited in an autosomal recessive pattern, is the most serious. Children are born with blisters or patches of missing skin as a result of passing through the birth canal. Blisters cover the entire body and are in the mucous membrane linings of the mouth and the moist areas of the intestinal tract. When the blisters heal, scars remain. The child then may have difficulty eating and swallowing, problems that can lead to malnutrition.

Scarring can also lead to fusion of the fingers and toes, loss of fingernails and toenails, joint deformities, and vision difficulties. Young adults with the condition are at risk for a very aggressive form of squamous cell carcinoma, which is life-threatening.

- Non-Hallopeau-Siemens Dystrophic epidermolysis bullosa (non-HS-RDEB): This type, inherited in an autosomal recessive pattern, is less severe than the classic form but includes many subtypes. Blistering is usually on the hands, feet, knees, and elbows. Scarring may be present, but it is not as widespread as in the classic type. Individuals may have malformed fingernails and toenails.
- Dominant Dystrophic Epidermolysis Bullosa (DDEB): This autosomal dominant condition tends to be milder, with blisters still appearing on hands, feet, knees, and elbows. The person may have malformed fingernails. In some of the mildest cases, this may be the only symptom of the disorder.

What Is the Genetic Cause of Dystrophic Epidermolysis Bullosa (EB)?

Mutations in the *COL7A1* gene, officially known as the "Collagen, type VII, alpha 1" gene, cause all three forms of dystrophic epidermolysis bullosa. Normally, *COL7A1* instructs for proteins that make type VII collagen. The collagens are a family that stabilize the skin and strengthen and support all connective tissues such as bone, tendons, and ligaments. The proteins produced by this gene are called pro-1(VII) chains and make up type VII collagen. Three of the chains twist together to make a tough rope-like molecule called procollagen, which is processed by the cell to remove extra protein segments from the ends. They then arrange themselves into long, thin bundles to make type VII collagen. This collagen then become anchoring fibrils that hold the two layers of skin—the dermis and epidermis—together.

Mutations in *COL7A1* gene disrupt the process of collagen formation. More than 400 mutations have been identified. These mutations alter the structure of the anchoring fibrils so that any friction or trauma can cause the layers to separate. Fluid then floods into the area, causing blisters. *COL7A1* can be inherited in two ways: autosomal recessive and autosomal dominant. *COL7A1* is located on the short arm (p) of chromosome 3 at position 21.1.

What Is the Treatment for Dystrophic Epidermolysis Bullosa (EB)?

Today, there is no cure for dystrophic epidermolysis bullosa. Living with the condition is like living consistently with third-degree burns. At present, the only treatment is daily wound care, which may involve different strategies for controlling the blistering and scarring.

Further Reading

"*COL7A1*." 2010. Genetics Home Reference. National Library of Medicine (U.S.). http://ghr.nlm.nih.gov/gene/COL7A1. Accessed 5/11/12.

DEBRA Foundation. http://www.debra.org. Accessed 5/11/12.

"Dystrophic Epidermolysis Bullosa." 2010. Genetics Home Reference. National Institutes of Health (U.S.). http://ghr.nlm.nih.gov/condition/dystrophic-epidermolysis-bullosa. Accessed 5/11/12.

E

Early-Onset Glaucoma

Prevalence Affects about 1 in 10,000 people; higher frequency in Middle East; juvenile angle glaucoma about 1 in 50,000 people

Other Names hereditary glaucoma

Glaucoma is a condition in which the optic nerves begin to malfunction and die. Usually it develops in older adults who have risky medical conditions, such as high blood pressure and diabetes. However, if the condition occurs before the age of 40, it is called early-onset glaucoma.

What Is Early-Onset Glaucoma?

The symptoms of glaucoma, a serious eye disease, begin with spots of floating particles of light within the field of vision and then progress to reduced peripheral vision and eventual blindness. Other signs may be bulging eyes, tearing, and great sensitivity to light. The basic cause is the pressure that is built up within the fluid of the eyes.

In early-onset glaucoma, inherited structural abnormalities at birth may interrupt the fluid pressure. If the condition occurs before the age of five, it is referred to as primary congenital glaucoma. These conditions can occur in newborns, appear in early infancy, or can appear as late-recognized congenital glaucoma.

However, if glaucoma occurs without the structure abnormalities, it is called juvenile open-angle glaucoma (JOAG). These abnormalities may be part of a syndrome. Juvenile-onset open angle glaucoma is often found with nearsightedness or myopia. JOAG may develop after the third year of life and even can appear in older children and adolescents. In this condition, the fluid or aqueous humor, located behind the iris, does not drain properly through a sieve-like structure called the trabecular meshwork back into the bloodstream. Thus, the pressure behind the

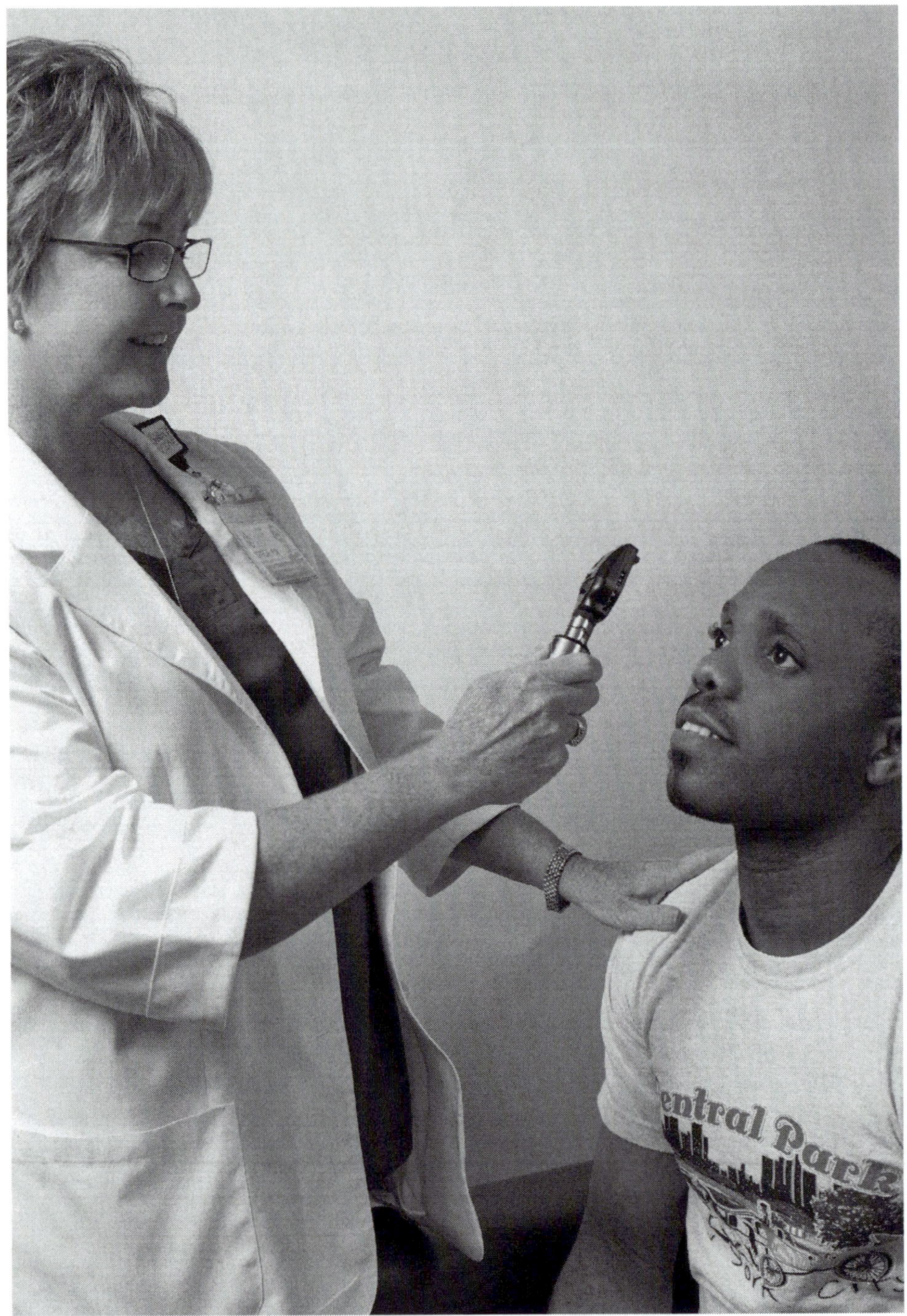

Eye examinations are important for people with glaucoma. (CDC / Amanda Mills)

iris is raised, and pressure will slowly damage the optic nerve. Unlike the congenital form, the problem may go undetected until major vision difficulties occur.

What Is the Genetic Cause of Early-Onset Glaucoma?

Mutations in two genes cause early-onset glaucoma. The genes are *CYP1B1*, officially called the "cytochrome P450, family1, subfamily B, polypeptide 1" gene, and *MYOC*, officially known as the "myocilin, trabecular meshwork, inducible glucocorticoid response" gene. The two genes are related to the two kinds of early-onset glaucoma.

CYP1B1

Mutations in *CYP1B1* are found in people with primary congenital glaucoma. Normally, *CYP1B1* instructs for the production of the cytochrome P450 protein. This protein, a member of a large family of enzymes, is involved in many body processes related to breaking down drugs in the body and producing fats. The protein is a part of many biochemical reactions that involve adding oxygen to molecules. Active in many body tissues, *CYP1B1* is present in the trabecular meshwork, which drains fluid from the back of the eyes, and other eye structures.

Mutations in the *CYP1B1* gene disrupt the protein that is involved in the formation of the trabecular meshwork, causing excess fluid buildup to put pressure on the intraocular fluid. The symptoms of glaucoma develop. Some researchers think that *CYP1B1* may interact with the *MYOC* gene to produce a more severe case of glaucoma at an earlier age. *CYP1B1* is thought to be inherited in an autosomal recessive pattern. The gene is located on the short arm (p) of chromosome 2 at position 21.

MYOC

Mutations in the *MYOC* gene are found in children with juvenile open-angle glaucoma (JOAG). Normally, *MYOC* instructs for producing the protein myocilin, found in the trabecular meshwork and the ciliary body (muscle tissues that move the eye). Myocilin may interact with a number of other proteins like a form of cytochrome P450. The mutations of *MYOC* may disrupt the normal function of these proteins and cause the symptoms of glaucoma. *MYOC* is inherited in an autosomal dominant pattern and is located on the long arm (q) of chromosome 1 at position 23-q24.

What Is the Treatment for Early-Onset Glaucoma?

It is important to check and treat ocular pressure in children. Several types of surgery are available for correcting some of the disorders of the eye. One of the most effective appears to be the trabeculectomy, in which the surgeon creates a new drainage channel for the fluid to pass through. Drops may reduce the intraocular pressure.

Further Reading

"*CYP1B1*." 2010. Genetics Home Reference. National Library of Medicine (U.S.). http://ghr.nlm.nih.gov/gene/CYP1B1. Accessed 5/11/12.

"Early-Onset Glaucoma." 2012. Genetics Home Reference. National Library of Medicine (U.S.). http://ghr.nlm.nih.gov/condition/early-onset-glaucoma. Accessed 5/11/12.

"Early-Onset Glaucoma." Medic8.com. http://www.medic8.com/genetics/early-onset-glaucoma.htm. Accessed 5/11/12.

"*MYOC*." 2010. Genetics Home Reference. National Library of Medicine (U.S.). http://ghr.nlm.nih.gov/gene/MYOC. Accessed 5/11/12.

Early-Onset Primary Dystonia

Prevalence One of the most common forms of dystonia; found especially in people of Ashkenazi Jewish heritage, affecting about 1 in 3,000 to 9,000 people; less common in other populations, affecting 1 in 10,000 to 30,000 non-Jewish people.

Other Names dystonia musculorum deformans 1; DYT1; early-onset generalized torsion dystonia; Oppenheim dystonia; Oppenheim's dystonia; primary torsion dystonia

Jews of Ashkenazi or central and eastern European descent have a set of genetic disorders that are much more prevalent in that population. Two effects may reflect the reason these disorders exist at a much higher proportion. One is the theory called the "Founder Effect," the idea that these ancestors who spread throughout Europe at the time of the Diaspora in about 70 AD carried the original mutation. The second theory is that "Genetic drift" Jews tended to live in specific communities and did not marry outside their faith. This concentrated the genes in that ethnic group.

In 1908, Schwalbe described an unusual torsion or twisting of limbs in a Jewish family; in 1911, Oppenheim termed the disorder dystonia musculorum deformans and surmised that it was caused by hysteria. Later, it was determined that the condition was a neurological disorder and one of the genetic disorders prevalent in the Jewish community. Early-onset primary dystonia is now recognized as one of the most common dystonias.

What Is Early-Onset Primary Dystonia?

Early-onset primary dystonia is a progressive movement disorder characterized by involuntary contractions of muscles, twisting of certain body parts, and shaking or tremors. Early-onset means that the conditions appear before the age of 12. The term primary means that the symptoms occur without other neurological occurrences such as seizures or dementia. Early-onset primary dystonia does not affect the person's intellectual abilities.

The signs and symptoms vary from person to person. However, once the symptoms start, they last for the life span of the person. The mildest symptoms may be similar to a writer's cramp. Severe cases may affect many parts of the body.

What Are the Genetic Cause of Early-Onset Primary Dystonia?

Mutations in the *TOR1A* gene are related to this condition. This gene is also called *DYT1*. Normally, *TOR1A* instructs for making the protein torsinA. Looking within the cell of an individual, there is the nuclear envelope, which surrounds the nucleus, and the endoplasmic reticulum, a maze that transports material in and out of the cell. In the cell space between these two structures, the protein torsinA is found. The function of torsinA appears to be to process and transport other proteins; it may also be related to movement of the nuclear membrane and endoplasmic reticulum. TorsinA is found in many body tissues and is especially important for a part of the brain called the substantia nigra, an area that produces the chemical messenger dopamine essential for smooth muscle function.

The mutation in *TOR1A* appears to be a GAG (guanine-adenine-guanine) deletion from the gene. The critical missing part disrupts the protein's effect on the nerve cells that control movement. *TOR1A* is inherited in an autosomal dominant pattern, which means that one of two copies of the gene is changed in each cell. Many people with the mutation are not affected but may pass the defective gene to offspring. *TOR1A* is located on the long arm (q) of chromosome 9 at position 34.

What Is the Treatment for Early-Onset Primary Dystonia?

At this time no cure exists for the disorder, but the symptoms can be treated with medications and surgical options. Hypnosis, sleep, and relaxation techniques may be recommended to manage stress, which appears to aggravate contractions.

Further Reading

"Early-Onset Primary Dystonia." 2010. Genetics Home Reference. National Library of Medicine (U.S.). http://www.ghr.nlm.nih.gov/condition/early-onset-primary-dystonia. Accessed 5/11/12.

Patil, Vijaya. 2010. "Primary Torsion Dystonia." Medscape. http://emedicine.medscape.com/article/1150643-overview. Accessed 5/11/12.

"*TOR1A*." 2010. Genetics Home Reference. National Library of Medicine (U.S.). http://ghr.nlm.nih.gov/gene/TOR1A. Accessed 5/11/12.

"Torsion Dystonia." 2008. Chicago Center for Jewish Genetic Disorders. http://www.jewishgenetics.org/?q=content/torsion-dystonia. Accessed 5/11/12.

Ehlers-Danlos Syndrome (EDS)

Prevalence 1 in 5,000 births; hypermobility as many as 1 in 10,000 to 15,000 people; elasticity type 1 in 20,000 to 40,000 people

Other Names cutis hyperelastica; EDS

Always fascinating is the contortionist who can bend and stretch into unusual positions. Such people may be dubbed "Rubber Boy," "Rubber band Man," or "Bendy Girl." The hyperflexible person may have a condition known as Ehlers-Danlos syndrome, or EDS. Television programs, such as *CSI: New York* and *House*, have featured EDS in their plots, and it is even thought that EDS helped the violinist Paganini do feats with the violin that others could not do. Even the schoolyard child with EDS may entertain his fellow classmates with his talents.

However, depending on the type, EDS can be a serious life-threatening disorder. The condition was identified around the beginning of the twentieth century by Danish physician Edvard Ehlers and French doctor Henri-Alexander Danlos. The condition now bears the name Ehlers-Danlos syndrome.

What Is Ehlers-Danlos Syndrome (EDS)?

EDS is a group of disorders that affect the connective tissue or collagen. It is collagen that holds skin, bones, blood vessels, and other organs in place. Without collagen, a person would just fall apart. Defects in collagen play a role in EDS, which can range from just mildly loose joints to severe complications.

Although there are six recognized types of EDS, the following general symptoms may include:

- Very flexible fingers and toes
- Loose joints that may be prone to dislocation and sprain
- Flat feet
- Easy bruising
- Crowded teeth as a result of a high, narrow palate
- Velvety smooth, stretchy skin
- Low muscle tone
- Cardiovascular problems
- Fibromyalgia

Often in childhood, symptoms of EDS may be diagnosed as child abuse. Pain associated with the condition is a serious complication. The soft, velvety skin may cause the individual to bruise easily, and the skin may split open leaving small scars.

What Are the Types of Ehlers-Danlos Syndrome (EDS) and Their Genetic Causes?

Although all types affect the joints and some affect the skin, six recognized types of EDS have been named:

- Hypermobility: This type is characterized by an unusual range of motion in joints and excessive flexibility. Infants may have delayed sitting, standing, and walking due to weak muscle tone. The loose joints are very unstable and cause pain, dislocations, and early-onset arthritis. This type involves two separate genes: the collagen, type III alpha 1, or *COL3A1*, inherited as an autosomal dominant trait and located on the long arm (q) of chromosome 2 at position 31; and "tenascin XB," or *TNXB*, an autosomal recessive trait located on the short arm (p) of chromosome 6 at position 21.3. Both of these genes instruct for making a component of collagen. The presence of the mutation in the gene disrupts function of the protein and causes the extreme flexibility related to EDS. This type affects 1 in 10,000 to 15,000 people.
- Classical—types 1 and 2: The classical types vary also; type 1 presents severe skin involvement and type 2 presents milder and more moderate symptoms. Three genes that affect type V collagen are involved: *COL5A1*, located on the long arm (q) of chromosome 9 at position 34.20q34.3; *COL5A2*, located on the long arm (q) of chromosome 2 at position 14-q32; *COL1A1*, located on the long arm (q) of chromosome 17 at position 21.33. All three of these genes are inherited in an autosomal dominant pattern and disrupt the production of type V collagen.
- Vascular: This more serious form of EDS involves vascular disorders in which the blood vessels and organs are prone to tearing, causing internal bleeding, stroke, and shock. During pregnancy, the uterus may rupture. This type is inherited in an autosomal dominant pattern involving Type III collagen and caused by a mutation in the *COL3A1* gene. People with this type often have characteristic facial appearance of large eyes, a thin build, and translucent skin. About one in four experience serious health problems by age 20, and more than 80% have life-threatening conditions by age 40.
- Kyphoscoloisis: This very rare form is due to a mutation in the *PLOD1* gene located on the short arm (p) of chromosome 1 at position 36.22. The autosomal recessive gene has about 20 mutations that cause a deficiency of an enzyme, lysyl hydroxylase. It is characterized by progressive curvature of the spine or scoliosis, very weak eyes, and muscle weakness. Fewer than 60 cases of this type have been reported.
- Arthrochalasis: This very rare type affects about 30 known cases. It is caused by mutations of *COL1A1* and *COL1A2* and inherited in an autosomal recessive pattern. This type is characterized by very loose joints and dislocations of the hips.

- Dermatosparaxis: This type is an extremely rare, with only 10 cases reported and inherited in an autosomal recessive pattern. Characterized by extremely fragile and sagging skin, mutations in the gene disrupt the collagen process and cause the characteristic symptoms.

There are several other types of EDS that do not fit into a neat category.

What Is the Treatment for Ehlers-Danlos Syndrome (EDS)?

Because EDS is a genetic disorder, no cure exists for it. EDS is a lifelong condition. The treatment must be supportive and related to the symptoms. Careful monitoring of the heart as well as physical and occupational therapy may be helpful. In general, medical intervention depends upon the symptoms. Children with EDS should be given information and taught early not to demonstrate the flexibility of their joints to classmates, as this can lead to early deterioration of the joints.

Further Reading

Ehlers-Danlos National Foundation. 2010. http://www.ednf.org. Accessed 5/11/12.

"Ehlers-Danlos Syndrome." 2010. Genetics Home Reference. National Library of Medicine (U.S.). http://ghr.nlm.nih.gov/condition/ehlers-danlos-syndrome. Accessed 5/11/12.

Ellis–Van Creveld Syndrome

Prevalence	Rare; about 1 in 60,000 to 200,000 newborns in the general population; more common among the Amish population of Lancaster County, Pennsylvania, and in the indigenous native population of Western Australia
Other Names	chondroectodermal dysplasia; mesoectodermal dysplasia

In the late 1930s, two pediatricians, Dr. Richard W. B. Ellis (1902–1966) of Edinburgh and Simon van Creveld (1895–1971) of Amsterdam, were riding on a train to a conference in England. In talking, they found they had patients with similar conditions—short limbs and extra fingers on the hand. Studying additional cases, they wrote a paper on the topic that was included in a 1933 medical textbook. The condition became known as Ellis–van Creveld syndrome.

About 20 years later, in 1964, Victor McCusick studied the Old Order Amish population in Lancaster County, Pennsylvania, and reported the presence of six-digit dwarfism in that population. The gene was possibly present in Samuel King and his wife, the founders of the group who immigrated to the area and established an Old Amish colony. Also, McCusick found that years of inbreeding had produced more cases of Ellis–van Creveld syndrome than was reported in all the years of medical literature at that time. In the Amish population, the prevalence is about 5 in 1,000 births; it is estimated that the frequency of carriers is about 13%.

What Is Ellis–Van Creveld Syndrome?

Ellis–van Creveld syndrome (EVC) is a type of dwarfism characterized by very short limbs and an extremely narrow chest at birth. The following physical disorders may also be present:

- Polydactyly, or extra fingers and toes: These digits are displayed in a post-axial position.
- Short ribs and a small chest
- Thoracic dysplasia: This can lead to respiratory deficiency and death.
- Congenital heart defects: One defect is the absence of the atrial septum that divides the two upper chambers of the heart; about 60% of individuals with heart defects have this disorder. This can also cause death.
- Prenatal tooth eruption: Individual is born with teeth.
- Fingernail dysplasia
- Cleft palate
- Malformation or fusion of the wrist bones

The disorder appears equally in both males and females. Individuals with the disorder have normal intellectual functioning.

What Is the Genetic Cause of Ellis–Van Creveld Syndrome?

The official name of the genes that cause the syndrome is "Ellis van Creveld syndrome," or *EVC* and *EVC2*. Normally, these genes instruct for a protein responsible for normal growth and development. The two genes probably work together especially before birth to develop the heart, bones, kidneys, and lungs.

Mutations in *EVC* and *EVC2* cause Ellis–van Creveld syndrome. The mutations are inherited in an autosomal recessive pattern, meaning that both copies of the mutation occur in every cell. Parents carry the mutated gene without having the condition themselves. These mutations disrupt the normal function of the EVC protein and produce the symptoms of dwarfism, heart defects, and the other features of *EVC*. The *EVC* gene is located on the short arm (p) of chromosome 4 at position 16. The *EVC2* gene is located on the short arm (p) of chromosome 4 at position 16.2-p18.1.

What Is the Treatment for Ellis–Van Creveld Syndrome?

Because EVC is a genetic disorder, there is no cure. Basic treatment includes treating the symptoms of the condition, such as malfunction of the heart, lung, or kidney.

Further Reading

"Ellis–Van Creveld Syndrome." 2010. Medscape. http://emedicine.medscape.com/article/943684-overview. Accessed 5/11/12.

"Ellis–Van Creveld Syndrome." 2010. National Center for Biotechnology Information (U.S.). http://www.ncbi.nlm.nih.gov/books/NBK22264. Accessed 5/11/12.

"*EVC*." 2010. Genetics Home Reference. National Library of Medicine (U.S.). http://ghr.nlm.nih.gov/gene/EVC. Accessed 5/11/12.

Emanuel Syndrome

Prevalence Rare with unknown prevalence; more than 100 cases have been reported

Other Names der(22) syndrome due to 3:1 meiotic disjunction events; supernumary der(22) syndrome; supernumary der(22)t(11;22) syndrome; supernumary derivative 22 chromosome syndrome

This syndrome has had numerous names, most of which are confusing to families of children with the disorder. For example, calling the disorder supernumerary der (22) syndrome or the incorrect term "cat-eye syndrome" has little meaning to parents trying to understand what is happening to their children. Even referring to the syndrome as t(11;22) did not explain that this syndrome was a translocation between chromosomes 11 and 22.

In 2004, the support group named the syndrome after a researcher at the Children's Hospital of Pennsylvania Medical Center who had worked to describe the mechanism of how chromosomes 11 and 22 get crossed. Dr. Beverly Emanuel, a cytogeneticist at the hospital, and colleagues have studied the disorder since the early 1970s and have continued to do so. Her research on this translocation inspired the group in the parent support system to request her name be given to this unique and rare syndrome. The term "Emanuel syndrome" is slowly being adopted by the medical community. The parents in the support site feel this name gives unity and identity to the disorder.

What Is Emanuel Syndrome?

Emanuel syndrome is a condition in which genetic material from two chromosomes are crossed, causing an extra chromosome and numerous health problems. The term "translocation" is used to describe what happens when two chromosome exchange material and created a derivative chromosome that is present in many of the cells. The term "der" is used for derivative.

The presence of the extra genetic material disrupts normal development and causes many of the following symptoms:

- Poor muscle tone at birth.
- Distinct facial features. The infant has a very small lower jaw and very small head.
- Ear disorders: Abnormalities include small holes in the skin in front of the ears. These holes are called preauricular pits or sinuses.
- Cleft palate: About half of the infants have a hole in the roof of the mouth or an abnormally high and arched palate.
- Failure to thrive: The child does not gain weight at the expected rate.
- Developmental delay.
- Heart defects.
- Kidney issue: Some children have abnormally small, called hypoplastic kidneys, or missing kidneys. This issue can be life-threatening in infancy.
- Sex organ abnormalities: Males with the condition have underdeveloped genitals
- Severe intellectual disabilities.

What Are the Genetic Causes of Emanuel Syndrome?

Extra genetic material from both chromosome 11 and chromosome 22 cause Emanuel syndrome. Normally, people have 46 chromosomes or 23 pairs. But the children with Emanuel syndrome have an extra chromosome, with material from each of the chromosomes. The term "supernumerary" is used to describe this extra chromosome. The word "derivative" also refers to this additional genetic material. Thus, the new chromosome is known as derivative 22 or the der(22) chromosome. People with the disorder have three copies of some genes in each cell, which disrupt normal development.

Unlike many chromosomal aberrations that happen during egg and sperm formation or in early embryonic development, der(22) appears to come from a parent who does not have the symptoms of the disorder but does have an arrangement of the chromosome called a balanced translocation between chromosomes 11 and 22. This means that no genetic material is lost and no health problems result. However, these translocations can become unbalanced when they are passed to offspring. The children with Emanuel syndrome inherited the unbalanced translocation that creates the derivative 22 chromosome.

Chromosome 11

This chromosome has about 134 million amino acid base pairs and about 1,500 genes, which have different roles in the body. About 150 genes relate to the sense of smell. Extra genetic material from this chromosome is in each cell. This extra material leads to the many defects that are present at birth and to profound intellectual disability.

Chromosome 22

Chromosome 22 is the second-smallest human chromosome, with about 50 million amino acid base pairs and between 500 to 800 genes, which perform a variety of functions. Scientists are still working on which genes are replicated in the extra derivative 22 chromosome.

What Is the Treatment for Emanuel Syndrome?

Emanuel syndrome is a very serious genetic disorder with so many conditions affecting various body areas. Treating the most life-threatening conditions, such as kidney and heart defects, is of first importance. Care by a multidisciplinary team is usually necessary. Standard treatment for cleft palate and other defects may be used. The child qualifies for services, such as speech, physical, and occupational therapy, under the Americans with Disabilities Act.

Further Reading

"Emanuel Syndrome." 2012. Genetics Home Reference. National Library of Medicine (U.S.). http://ghr.nlm.nih.gov/condition/emanuel-syndrome. Accessed 1/8/12.

"Emanuel Syndrome." 2012. National Center for Biotechnology Information (U.S.). http://www.ncbi.nlm.nih.gov/books/NBK1263. Accessed 1/8/12.

Emanuel Syndrome.org. 2011. http://www.emanuelsyndrome.org. Accessed 1/8/12.

Embryology: A Special Topic

Human Embryology

One cannot understand human genetic disorders without understanding something about how the embryo develops. Many of the genetic disorders caused by genetic mutations occur early in embryonic development. A second topic on human embryopathies follows this section.

Throughout history, people could only imagine what happens before birth. In the eleventh century, Trotula, a female physician at Salerno, Italy, wrote a treatise called *The Diseases of Women* that was illustrated with drawings of the child before birth as a miniature adult floating in water. Centuries later, Hieronymus Fabricus (1533–1609) used the same childlike drawings to write one of the first books on the formation of the developing human being. The word "embryo"

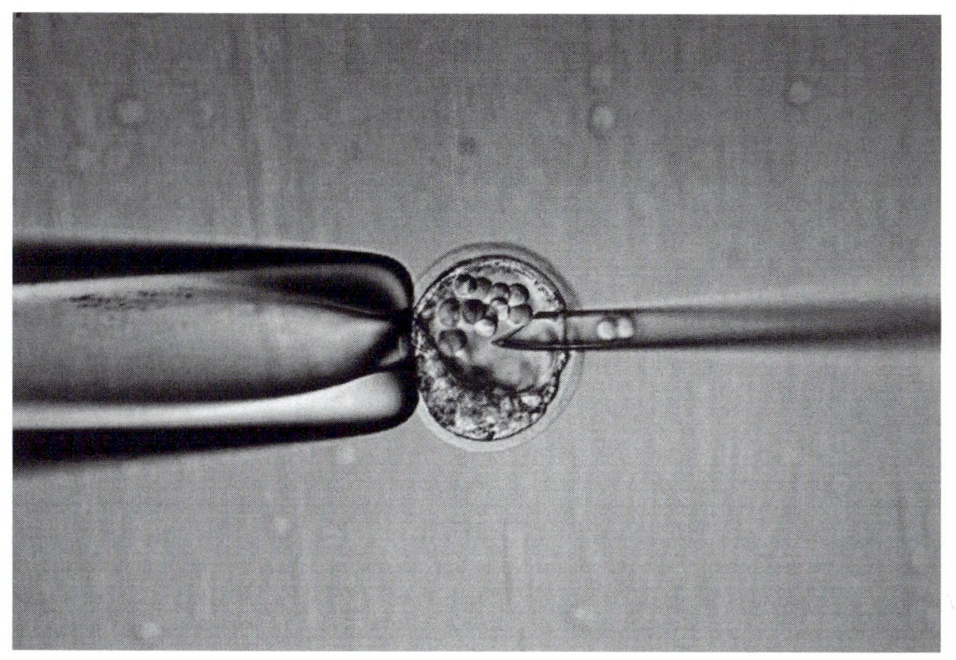

A fine needle injects embryonic stem cells. (ANNE-CHRISTINE POUJOULAT/ AFP/Getty Images/)

comes from the Greek words with the literal meaning "that which grows." The Greek roots are: *en*, meaning "in," and *bryein*, "to swell or be full."

The sciences of genetics and embryology were late developing, and it was not until the early twentieth century that they began to expand. Although the fields developed separately, some areas of current research combine genetics and embryology to understand why birth defects and genetic disorders may develop. Scientists consider the interplay between genetics and the environmental factors of the uterus critical. The human embryo is exceedingly vulnerable to drugs, viruses, and radiation during the first several months when systems are forming.

The preembryonic period entails the first two weeks of human development beginning at fertilization. At this point, the fertilized egg is called a zygote, which then undergoes division to become a blastocyst. Some of these cells undergo division to form three layers that become the cells of the embryo. These cells are called primary germ layers or stem cells and will become the following body tissues:

- Ectoderm: The outer layer that gives rise to the outer epithelium, such as hair, nails, skin, the sense organs, the brain, and the spinal cord. The epithelium forms the epidermis and other structures.
- Mesoderm: The middle layer that gives rise to the bones, muscle, connective tissue, circulatory system, and most of the excretory and reproductive systems.

- Endoderm: The inner layer gives rise to the epithelial linings, the cells that from the linings of the body cavity such as the digestive tract, most of the respiratory tract, bladder, liver, pancreas, and some of the endocrine glands.

Embryonic stem cell research is based at this stage of development.

The Human Embryo

Following is a week-by-week development of the human embryo:

- Week 1: By the end of the first week, the fertilized egg in the Fallopian tube reaches the uterus but does not increased in size. The blastocyst or blastula attaches to the wall of the uterus, called the endometrium and implants. A connection between the mother begins to develop that will be the umbilical cord.
- Week 2: Around 12 days, the primitive streak begins to form. This band appears along the back of the mass and is a temporary structure that will form other structures.
- Week 3: Around 21 days, the notochord forms the central axis that will become the vertebral column. Somites, or little cube-like bodies, project from each side of the neural groove; these somites will become the 31 spinal nerves that extend control to the entire body. The brain, spinal cord, heart, and gastrointestinal tract begin to form.
- Week 4: Now a period of spectacular growth occurs. Somites begin to differentiate into the spinal nerves, and a part of the sclerotoma forms the vertebral column; the neural tube closes; and the tail or caudal extremity is prominent.
- Week 5: First traces of the hand appear; the feet, clavicle, and lower jaw begin; a primitive palate appears; and the general form of the human body now becomes evident. Neurogenesis begins with noticeable brain activity; the brain develops into five areas. The heart begins to beat about the same time, and blood begins to flow. Organs begin to form. The head represents about one-half of the embryo's length and more than half of the embryo's mass.
- Week 6: The vertebral column, primitive cranium, and ribs appear as cartilage; furrows mark the three divisions of limbs; the general form of the human skeleton is evident; primitive germ cells that will become teeth appear. The muscles and nervous system have progressed to allow for motion, and the eyes begin to form.
- Week 7: The ribs, shoulder blade, shaft of the humerus or upper arm, femur, tibia, palate, and upper jaw begin to form bone or ossify; 10 finger rays appear.
- Week 8: Distinction of arm and forearm occurs; all vertebral bodies are mode of cartilage, but bone is beginning to form in the extremities and ilium in the hip; two halves of the hard palate unite; the head and nose are formed. Organs

and growth continue, and hair has started to form with all essential organs. Facial features begin to develop. The embryo is now considered a fetus.

Embryologists use a system called Carnegie stages to describe the growth of the embryo and fetus. The stages are based on arbitrary levels of maturity and the way the average develops. Carnegie stages are numbered from 1 to 23 based on external features.

The Human Fetus

The stage of the human fetus begins at about nine weeks. The word "fetus" with the plural "fetuses" comes from the Latin *fetus*, meaning "offspring or bringing forth."

Beginning with this stage, development may vary.

- Weeks 9 to 25: Week 9 shows a human being beginning to take shape and form, although it may be only 1.2 inches or 30 mm in length and weigh about 8 grams. To stimulate lung development, the fetus may show breathing-type movements but this motion is not to get oxygen. The heart, hands, feet, brain, and other organs are present but are only beginning development and are not functioning. Arches of the vertebrae, the frontal and cheek bones, the shafts of the metatarsal bones, and all the toes and fingers appear. Union of the hard palate is complete and hands appear. At about 12 weeks, the fetus is noticeably attached to the placenta via the umbilical cord. This is the lifeline of the fetus. It is through the umbilical cord that the young offspring gets its oxygen and nourishment and gives off carbon dioxide and waste products. All basic processes for living are carried on through the umbilical cord. The tail disappears and the human shape takes form. Bones continue to ossify.
- Weeks 16 to 25: Around the 21st week, the woman will begin to feel movement. Closure of the cartilage arches of the spine occurs, and bones continue to develop. Some of the tiny bones in the ear develop; ossification of the bones of teeth occurs. By the end of the fifth month, the fetus is about 20 cm (8 inches). Development of bones in the vertebrae continues.
- Weeks 26 to 42 (seventh and eighth months): Bones continue to ossify but are still soft and pliable. Final points of development continue; the fetus gains weight and then normally assumes the birth position of head first. By the end of 40 weeks after fertilization, the fetus is about 20 inches ling (51 cm) from head to toe. Body fat increases. Through the umbilical cord, the fetus gets iron, calcium, and phosphorus. Fingernails develop. Head hair becomes coarse and thicker. The fetus is considered full term between 38 and 42 weeks, which means the baby is sufficiently equipped for life outside the uterus.

If given postnatal care, fetuses weighing less than 500 gm may live and are referred to as extremely low birth weight or immature infants.

At any point from conception to birth, things can go wrong. Disorders may arise from genetic conditions and disruption in the functions of genes. The mother may expose the embryo or fetus to something that affects development.

See also Embryopathies: A Special Topic

Further Reading

"Atlas of Human Embryology." 2009. Chronolab. http://www.embryo.chronolab.com. Accessed 2/10/12.

"Human Embryology Animations." 2009. Indiana University. http://www.indiana.edu/~anat550/embryo_main. Accessed 2/10/12.

"The Multi-Dimensional Human Embryo." 2009. http://embryo.soad.umich.edu. Accessed 2/10/12.

Standring, Susan, ed. 2009. *Gray's Anatomy*. 40th ed. New York: Elsevier. For a detailed presentation on embryology, see *Gray's Anatomy* (1918 edition) online—http://education.yahoo.com/reference/gray—or the 40th edition from 2009.

Embryopathies: A Special Topic

Embryopathies are birth defects, problems that happen while the baby is developing during pregnancy. According to the March of Dimes, about 120,000 babies are born each year with physical or mental birth defects. Understanding human embryology and development *in utero* is crucial to knowing about human genetic disorders and birth defects. Most of the genetic mutations show up in early formative days. Some birth defects such as Huntington disease do not appear until many years later. In addition to genetic disorders, a group of environmental conditions may occur that affect the development of the embryo and fetus.

The term "embryopathies" refer to congenital abnormalities that occur during the embryonic period. The development of the offspring can be thought of in several stages:

- Weeks 1–2 or the first 14 days: The first 14 days after fertilization, an insult or injury generally does not result in an abnormality. The problem generally repairs itself or the fertilized egg dies, a condition known as spontaneous abortion.
- Third week to eighth week: The embryo is very vulnerable. During these weeks, numerous divisions occur, and organs begin to form. The embryo grows at a relatively enormous rate. The term "noxae" is used to describe anything that is harmful to health. The word has the same root and meaning as the word noxious, meaning harmful or not healthful.
- Fetal period beginning at 9 weeks to birth: The term "fetopathy" is used to describe abnormalities that happen during this period. The sensitivity to

injury now is greatly reduced. Most of the organs are formed. The one exception is the cerebral cortex, the forward and thinking part of the brain. This organ continues to be susceptible to noxae between the 8th and 15th weeks of development.

Embryopathies are divided into primary abnormalities and secondary abnormalities.

Primary Abnormalities

A primary abnormality is one that can be traced to a problem in the development of a structure or organ. Most of the genetic conditions presented in this encyclopedia are primary abnormalities traced to a gene or family of genes. Some are related to chromosome abnormalities, which may be genetic or occur at random. Another group is a combination of many factors, including genetics and environment. Following is the breakdown of the three subgroups:

- Genetic aberrations: Mutations in specific genes cause certain disorders, and many of those presented in this encyclopedia. About 7.5% of all birth defects are caused by gene mutations. These defects can be caused by a single gene or by several genes interacting. Conditions relating to a single gene include cystic fibrosis, sickle cell disease, and osteogenesis imperfecta. The conditions are inherited in a dominance or recessive pattern according to Mendelian law.
- Chromosomal aberrations: These defects are related to repeats in the chromosomes, such as the extra chromosome 21 in Down syndrome. These defects comprise about 0.5% of birth defects.
- Multifactorial anomalies: Several genes are involved, and these susceptibility genes may also respond to conditions in the environment such as medicines or chemical products. This group includes conditions such as spina bifida and other neural tube defects, cleft palates, and harelips. All sorts of heart and circulation disorders, dysplasia, or abnormal organization of cells in the tissue are in this group.

Secondary Abnormalities

In these conditions, the individual was developing normally until an outside factor exposed the embryo to some agent. The child may have the genetic propensity for a disorder, but it would not have occurred without the exposure to the noxae. The agents responsible are called teratogenic factors, from the Greek words *teratos*, meaning "monster," and *gen*, meaning "giving rise to."

Many secondary abnormalities depend on the health and behavior lifestyle of the mother. The noxae or insult must occur at the time when the organ or system is developing, so timing is important here. Teratogenic agents include drugs and medicines, environmental chemicals, ionizing radiation, infections, and metabolic imbalance.

Drugs and Medications

Many drugs are teratogens. Smoking and excessive amounts of caffeine and alcohol are teratogens under the right conditions and at the right time. Fetal alcohol spectrum disorder causes the young infant to have certain mental and physical defects. (*See also* Fetal Alcohol Spectrum Disorder.) Another serious chemical that causes birth defects is isotretinoin (Accutane). (*See also* Fetal Isotretinoin Syndrome). The following drugs have been associated with embryopathies:

- Temazepam (Restoril, Normisson): Prescribed to combat sleeplessness, this drug may harm the fetus.
- Nitrazepam (Mogadon): This hypnotic drug, used as a treatment for insomnia, causes abnormalities in 30% of the newborns of mothers who have taken this drug.
- Nimetazepan (Ermin): This drug is a hypnotic agent used also as an anticonvulsant.
- Aminopterin: An analog of folic acid, Aminopterin has immunosuppressive properties used in chemotherapy. Because it binds certain proteins, it depletes nucleotide precursors and inhibits DNA, RNA, and protein.
- Androgenic or male hormones: Androgen is the generic term for any natural or synthetic compound that stimulates male characteristics. Usually a steroid, testosterone is the best-known androgenic hormone. Warning labels to pregnant women not to touch the compound indicate the seriousness of this teratogen.
- Busulfan: Busulfan was the mainstay of the chemotherapy treatment of chronic myeloid leukemia (CML). It has now been replaced by another drug for this treatment but may still be prescribed because of its low cost.
- Captopril: This is one of the popular angiotensin-converting enzyme inhibitors or ACE inhibitors used for treating high blood pressure and some types of congestive heart failure.
- Enalapril: This drug is another ACE inhibitor used for treating high blood pressure and heart failure.
- Cyclophosphamide: This drug is used to treat certain cancers and immune disorders. Considered a prodrug, it is converted in the liver to an active form to fight cancer.
- Diethylstilbestrol (DES): This synthetic form of nonsteroid estrogen achieved notoriety in the late 1970s for causing a rare vaginal cancer for young women who were exposed to DES during their mother's pregnancy. DES was used in cattle feed and treatment for some medical conditions.
- Phenytoin (Dilantin): The drug is commonly used for treatment of epilepsy.
- Etretinate: This retinoid product is used for treatment of psoriasis and other skin conditions. It can cause birth defects long after use.

- Lithium: This drug is used in the treatment of bipolar disorder.
- Methimazole: This drug is used to treat hyperthyroidism, a condition that occurs when the thyroid produces too much thyroid hormone.
- Penicillin and tetracyclines: These antibiotics are used for a variety of treatments of infections.
- Trimethadione: This drug is used to treat epilepsy. However, if taken during pregnancy, the fetus may develop fetal trimethadione syndrome, resulting in facial dysmorphies, heart defects, growth retardations, and mental retardation. Fetal loss has been reported as high as 87%.
- Methoxyethyl ethers: At one time, ethers were used as anaesthetics.
- Valproric acid (VPA): This compound is used as an anticonvulsant and mood stabilizer especially in treating epilepsy and bipolar disorder.

Especially noted are the blood thinner coumadin or and the drug thalidomide, which took on a life of its own in the 1960s.

Environmental Chemicals

Exposure to certain chemical products can be teratogenic. The polychlorinated biphenyls (PCBs), polychlorinated dibenzodioxins (the infamous dioxin), polychlorinated dibenzofurans (PCDFs), hexachlorophene, organic mercury, ethidium bromide, and others have been shown to be harmful to developing embryos. Many of these chemicals are used in industrial plants and may be by-products of waste. Ethidium bromide is used in biological and genetic studies.

Ionizing Radiation

Atomic weapons fallout from explosions or tests affect the fetal development. Also background radiation, diagnostic X-rays, and radiation therapy are known teratogens.

Pathogens

The embryo's connection to the world is through the mother's placenta. The umbilical cord connected to the placenta is the lifeline for receiving nourishment and oxygen as well as the vehicle for giving off waste products and carbon dioxide. If the mother has a viral infection, the embryo may get the virus via the placenta and be affected in several ways. A vaginal infection of the mother may affect the amniotic fluid and transmit the virus to the fetus.

The mnemonic device TORCH is used to help medical students remember the pathogenic agents: toxoplasmosis, other agents, rubella, CMV, HSV. Following are pathogenic agents:

- T—Toxoplasmosis: One of the most serious teratogenic agents is commonly called cat-scratch fever. The condition is caused by a one-celled protozoan *Toxoplasma Gondii* and is carried by many animals, with the cat being the

Thalidomide: A Medical Disaster

The drug thalidomide was developed in 1953, but the company stopped its production and gave the rights to a German drug company, Grünenthal, in 1957. The drug was prescribed to women with severe morning sickness. No one paid much attention when, in 1956, a baby with an unusual defect was born. The child had flipper-like arms and other birth defects. The mother had taken thalidomide during her pregnancy, but the connection was not made between drug and defect.

During the next few years in Europe, a dramatic increase in birth defects occurred, including phocomelia, a shortened-limb defect in which the hands, feet, or both appear like small flippers. An Australian researcher made the connection between thalidomide and the birth defects. He found that thalidomide exposure blocked the middle stem cell layer in the mesenchyme during the time that the intermediate limb bud limb was developing. The entire limb bud then formed only the most distal elements at the finger ends. The limb defect was the only one caused by the drug. Trunks lacking arms and legs or both, deafness, cleft palate, and malformed internal organs are only a few of the disabilities experienced by over 12,000 children in 46 countries. These disorders were traced to use of thalidomide during pregnancy. In 1962, a Phoenix, Arizona, mother of four, Sherri Chessen Finkbine, who had been taking thalidomide for stress during her pregnancy, tried to get an abortion at a local hospital when she learned that the drug would probably deform her child severely. She was denied but traveled to Sweden with her husband and had the abortion there.

In the United States, Frances Kelsey, MD, had taken a job in 1960 as medical evaluator for the Food and Drug Administration (FDA). The German company wanted to market the drug in the United States, but Kelsey insisted more tests be done to show safety. She was under great pressure to approve, but stalled. Kelsey had saved many Americans the heartbreak of these severe deformities. As a result, the FDA significantly tightened its drug approval protocols. Now, the agency is one of the strictest in the world.

most common. Cases of toxoplasmosis are contracted by ingestion of feces from an infected cat. Emptying the family cat litter box is one of the chief sources of contamination. The condition causes mild flu-like symptoms in healthy adults, but people with compromised immune systems or pregnant women can become seriously ill.

- O—others: Others include syphilis, varicella virus, and Venezuelan equine encephalitis.

- R—Rubella virus: This virus causes German measles and is related to severe birth anomalies. Symptoms include heart defects, cataracts, deafness, and sometimes mental disability. It is recommended that pregnant women be vaccinated against this disease.
- C—CMV or cytomegalovirus: The human herpes virus-5 (HHV-5) is often contracted from the genital infection of the mother, which affects about 3% of pregnant women. During the embryonic period, the infection may lead to death or spontaneous abortion of the embryo. However, if contracted during the second trimester, an infection could lead to microcephaly, retarded growth, central nervous system disorders, problems with the senses, and enlarged liver.
- H—HIV or Human Immunodeficiency virus: HIV in pregnant women has grown to be a serious worldwide problem with over 33.4 million people now carrying the virus. When the mother carries the virus, about one-third of the children become infected. The anomalies include retarded growth, microcephaly, and mental deficiency.

Metabolic Imbalance

Certain conditions that the mother has may cause problems with the fetus. These include alcoholism (*see also* Fetal Alcohol Spectrum Disorder), endemic cretinism, diabetes (*see also* Diabetic Embryopathy), folic acid deficiency, iodine deficiency, hyperthermia, phenylketonuria, rheumatic disease, and congenital heart defects.

Although the status of some of the teratogens has been established through observation at birth, many of the substances have been shown to cause major problems in animal testing. Other compounds, including aspirin, Agent Orange, and NSAIDs could be added to this list. Some drugs have not been studied in humans.

Further Reading

"Embryopathies." *Human Embryology* [Online course for medical students]. http://www.embryology.ch/anglais/iperiodembry/patholperiod01.html. Accessed 2/10/12.

The Teratology Society: Birth Defects Research, Education, Prevention. 2012. http://www.teratology.org. Accessed 5/12/12.

Emery-Dreifuss Muscular Dystrophy (EDMD)

Prevalence X-linked form affects about 1 in 1,000,000 people; autosomal recessive type very rare, with only a few cases reported worldwide

Other Names benign scapuloperoneal muscular dystrophy with early contractures; EDMD; Emery-Dreifuss syndrome; muscular dystrophy, Emery-Dreifuss

In 1902, French neurologists Raymond Cestan and N. J. LeJonne noted a condition in young boys who had very weak muscles and a tendency to walk on their toes. Later, two researchers, Professor Emery from the UK and Professor Dreifuss from the United States, worked together to study eight affected males in Virginia in three generations. The study indicated an X-linked pedigree pattern. However, later studies in 1994 showed that there was another possible inheritance pattern that was autosomal dominant but also consistent with X-linked patterns. Thus, it was concluded that two genes displaying different symptoms are involved in this condition.

What Is Emery-Dreifuss Muscular Dystrophy (EDMD)?

All forms of muscular dystrophy are characterized by the progressive wasting of muscles. Like many other types, Emery-Dreifuss muscular dystrophy begins in childhood or early adolescence and affects the skeletal muscles that are used for movement. However, the major differences include early muscle contractures, muscle weakness that is distributed throughout the body, and deterioration of the heart muscle.

Contractures are tightening and shortening of certain muscle groups so that it is difficult to move those muscles. The contractures occur at certain joints and most often occur in the elbows, ankles, and neck. The affected individuals experience progressive weakness, beginning in the upper arms, then to the lower legs, and to the muscles in the shoulders and hips. Contractures may limit the elbow so that the arms always appear in a semi-flexed position. Affecting the muscles in the ankle make the child always appear to walk on his or her toes. Also, the child may not be able to bend the neck.

Boys with this EDMD almost always have heart problems, at least by young adulthood. However, rather than affecting the heart muscle, EDMD disturbs the electrical wiring of the heart, called the cardiac conduction system. This system controls the rate of the heartbeat, which is usually abnormally slow. The person may feel palpitations or fluttering of the heart as well as attacks of fainting spells and weakness. A pacemaker may correct this problem.

What Are the Genetic Causes of Emery-Dreifuss Muscular Dystrophy (EDMD)?

Three types of inheritance patterns cause EDMD: an X-linked pattern, an autosomal dominant pattern, and an autosomal recessive pattern. Two genes appear to be involved: the "lamin A/C" or *LMNA* gene, and "emerin" or *EMD* gene. These two genes instruct for the making of proteins that are part of the nuclear membrane surrounding the nucleus and appear to play a part in the movement of proteins in and out of the nucleus.

EMD

The *EMD* gene also provides instructions for the protein that is essential for the normal function of skeletal muscles and heart muscle. About 100 mutations of the *EMD* gene have been reported to disrupt the instructions for the protein emerin, which may lead to the abnormalities of the muscles and heart. This gene is inherited in an X-linked pattern and is located on the long arm (q) of the X chromosome at position 28.

LMNA

The second gene, *LMNA*, instructs for a group of proteins called lamins. Lamins A and C are also related to the nuclear envelope and appear to also regulate movement of certain proteins in and out of the nucleus. More than 100 mutations in the *LMNA* gene have been identified. Researchers are unclear how the protein affects muscles and heart, which are the characteristics of EDMD. This gene is inherited in an autosomal dominant pattern and occurs generally in people with no history of the disorder. A rare *LMNA* mutation is inherited in an autosomal recessive pattern. *LMNA* gene is located on the long arm (q) of chromosome 1 at position 22.

What Is the Treatment for Emery-Dreifuss Muscular Dystrophy (EDMD)?

In general, EDMD is less severe than other forms of muscular dystrophy and many affected individuals can live to middle age or longer. However, it is essential that individuals with EDMD be under a doctor's care and have checkups every year Maintaining good, healthy lifestyle is very important in any form of muscular dystrophy.

See also Duchenne/Becker Muscular Dystrophy; Fascioscapulohumeral Muscular Dystrophy; Ullrich Congenital Muscular Dystrophy

Further Reading

"*EMD*." 2010. Genetics Home Reference. National Library of Medicine (U.S.). http://ghr.nlm.nih.gov/gene/EMD. Accessed 5/12/12.

"Emery-Dreifuss Muscular Dystrophy." 2010. Genetics Home Reference. National Library of Medicine (U.S.). http://ghr.nlm.nih.gov/condition/emery-dreifuss-muscular-dystrophy. Accessed 5/12/12.

"Emery-Dreifuss Muscular Dystrophy." 2010. Muscular Dystrophy Campaign. http://www.muscular-dystrophy.org/about_muscular_dystrophy/conditions/98_emery-dreifuss_muscular_dystrophy. Accessed 5/12/12.

"*LMNA*." 2010. Genetics Home Reference. National Library of Medicine (U.S.). http://ghr.nlm.nih.gov/gene/LMNA. Accessed 5/12/12.

Epidermal Nevus and Epidermal Nevus Syndrome

Prevalence Epidermal nevi affect about 1 in 1,000 people

Other Names epidermal naevus

A nevus is an abnormal benign overgrowth of skin cells. The plural uses the Latin form "nevi." Epidermal nevus syndrome combines epidermal nevi with other symptoms that affect many systems. First, this article will describe the epidermal nevus and then discuss epidermal nevus syndrome or EVS.

What Is Epidermal Nevus?

An epidermal nevus is a noncancerous skin growth in certain areas. The areas will have an appearance that is distinct from the rest of the skin. They may appear as different colored and uneven scale-like patches. Sometimes they are smooth and soft to the touch, and other times they may be wart-like. A child may be born with epidermal nevi or may develop them during childhood. They may develop along the lines or paths that embryonic skin cells follow before birth.

Following are the three types of epidermal nevi:

- Keratinocyte: One of the types of skin cell found in the outer layer of the skin is called the keratinocyte. These are the tough, fibrous cells that can be hard as in the hair or nails, or soft like the cells in stratum corneum of the outermost layer of the skin or epidermis that form layers and consistently slough off. One type of nevi involves only these types. This kind is called nonorganoid epidermal nevi.
- Cells that make up the hair follicles: This type is one of a group called organoid epidermal nevi.
- Cells that make up the sebaceous glands: These glands produce the fluid sebum or oil that protects the skin and hair. This type is also one of a group called organoid epidermal nevi.

The nevi that develop vary according to the type of skin cell that is affected.

What Is the Genetic Cause of Epidermal Nevus?

Changes in the *FGFR3* gene, officially known as the "fibroblast growth factor receptor 3" gene, cause epidermal nevus. Normally, *FGFR3* provides instructions for the FGFR3 protein that lives both inside the cell and outside the cell. The one outside the cell looks for growth factors that are in the bloodstream. When these factors are found, the two connect like a key that fits into a lock. The other end of the protein extends into the cell telling the cell itself what to do when the signals are detected. There are several different forms called isoforms of the FGFR3 protein found in the cells of different tissues, which also interact with different growth factors. The form found epidermal layer of the skin is most likely is responsible for controlling the processes of cell growth and division, which is important for the renewal of skin cells but also can lead to the overgrowth of the cells, causing birth marks and the condition, epidermal nevus.

Mutations in the *FGFR3* gene have been associated with about 30% of the people with the keratinocytes or the type made with the outer layer of the skin cells. The mutation results in a change in only one amino acid building block, in which cysteine replaces arginine. The abnormal protein disrupts the normal control of the growth factor and therefore is consistently turned on, resulting in the buildup of abnormal skin cells. The gene has not been found in the other types. This gene does not appear to be inherited but is somatic meaning that it changes after fertilization during embryonic development. *FGFR3* is located on the short arm (p) of chromosome 4 at position 16.3.

What Is Epidermal Nevus Syndrome?

Although some people with epidermal nevus have only skin growth, others may also have conditions that affect many systems. This is called epidermal nevus syndrome. About one-third of the people with EVS have some type of bone, eye, or brain involvement.

In 1957, Gustav Schimmelpenning wrote about a case of sebaceous nevus involving the heart, eye lesions, dense cranial bones, seizures, and mental retardation. Six different syndromes with epidermal nevi as part have been described. The age of diagnosis ranges from birth to 40 years.

What Is the Treatment for Epidermal Nevus?

Epidermal nevi can be treated according to the symptoms. Conventional medications, cyrosurgery, and other treatments can be used. The symptoms of the syndrome can be quite serious. Disorders must be addressed according to severity of the symptoms.

Further Reading

"Epidermal Nevus." 2008. Medicis: The Dermatology Company. http://www.dermatopedia.com/epidermal-nevus. Accessed 1/10/12.

"Epidermal Nevus." 2012. Genetics Home Reference. National Library of Medicine (U.S.). http://ghr.nlm.nih.gov/condition/epidermal-nevus. Accessed 1/10/12.

"Epidermal Nevus Syndrome." 2012. Medscape. http://emedicine.medscape.com/article/1117506-overview#showall. Accessed 1/10/12.

Epilepsy

The word epilepsy is probably as misunderstood as using the term cancer. For many people, these two words evoke a unilateral condition that has negative

An woman prepares to take medications to control her epilepsy. (Jeff Roberson/ Associated Press)

connotations. But like cancer, there are many forms of epilepsy and many different causes. Actually, epilepsy is a general term for a condition in which the person has recurring seizures. Likewise, there are many types of seizures. All involve abnormal electrical activity in the brain that causes an involuntary change in body movement, function, sensation, awareness, or behavior. The seizures can vary from brief moments of staring into space, to convulsions where the person experiences falling down, unconsciousness, and shaking.

Epilepsy has been documented in medical history. The word *epilepsia* comes from the Greek meaning "to take hold of." Many ancients associated epilepsy with religious experiences or even with demon possession. Visions experienced during the seizures were considered to come from the gods. However, in most cultures, the condition was shunned and stigmatized. Today, in developed countries, that is no longer the case.

Efforts are being made on many fronts to understand this disease. According to the National Institutes of Health, about 107 of the conditions have genetic causes. Following are some of the conditions.

1. 17q21.31 microdeletion syndrome: This deletion of only a small piece of chromosome 17 causes a syndrome in which 50% of the individuals have seizures.
2. Aicardi syndrome: This syndrome found only in girls is related to a mutation on the X chromosome. Seizures begin in infancy and develop into a form of epilepsy that is very difficult to treat.
3. Autosomal dominant nocturnal frontal lobe epilepsy: This type of epilepsy runs in families and tends to occur at night. Mutations of the gene are on chromosome 8.
4. Autosomal dominant partial epilepsy with auditory features: In this form of epilepsy, the individual may hear buzzing or noises with the seizure. Mutations are on chromosome 10.
5. Denatorubral-pallidoluysian atrophy: This type of brain atrophy usually occurs before the age of 20 and involves involuntary jerking and twitching. The cause is a mutation on chromosome 12.
6. Hereditary hyperekplexia: This condition is characterized by periods of rigidity and seizures beginning in infancy. Mutation of the gene is located on chromosome 5.
7. Kufs disease: This disorder is characterized by seizures and muscle stiffness. Mutation is found on chromosome 1.
8. Lafora progressive myoclonus epilepsy: This very serious type of epilepsy involves sudden falling down and shaking all over. The gene is located on chromosome 6.
9. Lennox-Gestaut syndrome: This very serious form starts in infancy, and the child has tonic or stiffening seizures. The gene has not been located, although it is suspected that it runs in families.
10. Myoclonic epilepsy with ragged-red fibers: This type is characterized by stiffness and mental deterioration. It is related to mitochondrial DNA.
11. Pyridoxal 5'-phosphate dependent epilepsy: This condition of epilepsy is a metabolic condition involving use of certain chemicals. The gene is located on chromosome 17.

12. Pyridoxine-dependent epilepsy: The seizures may begin before birth and continue. This metabolic condition is difficult to treat. The gene is located on chromosome 5.
13. Ring chromosome 14 syndrome: The child has seizures and intellectual disability. The condition is formed by a ring-like structure in chromosome 14.
14. Ring chromosome 20 syndrome: This condition begins at birth and is characterized by seizure during both day and night. The mutation is formed by a ring-like structure in chromosome 20.
15. Unverricht-Lundborg disease: This rare disease affects children between 5 and 10. The mutation of the gene is located on chromosome 21.

According to the Centers for Disease Control and Prevention, epilepsy affects about two million Americans. In 2010 alone, nearly 140,000 new cases were diagnosed.

Further Reading

"Epilepsy." 2011. Centers for Disease Control and Prevention. http://www.cdc.gov/epilepsy. Accessed 1/27/12.

"Epilepsy." 2012. Genetics Home Reference. National Library of Medicine (U.S.). http://ghr.nlm.nih.gov/conditionGroup/epilepsy. Accessed 1/27/12.

Epilepsy Foundation. http://www.epilepsyfoundation.org. Accessed 1/27/12.

National Society for Epilepsy. 2012. http://www.epilepsynse.org.uk. Accessed 5/12/12.

Erythromelalgia (EM)

Prevalence 1.3 per 100,000; more in women than men; median age at diagnosis is 61

Other Names acromelalgia; erythromalgia; Gerhardt disease; Mitchell's disease; primary erythromelalgia; red neuralgia; Weir-Mitchell disease

Silas Weir Mitchell was a physician, psychiatrist, novelist, actor, and poet; he is noted as one of the most versatile Americans since Benjamin Franklin. His career began in the American Civil War when he was one of the first to pursue causes of nervous system conditions. In the early 1870s, he was appointed to the Philadelphia Orthopaedic Hospital and Infirmary for Nervous Diseases. Here, he first described a disorder in which the patient had periodic episodes of a skin condition that was very painful, showed elevated skin temperatures, and exhibited terrible burning sensations. The condition at that time was named after him—Mitchell's disease. Now it is called erythromelalgia.

What Is Erythromelalgia (EM)?

Erythromelalgia is a rare disorder characterized by burning pain, swelling, and redness in various parts of the body. The word comes from three Greek roots: *erythros*, meaning "red"; *melos*, meaning "limb"; and *algos*, meaning "pain." The hands and feet are usually affected, but the redness and swelling may extend to the upper legs and face. Typically, it affects both sides of the body but may involve only one side. This disease is extremely variable and may have a gradual onset or an acute onset. The pain may be so debilitating that the person cannot wear shoes or walk, and it may interfere with normal activities.

Erythromelalgia occurs when blood vessels in the extremities are blocked and inflamed. The episodes may be triggered by heat, alcohol consumption, exercise, or spicy foods. Even wearing socks that are too tight may evoke an attack. The condition, which is more frequent in women, may be primary or a secondary disorder related to another condition. The secondary conditions include high blood pressure, neuropathy, some autoimmune disorders, and mushroom or mercury poisoning.

What Is the Genetic Cause of Erythromelalgia (EM)?

Mutations in the *SCN9A* gene, officially called the "sodium channel, voltage-gated, type IX, alpha subunit" gene, cause EM. *SCN9A* is a member of a large family of genes that instruct for making sodium channels, which transport positive-charged sodium ions (atoms) into cells. These atoms are important in the cell's generating and transmitting electrical signals. Normally, *SCN9A* instructs for making one part of a sodium channel called the alpha unit. These channels are found in areas near the spinal cord designated as nociceptors, nerve cells that receive and transmit pain.

About 10 mutations in the *SCN9A* gene have been related to EM. All mutations involve changing just one building block or amino acid. The mutation causes the sodium channel to open more easily and stay open longer. The increased flow of sodium atoms produce nerve impulses within the pain receptors (nociceptors), and lead to the symptoms of EM. *SCN9A* is inherited in an autosomal dominant pattern and is located on the long arm (q) of chromosome 2 at position 24.

What Is the Treatment for Erythromelalgia (EM)?

EM is a very difficult disorder to diagnose, and the patient may take months and visits to as many as six to seven specialists to diagnose the conditions. If it is related to a secondary cause, that condition must be treated first. The treatment is mostly controlling the symptoms, which may involve painkillers and avoiding the things that may trigger the conditions. The Erythromelalgia Association seeks to identify, educate, and support those people with the painful condition. One mission is to help general practitioners recognize and diagnose EM.

Further Reading

"Erythromelalgia." 2010. Genetics Home Reference. National Library of Medicine (U.S.). http://ghr.nlm.nih.gov/condition/erythromelagia. Accessed 5/12/12.

The Erythromelalgia Association. 2010. http://www.erythromelalgia.org. Accessed 5/12/12.

"*SCN9A*." 2010. Genetics Home Reference. National Library of Medicine (U.S.). http://ghr.nlm.nih.gov/gene/SCN9A. Accessed 5/12/12.

Essential Tremor (ET)

Prevalence A common disorder affecting millions of people; estimates vary because several disorders have the symptoms; may affect as many as 14% of those over 65.

Other Names benign essential tremor (the benign has been dropped because some of the conditions are quite serious); familial tremor; hereditary essential tremor

What do Archimedes, the third-century Greek mathematician, and essential tremor have in common? Archimedes developed a spiral that bears his name. The spiral begins at a fixed point and then moves from that point with a constant speed along a line rotating at a constant angular velocity. So what is the connection? Neurologists ask people who are suspected of having essential tremor to draw the spiral; this tool is used to diagnose the condition. The logo of the International Essential Tremor Foundation is the Archimedes spiral.

What Is Essential Tremor (ET)?

Essential tremor (ET) is the most common neurological disorder characterized by involuntary, rhythmical shaking. The shaking can affect the following body parts:

- Hand and arm tremor: This area is the most common and is evident when the person extends the limb or hand for eating, writing, sewing, or shaving. Unlike other tremors such as those of Parkinson disease, ET appears to occur when the limb defies gravity and does not usually occur when the limb is at rest.
- Head tremor: The head may oscillate in a "yes-yes" or a "no-no" pattern.
- Tongue and voice: The voice may appear shaky.
- Legs and feet: This type of tremor may affect walking and give an awkward-appearing gait.

Tremor often begins in early adulthood but becomes more noticeable as people get older; however, the symptoms can appear at any age. The condition may be progressive and begin in one area, then progress to another. Late-onset ET—over age 65—is often associated with dementia. There are more than 20 different kinds of tremor, with essential tremor the most common.

Several things may things may trigger ET, including stress, fever, exhaustion, cigarette smoking, hunger, caffeine, extremes of temperature, or low blood sugar. ET is not considered a life-threatening condition or even a debilitating condition, but it can disrupt certain activities that one needs to do in life.

What Are the Genetic Causes of Essential Tremor (ET)?

Essential tremor is a very complex disorder. Children of a person with ET may have a 50-50 chance inheriting the disorder. This statistic indicates an autosomal dominant inheritance pattern. At least two genes appear to be involved in ET: *DRD3* and *HS1BP3*.

A mutation in *DRD3* has been associated with ET. The official name of this gene is "dopamine receptor D3." Normally, *DRD3* instructs for making a protein called dopamine receptor D3. Dopamine is a neurotransmitter or chemical messenger that signals the nervous system to produce physical movement. Changes in *DRD3* appear to be the result of a protein building-block replacement error. The amino acid glycine replaces serine at position 9. The changed protein then evokes a stronger response to dopamine and causes the involuntary shaking of ET. *DRD3* is located on the long arm (q) of chromosome 3 at position 13.3.

HS1BP3 is the official symbol for "HCLS1 binding protein 3." Normally, *HS1BP3* instructs for the making of a protein called hematopoietic-specific protein 1 binding protein 3. This important protein regulates chemical signaling in an area of the brain called the cerebellum, which controls motor functions of the body. A mutation in *HS1BP3* disrupts the communication between the cerebellum and messages to motor neurons. The mutation is caused by a substitution when the protein building block glycine replaces alanine at position 265. *HS1BP3* is located on the short arm (p) of chromosome 2 at position 24.1.

What Is the Treatment for Essential Tremor (ET)?

Although no cure exists for ET, some treatments may help. Eliminating certain "triggers" or stressors, such as caffeine or stimulants, may help. Two medications are prescribed: beta blockers and antiepileptic or anticonvulsants. Some people may require surgical treatments with deep brain stimulation or thalamotomy (surgically cutting part of the thalamus). Minor cases may be treated with physical therapy. The International Essential Tremor Foundation (IETF) is a strong support group for individuals and families affected by ET.

Further Reading

"Essential Tremor." 2010. Medscape. http://emedicine.medscape.com/article/1150290-overview. Accessed 5/12/12.

International Essential Tremor Foundation (IETF). 2010. http://www.essentialtremor.org. Accessed 5/12/12.

"Treatments for Essential Tremor." 2010. American Academy of Neurology. http://www.aan.com/globals/axon/assets/2585.pdf. Accessed 5/12/12.

Eugenics: A Special Topic

The topic of eugenics was very controversial from its conception and evoked strong feelings from those who thought the concept an answer to societies' problems and those who felt it a violation of human rights, if not evil. This article presents a brief history of the movement, describing its roots and present standing.

What Is "Eugenics"?

The word "eugenics" comes from two Greek words: *eu*, meaning "good" or "well," and *gen*, meaning "birth," "seed" or "born." This root is also used in the word "genetics," the topic of this encyclopedia. In 1883, Sir Francis Galton, a British social scientist, coined the term to mean a study of all agencies under human control that can improve the quality of future generations. In other words, he was asking how society can improve itself.

At the end of the nineteenth century and beginning of the twentieth century, the idea of eugenics became popular, developing many meanings. Some perceived it as meaning good prenatal care and personal responsibility for mother and child. Others perceived it to mean to discourage inbreeding among close relatives. At one time, when villages were quite isolated, marrying cousins or other relatives were accepted. People began to note that disabilities were more frequent in the children born from these marriages. Someone said that the motorized truck, which enabled people to go from one village to another, was a great tool for eugenics. However, the extreme meaning included forced sterilization and even euthanasia of those with disabilities.

Social Engineering of Population

Controlling the population is an old idea that was practiced in many civilizations. The ancient Greek philosopher Plato suggested a method to evaluate a couple's quality before they could have children, resulting in a scoring by number. Those with a high number could have children, and those with a low number could not. The idea was not popular and was never adopted. In Sparta, in ancient Greece, a group of bureaucrats examined all newborns. If the boy was considered desirable, he was covered with wine and left in the mountains. If he lived, he would be strong enough to fight and have children. This was the survival of the fittest.

Charles Darwin did not originally apply his theory of the survival of the fittest to humans, but later philosophers combined Mendelian genetics and evolution to

Sir Francis Galton (1822–1911), considered the father of eugenics. (National Institutes of Health)

make a set of ideas called social Darwinism. These people believed that genetics could improve populations, and in 1912, they convened the first International Congress of Eugenics. Several prominent people were there, including Winston Churchill, Alexander Graham Bell, Margaret Sanger of Planned Parenthood, and Leonard Darwin, the son of Charles Darwin.

Francis Galton

Sir Francis Galton was a British anthropologist, with many other accomplishments, and half-cousin of Charles Darwin. When he read *The Origin of Species*, he immediately began to apply this knowledge to human beings using social statistics. He determined that civilization thwarted natural selection by protecting the weak members of society, and only by changing social policies could society be changed. He reasoned that if selective breeding could be used to improve animals, human intelligence and talent could also be improved by applying similar methods. In his book *Hereditary Genius*, he stated that the less intelligent had larger families, and he encouraged people to improve the bloodline by human selective breeding, such as more intelligent people having more children. Thus, eugenics became defined as selective reproduction that creates children with desirable traits.

Eugenics in the United States

The idea of eugenics—promoting the idea of high reproduction in certain people with more desirable traits and reducing the reproduction of those with undesirable traits—caught on and was developed by governments around the world. But nowhere was it more strongly embraced than among the progressives in the United States as an answer to social ills.

American scientist Charles Davenport was one of the major proponents of eugenics. He received funds to establish the Biological Experiment Station at Cold Spring Harbor in Maine in 1904 and kept statistics that became the basis for enforced sterilization studies. He brought in distinguished scientists to promote the idea of eugenics and had even had support among progressive thinkers, although not all went along with his racist beliefs.

The end of the nineteenth century and the early decades of the twentieth century were a breeding ground for the eugenics movement. Along with many other countries, some laws and policies in the United States were enacted that promoted eugenics. The height of the eugenics periods was the year 1923, which coincided with concern about immigration from other countries. Communities were encouraged to have "best baby" contests.

The "science of eugenics" that developed was really a pseudoscience. Laws were passed to wipe away the "unfit" and preserve the northern European Nordic stereotype. In 1909, California became the first to enact sterilization of those individuals with mental and physical disabilities. Laws were subsequently enacted in 27 states. Some 60,000 people were sterilized, with over one-third in California, the epicenter of the movement. Prestigious universities offered courses in eugenics, using a textbook called *Applied Eugenics*, which devoted a chapter to "lethal selection."

Eugenicists believed in the identification of people, which included the poor, deaf, mentally ill, blind, developmentally disabled, promiscuous women, homosexuals, and certain racial groups like gypsies, as unfit. Nazi Germany took it a step further and added most from this group, along with all Jews, and marked them for extermination.

After World War II, the eugenics movement lost favor with the revelation of the Nazi regime's cruelty in murdering approximately 11 million people, including children, in order to create and strengthen a "master race." Even academics who had earlier embraced the idea began to distance themselves from the movement. Eugenics became associated with racial hatred, human experimentation, and extermination of the "unfit." The German movement was seen as a gross and brutal violation of human rights, the concept that all people have rights.

Modern Eugenics

With the cracking of the human genome and advances in genetic technology, it may be possible to produce genetically modified humans. The difference between the former times and today is that today, the idea of improving the species is a concept to be feared and guarded against. Individuals with disabilities have rights, and the idea of sterilization of certain people with mental or physical disabilities is not accepted. However, reproductive technologies are such that today, many are raising questions about creating "designer babies" and "reprogenetics." The interest here seems to arise from the desire to improve the individual child, rather than the improvement of society.

This issue is an important moral and ethical issue. One of the differences is that the concern today is not like the compulsory nature of the one that developed during the early years of the twentieth century. The state is not involved in establishing laws. Today, geneticists are concerned about finding and possibly eliminating genetic diseases such as sickle-cell disease. Genetic screening under normal circumstances can greatly reduce dominant traits such as Huntington disease. Eliminating recessive genes is more difficult and could not be done unless all members of a genetic pool were screened. Nevertheless, the idea of improving individuals and even improving a society will continue to be a major bioethical issue in the following decades.

Further Reading

Black, Edwin. 2003. *War against the Weak: Eugenics and America's Campaign to Create a Master Race*. New York: Four Walls Eight Windows. http://www.waragainst theweak.com. Accessed 5/12/12.

Carlson, Elof Axel. 2001. *The Unfit: A History of a Bad Idea*. Cold Spring Harbor, NY: Cold Spring Harbor Press.

Engs, Ruth C. 2005. *The Eugenics Movement: An Encyclopedia*. Westport, CT: Greenwood Publishing Group.

Galton. David. 2002. *Eugenics: The Future of Human Life in the 21st Century*. Boothbay Harbor, ME: Abacus.

Kevles, Daniel. 1985. *In the Name of Eugenics: Genetics and the Uses of Human Heredity*. New York: Knopf.

Shakespeare, Tom, with Anne Kerr. 2002. *Genetic Politics: From Eugenics to Genome*. New York: New Clarion Press.

F

Fabry Disease

Prevalence Estimated 1 in 40,000 to 60,000; occurs in males more frequently than females; late-onset forms milder than the classic, severe form

Other Names Anderson-Fabry disease; angiokeratoma corporis diffusum; alpha-galactosidase A deficiency; Fabry's disease

In 1898, German dermatologist Johannes Fabry studied a boy who had tiny reddish-purple pimples or pustules that appeared around his navel, buttocks, and lower abdomen. He called the condition "purpura haemorrhagica nodularis." He studied several similar cases and continued interest in the disease until his death in 1930. At the same time, Anderson in England recorded that a young male aged 39 had similar eruptions from early childhood and that the disorder included several systems. He called the condition angiokeratoma. However, it was Fabry who was dedicated to investigating the disease, and who contributed his name to the disease.

What Is Fabry Disease?

Fabry disease is a condition that affects many parts of the body. A buildup of a type of fat or lipids in the body's cells causes the disorder. Fabry disease is classified as a problem of the lysosomes of the cells. Lysosomes are small structures called organelles (little organs) within the cell's cytoplasm, containing enzymes that are necessary for the cell's digestive process. Lysosomes use these digestive enzymes to recycle usable worn-out cell components.

A lysosomal disease occurs when these cells have too much or too little of an enzyme. In Fabry disease, there is little or no activity of enzyme A-galactosidase A, which breaks down a certain fatty substance called globotriaosylceramide. The lack of the enzymes allows the buildup of the fatty substance that can cause the following characteristics of Fabry disease:

- Skin: Small clusters of dark, red spots called angiokeratomas occur on the skin. These painless spots may first appear on the navel but can also be on the thighs, buttocks, lower abdomen, and groin.
- Pain: These episodes called acroparethesias occur in the hands and feet.
- Perspiring: The individual may not be able to sweat (anhydrosis) or may sweat very little (hypohydrosis).
- Kidneys: Renal complications are a common side effect of Fabry disease and may worsen throughout life. One of the first symptoms is very foamy urine, a condition called proteinuria. End-stage renal failure may occur in the last when the person is in the 30s.
- Heart: When the lipids build up in the heart cells, heart disease may result. High blood pressure, congestive heart failure, and cardiomyopathy are the most common heart-related problems.
- Eye: A clouding of the cornea may be one of the first indications of Fabry disease. Later, other eye problems such as cataracts and macular degeneration may occur.
- Stroke.
- Other conditions: Tiredness, ringing in the ears, vertigo, nausea, inability to gain weight, and gastrointestinal disturbances may occur.

Many of the symptoms of classic Fabry are experienced in early childhood and are often misdiagnosed. Milder forms of the disorder may appear later in life and be related to only one organ.

What Is the Genetic Cause of Fabry Disease?

Changes in the *GLA* gene, officially known as the "galactosidase, alpha" gene, cause Fabry disease. Normally, *GLA* instructs for making the lysosomal enzyme alpha-galactosidase A. This enzyme then breaks down the molecule globotriaosylceramide and acts as a part of the recycling of old red blood cells and other types of cells. Mutations in this gene usually involve changes in one building block, which of course changes the protein produced. The abnormal enzyme is not able to break down the lipids, allowing the buildup in the lining of blood vessels in the skin and cells of vulnerable organs. This progressive buildup damages the cells and causes the signs and symptoms of Fabry disease. The *GLA* gene is inherited in an X-linked recessive pattern and is located on the long arm (q) of the X chromosome at position 22.

What Is the Treatment for Fabry Disease?

For many years, the only recourse for treatment was targeting the symptoms and making the patients more comfortable. In 2001, a new therapy called enzyme replacement therapy (ERT) was released. These therapies attempt to replace the deficient enzyme. However, the treatments are very expensive and may cost as much as $250,000 a year.

Further Reading

"Genetics of Fabry Disease." 2010. Medscape. http://emedicine.medscape.com/article/951451-overview. Accessed 5/12/12.

"*GLA*." 2010. Genetics Home Reference. National Library of Medicine (U.S.). http://ghr.nlm.nih.gov/gene/GLA. Accessed 5/12/12.

National Fabry Disease Foundation. 2010. http://www.thenfdf.org. Accessed 5/12/12.

"NINDS Fabry Disease Information Page." 2010. National Institute of Neurological Disorders and Stroke (U.S.). http://www.ninds.nih.gov/disorders/fabrys/fabrys.htm. Accessed 5/12/12.

Facioscapulohumeral Muscular Dystrophy (FSHMD)

Prevalence About 1 in 14,000; most prevalent of the nine types of muscular dystrophy

Other Names facioscapulohumeral atrophy; facio-scapulo-humeral dystrophy; facioscapulohumeral type progressive muscular dystrophy; facioscapuloperoneal muscular dystrophy; FSHD; FSH muscular dystrophy; Landouzy-Dejerine dystrophy; muscular dystrophy, Landouzy Dejerine; muscular dystrophy-facioscapulohumeral

In 1885, Louis Landouzy, a French professor of medicine at Reims, and his friend Joseph Jules Dejerine reported the results of a careful study of a serious muscle condition that appeared to affect the upper body. The condition was named Landouzy-Dejerine dystrophy after the men. However, the preferred name using the affected body parts is commonly used now—facioscapulohumeral muscular dystrophy.

What Is Facioscapulohumeral Muscular Dystrophy (FSHMD)?

Facioscapulohumeral muscular dystrophy, or FSHMD, is the most prevalent type of the nine types of muscular dystrophy. The long term describes the body parts affected: *facio*, meaning "face"; *scapula*, meaning the "shoulder blade area"; and *humerus*, meaning the "bones of the upper arm." Thus, the muscle weakness and loss of muscle tissue is concentrated in the upper body, although it may extend to muscles around the pelvis, hips, and lower legs. This type of muscular dystrophy differs from Duchenne muscular dystrophy and Becker muscular dystrophy, which affect the lower body.

Symptoms of FSHMD, which usually do not appear until ages 10 to 26, begin in a mild form but then become progressively worse. The following symptoms may develop:

- Facial muscles: The weakness of the facial muscles makes it difficult to turn up the corner of the mouth when smiling, drinking from a straw, or whistling. The person may have difficulty with facial expressions and appear as always depressed or angry. He or she may have problems pronouncing words and talking.
- Eyes: The eyelids may droop. Muscle weakness around the eyes may make it difficult to fully close the eyes when sleeping and lead dry eye and other eye disorders.
- Shoulder area: Weak muscles cause the shoulder blades to stick out—a condition called scapular winging. This weakness makes it very difficult to raise the arms above the head or to throw objects such as a ball.
- Arms: The biceps, triceps, and deltoid muscles of the upper arm are first affected, then the weakness may progress to the lower arm muscles.
- Other body muscles: Loss of strength in the stomach muscles, which then may progress to the legs, eventually causing a situation known as foot drop.
- Senses: Hearing loss of high-tone hearing and other abnormalities including the retina at the back of the eye may occur.

However, rarely does FSHMD affect the heart muscles or breathing.

What Is the Genetic Cause of Facioscapulohumeral Muscular Dystrophy (FSHMD)?

A deletion in an area of chromosome 4 called *D4Z4* causes FSHMD. This area is located near the end of the chromosome and consists of 11 to more than 100 repeated DNA segments, each of which is about 3,300 base pairs. However, people with FSHMD have a very short segment in the area with the repeats of 1 to 10. The exact role and mechanism of how this causes the disorder is not yet known. It appears that the *D4Z4* area influences other genes located on chromosome 4. Other genetic factors may be involved. This gene is inherited in an autosomal dominant pattern and is found on the long arm (q) of chromosome 4 at positions 4–10.

There are two other forms of FSHMD: FSHMD 1B and infantile FSHMD or IFSHMD. FSHMD 1B is a less prevalent form that is caused by a gene possibly located on another area of chromosome 4 or on another gene. Infantile FSHMD is a very severe form of the disorder that is accomplished by early hearing loss, vision problems, and seizures.

What Is the Treatment for Facioscapulohumeral Muscular Dystrophy (FSHMD)?

Because FSHMD is a genetic disorder, no cure presently exists. Treatments may control some of the symptoms and improve the quality of life. Activity is encouraged because complete bed rest and inactivity cause the muscles to atrophy more quickly. Physical therapy may help.

See also Duchenne/Becker Muscular Dystrophy; Emery-Dreifuss Muscular Dystrophy (EDMD); Ullrich Congenital Muscular Dystrophy (UCMD)

Further Reading

"Facioscapulohumeral Muscular Dystrophy." 2010. MedlinePlus. National Library of Medicine (U.S.). http://www.nlm.nih.gov/medlineplus/ency/article/000707.htm. Accessed 5/13/12.

"Facioscapulohumeral Muscular Dystrophy: Making an Informed Choice about Genetic Testing." 2010. http://depts.washington.edu/neurogen/downloads/fshd.pdf. Accessed 5/13/12.

FSH Society. 2010. http://www.fshsociety.org/pages/about.html. Accessed 5/13/12.

Factor V Leiden Thrombophilia

Prevalence About 1 in 5,000; 5% of Caucasians in North America; less common among Hispanics and African Americans; extremely rare among Asians

Other Names APC resistance, Leiden type; Factor V; hereditary resistance to activated protein C

Naming of diseases and disorders varies so much. Some disorders are named after the people who first described or studied them. Some may be named after a famous person who had the disease. Others may be named after the area of the body affected. Factor V Leiden was named after a city. In 1994, Professor R. Betina and a team in Leiden, a city in the Netherlands, wrote about a mutation in blood coagulation called factor V in the publication *Nature*. The name of the city Leiden was added to the official name of the disorder.

What Is Factor V Leiden Thrombophilia?

Factor V Leiden thrombophilia is disorder of a person's blood clotting mechanism. The word "thrombophilia" is from two Greek words: *thrombus*, meaning "clot," and *philo*, meaning "lover of" So the word reflects that formation of clots happens readily and freely. This disorder is one of the most commonly inherited blood

Preventing Deep Vein Thrombosis

When taking long flights or even overland travel, take precautions by drinking plenty of water and moving around every 60 to 90 minutes. It is the lack of motion that causes blood to pool in the legs and cause the dreaded deep vein thrombosis.

Here are some movements to help improve the blood flow:

- Ankle circles: Draw circles with the toes in both clockwise and counter-clockwise manner, exercising both ankles.
- Seated calf raises: Start with feet on floor and slowly raise heels off the ground and press heels to the floor.
- Knee raises: Lift knees up to chest keeping them together and then one at a time.
- Reach: Raise the arms and reach to the ceiling.
- Neck stretch: Drop head to one side, then circle to the chin to the chest, and then to the other side.

And do not worry about what other people think—movement may save your life.

disorders among those living in Europe and of Caucasian ancestry in other parts of the world.

In order to understand this disorder, it is necessary to know the normal mechanism for clotting or coagulation. When an injury occurs to the blood vessel, the following things happen:

- Blood platelets or clotting cells move to the site and from a loose plug over the bleeding area.
- Enzymes on the surface of the platelets send messages to generate fibrin, strands of tough strings, which act as a biological adhesive tape to hold the platelets in place. Helping out in this process is a substance called Factor V.
- When the clot is in place, a substance called activated protein C (APC) begins to hold the clot-making process in check, inactivating Factor V and keeping the clot from growing larger than it should.

Anything disrupting the process can cause the clot to grow and endanger the life of the person.

Factor V is a protein in the blood that monitors the delicate balance of normal clotting. Most people do not realize that during just a normal day, the blood vessels that are not seen endure many scrapes and bruises. Factor V is called upon to repair the damage without the person realizing it.

What Is the Genetic Cause of Factor V Leiden Thrombophilia?

An abnormal mutation in the Factor V gene is called Factor V Leiden. The official name of the gene is "coagulation factor V (proaccelerin, labile factor)," or *F5*. Normally, this gene instructs for making a protein called coagulation factor V. Cells in the liver make that protein, and then it circulates in the bloodstream until called upon by an injury to the blood vessels. The Factor V then interacts with activated protein C (APC) to stop the coagulation by cutting or cleaving it at specific sites, shutting down the clot and keeping it from growing too large. Factor V Leiden turns off the process, and the result is a condition known as APC resistance.

The mutation in the Factor V Leiden gene is caused by the substitution of the amino acid arginine with glutamine. This mutated gene disrupts the protein causing the coagulation factor from working properly. The major damage of Factor V Leiden is developing a condition called deep vein thrombosis (DVT).

F5 is inherited in an autosomal recessive pattern and is located on the long arm (q) of chromosome 1 at position 23.

What Is the Treatment for Factor V Leiden?

If one develops DVT, the treatment is with a blood thinner like warfarin. However, this is not recommended for a long period of time. The individual who has been diagnosed with Factor V Leiden should take serious precautions by cutting down on the major risk factors of obesity, smoking, and generally living a healthy lifestyle. Flights on airplanes or long times of immobility increase the risk of blood clots. Women may have to forego hormone treatments (HR). There is also an increased risk of blood clotting during pregnancy.

Further Reading

"*F5*." Genetics Home Reference. National Library of Medicine (U.S.). http://ghr.nlm.nih.gov/gene/F5. Accessed 1/3/11.

"Factor V Leiden Thrombophilia." 2010. Genetics and Rare Diseases Information Center (GARD). National Institutes of Health (U.S.). http://rarediseases.info.nih.gov/GARD/QnASelected.aspx?diseaseID=6403. Accessed 5/12/12.

Ornstein, D., and M. Cushman. 2003. "Factor V Leiden." *Circulation* (American Heart Association). http://circ.ahajournals.org/cgi/reprint/107/15/e94.pdf. Accessed 5/12/12.

Fahr Disease

Prevalence Rare; about 30 families described in medical literature; may be underdiagnosed

Other Names cerebrovascular ferrocalcinosis; Fahr syndrome; idiopathic basic ganglia calcification; striopallidodentate calcinosis

In 1939, German neurologist Theodor Fahr noted a condition in which the person progressively lost control of motor movement and developed dementia. The condition had also been called Charany-Brunches syndrome and Fritche's syndrome after physicians who had expressed interest in a new disorder. However, Fahr's name emerged because of his extensive publication in the area.

What Is Fahr Disease?

Fahr disease is also known by the synonym familial idiopathic basal ganglia calcification or FIBGC, which describes a condition that is present in a family and involves abnormal deposits of calcium in an area of the brain called the basal ganglia. The word "idiopathic" refers to a condition of unknown origin. The basal ganglia are areas deep in the brain that control movement. Researchers believe these calcium deposits disrupt the normal functioning in different parts of the brain. The deposits may impair the pathways that connect the basal ganglia to other parts of the brain and may be responsible for the symptoms of Fahr disease.

The following symptoms usually begin to appear in the 40s or 50s but can occur at anytime during childhood or adolescence:

- Appearance: Person may develop a stiff, shuffling gait and a rigid, mask-like face.
- Movement: The person cannot control movement and experience involuntary tensing of the muscles called dystonia. They may have difficulty controlling movements or ataxia and experience uncontrollable movements of the limbs called choreoathetosis.
- Seizures: Affected persons may have seizures.
- Psychotic conditions: As the disease progress and the calcium deposits become more pronounced in the brain, the individual may have difficulty concentrating and experience memory loss. Later, changes in personality and a distorted view of the world may accompany dementia or loss of intellectual function. About 20% to 30% of people with Fahr disease have one or more of these psychotic conditions.

What Is the Genetic Cause of Fahr Disease?

In an idiopathic disease, the actual cause of the disorder is unknown. Researchers do suspect that Fahr disease appears to be associated with changes in chromosomes 2, 7, 8, 9, and 14 in some families. The extensive number of genes suggests that multiple genes may be involved. The condition is inherited in an autosomal dominant pattern.

What Is the Treatment for Fahr Disease?

The prognosis for this condition is not positive. There is no cure and no standard course of treatment. Medication may control some of the psychotic symptoms.

The nature of the disease with its progressive degeneration results in disability and death.

Further Reading

"Fahr Disease." 2010. Whonamedit? http://www.whonamedit.com/synd.cfm/451.html. Accessed 5/12/12.

"Fahr's Syndrome." 2010 Cleveland Clinic. http://my.clevelandclinic.org/disorders/fahrs_syndrome/hic_fahrs_syndrome.aspx. Accessed 1/3/11.

"NINDS Fahr's Syndrome Information Page." 2010. National Institute of Neurological Disorders and Stroke. http://www.ninds.nih.gov/disorders/fahrs/fahrs.htm. Accessed 5/12/12.

Familial Adenomatous Polyposis (FAP)

Prevalence Varies from 1 in 7,000 to 1 in 22,000

Other Names adenomatous polyposis coli; adenomatous polyposis of the colon; colon cancer—familial; familial intestinal polyposis; familial multiple polyposis syndrome; familial polyposis coli; familial polyposis syndrome; FAP; hereditary polyposis coli; MYH-associated polyposis; polyposis coli

Although genetic testing is important to diagnosing a disease, family health history perhaps provides the best insight into personal disease risk and even disease prevention. The emphasis on family health issues is especially important in studying the many diseases that tend to run in families. Looking at family health is not a new idea. In 2004, the U.S. surgeon general proclaimed National Family History Day to coincide with Thanksgiving Day each year. This would get families talking about their health background when they were gathered for dinner. Perhaps one of the most important family discussions should center on conditions that other members of the family have experienced. One of these is familial adenomatous polyposis, or FAP or colon cancer.

What Is Familial Adenomatous Polyposis (FAP)?

Familial adenomatous polyposis is a disorder of the gastrointestinal (GI) system that can lead to cancer. In fact, 1% of all cases of colorectal cancer are caused by FAP. The name describes the condition. "F" stands for familial, meaning that it runs in families. "A" stands for adenomatous, a type of growth of glandular origin that can turn into cancer. "P" stands for polyposis made of two Greek words: *polyp*, meaning "many growths," and *osis*, meaning "condition of."

Patients with classic FAP have thousands of these polyps in the GI system beginning in their teens. Unless these polyps are removed, the person will have cancer develop before the age of 40. In another form of the disorder, called attenuated FAP, the person develops a late-onset form around the age of 55. A milder form of FAP may occur. In this type the person may have fewer than 100 polyps.

The person with classic FAP has thousands of polyps occurring in the colon. Symptoms may include blood in the stools, anemia, weight loss, and altered bowel habits. The polyps may also be in the bottom of the stomach, in the duodenum (the first part of the small intestine). In addition, the genetic carriers of the FAP gene may also have the following other growths:

- Desmoid tumors: These growths are noncancerous and occur in the lining of the covering of the intestine. These tumors tend to reoccur even after they are surgically removed.
- Eye: Pigmented lesions may appear in the retina.
- Cysts: Cysts may form in the jaw, sebaceous glands, and in bone, called osteomatas. Gardner syndrome occurs with the combination of polyposis, osteomas, fibromas, and sebaceous cysts.

What Is the Genetic Basis for Familial Adenomatous Polyposis (FAP)?

Familial adenomatous polyposis is caused by mutations in two genes inherited in two different patterns: the *APC* gene and the *MUTYH* gene. The official name of the *APC* gene is "adenomatous polyposis coli" gene. Normally, *APC*, a tumor suppressor gene, instructs for making the APC protein that determines whether a cell will develop tumors. *APC* regulates cell division by keeping the cell growth in check and making sure that it does not multiply more than it should. In addition to cell division, APC protein helps insure that the number of chromosomes in a cell is correct following mitosis. More than 700 mutations of the *APC* gene lead to the shortened version of the APC protein, which is then responsible for the development of classic or attenuated FAP, depending on the mutation. *APC* gene is inherited in an autosomal dominant pattern and is located on the long arm (q) of chromosome 5 at positions 21-22.

A second gene that may be involved in FAP is a mutation in the *MUTHYH* gene or "mutYhomolog (E.coli)." Normally, *MUTHYH* instructs for the making of an enzyme called MYH glycosylase that helps repair DNA and correct mistakes when DNA is replicated. When the DNA is repaired properly, cell growth is normal and does not lead to tumor formation. This type of repair is called base excision repair. However, mutations in the gene keep the repair from taking place properly and the enzyme from not functioning to suppress tumor formation. This gene is more common in people of European descent. It is inherited in an autosomal recessive pattern and located on the short arm (p) of chromosome 1 at position 34.1.

What Is the Treatment for Familial Adenomatous Polyposis (FAP)?

Taking a family history and genetic testing are the first two steps in treatment. The colon is then examined using a thin, flexible tube called a scope or by an X-ray examination. If the polyps are found, surgery may be recommended. Because there are a thousand or more polyps, the surgeon may have to remove the colon leaving the small intestine joined to about five inches of the person's rectum. In this way the person may have normal bowel movements. Other procedures may lead to use of a bag for the bowel function. A variety of medications are being tried to shrink the polyps, but these—as do any other medications—may cause unwanted side effects.

Further Reading

"Familial Adenomatous Polyposis." 2010. Cleveland Clinic. http://www.clevelandclinic.org/registries/inherited/fap.htm. Accessed 5/12/12.

"Familial Adenomatous Polyposis." 2010. Medscape. http://emedicine.medscape.com/article/175377-overview. Accessed 5/12/12.

"Genetics of Colorectal Cancer." 2010. National Cancer Institute. http://www.cancer.gov/cancertopics/pdq/genetics/colorectal/healthprofessional. Accessed 5/12/12.

Familial Atrial Fibrillation (AF)

Prevalence	Affects about 3 million people in the United States; incidence of familial disorder unknown, but studies suggest AF may run as high as 30% of those with family member with the condition
Other Names	auricular fibrillation

When a condition occurs in several members of a family, it is said to be familial. Atrial fibrillation is a common disorder affecting more than three million people. Many of those affected can trace the condition to a grandmother or great-uncle. Although no exact studies of prevalence exist, studies suggest that more than 30% of those who have atrial fibrillation have familial atrial fibrillation.

What Is Familial Atrial Fibrillation (AF)?

Several heart disorders are caused by disruption of the electrical system of the heart, which then affects the heart function. In familial atrial fibrillation, the condition runs in the family. The word "atrium" comes from the Latin meaning "room" and refers to the upper chambers of the heart, which receives the blood from the various places and then passes it the lower chambers or ventricles to be pumped to the specific destinations. Auricle is another name for atrium. The word "fibrillation" refers to fine, rapid movements of the individual fibers of the heart muscle with little or no movement as a whole. Thus, in atrial fibrillation, the muscle fibers of the upper chamber of the heart are not working together. The electrical activity is not coordinated, and the heartbeat then becomes fast and irregular.

The following things can happen when the atrium area function is disrupted:

- A fluttering sensation in the chest area
- Pounding or palpitations in the chest
- Dizziness
- Chest pain
- Shortness of breath
- Fainting, a condition known as syncope
- Increase in the risk of stroke
- Death

The symptoms can occur at any age, and these complications of familial atrial fibrillation may be quite serious. Some people with AF never have any of the related health problems. However, other factors, such as high blood pressure, diabetes, a previous stroke, and hardening of the arteries, may increase the risk for atrial fibrillation.

What Are the Genetic Causes of Familial Atrial Fibrillation (AF)?

Three genes appear to be involved in this condition: *KCNE2*, *KCNJ2*, and *KCNQ1*. These three genes are members of a large family of *KCN* genes. *KCNE2* is the genetic symbol for "potassium voltage-gated channel, ISK-related family, member 2." This gene instructs for making a protein that regulates the activity of potassium channels. Channels are pathways that transport positively charged atoms or ions of potassium in and out of the cells and are important in generating and transmitting electrical signals. The protein creates channels to the heart muscle, where potassium is transmitted in and out. The protein is also responsible for recharging the heart muscle after each heartbeat to maintain the rhythm. A mutation in this gene occurs when the amino acid cysteine replaces arginine. This gene is inherited in an autosomal dominant pattern and is located on the long arm (q) of chromosome 21 at position 22.12.

The second gene is *KCNJ2* whose official name is "potassium inwardly-rectifying channel, subfamily J, member 2." This gene is another of the large family of genes (*KCN*) that produce potassium channels. These channels are active in both the heart and skeletal muscles. A mutation occurs in a single amino acid exchange; valine replaces the amino acid isoleucine. The mutation appears to increase the flow of potassium ions through the channel created by the KCNJ2 protein, thus disrupting the normal heart rhythm. *KCNJ2* is inherited in an autosomal dominant pattern and located on the long arm (q) of chromosome 17 at position 24.3.

The third gene, *KCNQ1*, is the official name of symbol is "potassium-voltage-gated channel, KQT-like subfamily, member 1." This gene instructs for making potassium channels active in the inner ear and in heart or cardiac muscles. The heart channels are involved in recharging the cardiac muscle after each heartbeat. The mutation in this gene appears to increase the flow of potassium ions formed by the *KCNQ1* protein resulting in atrial fibrillation. The gene is inherited in an autosomal dominant pattern and is located on the short arm (p) of chromosome 11 at position 15.5.

What Is the Treatment for Familial Atrial Fibrillation (AF)?

Because this disorder is a genetic disorder, there is no cure. However, treatment of the symptoms is conducted under the care of a cardiologist.

Further Reading

Darbar, D., et al. 2003. "Familial Atrial Fibrillation Is a Genetically Heterogeneous Disorder (Abstract)." *Journal of the American College of Cardiology*. June 18. 41(12): 2185–2192. http://www.ncbi.nlm.nih.gov/pubmed/12821245. Accessed 5/12/12.

"Familial Atrial Fibrillation." 2007. Genetics Home Reference. National Library of Medicine (U.S.) http://ghr.nlm.nih.gov/condition/familial-atrial-fibrillation. Accessed 1/29/12.

"*KCNE2*." 2007. Genetics Home Reference. National Library of Medicine (U.S.). http://ghr.nlm.nih.gov/gene/KCNE2. Accessed 5/12/12.

Familial Cold Autoinflammatory Syndrome

Prevalence Very rare condition; affects less than 1 per million people

Other Names cold hypersensitivity; familial cold-induced autoinflammatory syndrome; familial cold urticaria; FCAS; FCU

What Is Familial Cold Autoinflammatory Syndrome?

Familial cold autoinflammatory syndrome is a condition that appears to run in families in which people react violently to exposure to cold temperatures. The infant experiences the reaction early, and it continues throughout life. Following are some of the symptoms of familial cold autoinflammatory syndrome:

- Time of exposure; the people begin to react within an hour of exposure to cold
- Time of episodes; the episodes usually last an average of 12 hours; some may continue up to three days

- Rash; an itching and burning rash begins on the face or extremities and then spreads to other parts of the body
- Swelling of extremities
- Intermittent fever and chills
- Joint pain
- Eye pain and redness
- Fatigue
- Headache
- Thirst
- Nausea

This condition should not be confused with acquired cold urticaria, which develops later in life and is not inherited.

What Are the Genetic Causes of Familial Cold Autoinflammatory Syndrome?

Two genes—*NLRP3* and *NLRP12*—are associated with familial cold autoinflammatory syndrome.

NLRP3

The *NLRP3* gene is officially known as the "NLR family, pyrin domain containing 3" gene. Normally, *NLRP3* provides instruction for making the protein cryopyrin. The word "cryopyrin" comes from two Greek words: *cryo*, meaning "cold," and *pyr*, meaning "fire." This reflects the intermittent cold and fever that are the symptoms of the disorder. The protein cryopyrin is a member of the nucleotide-binding domain and leucine-rich repeat containing (NLR) proteins and is found inside the cytoplasm of the cells, especially the white blood cells. Its action is involved in the immune system where it assists in responding to injury, toxins, and invasion by outside organisms. When it is activated, groups of cryopyrins work along with other structures called inflammasomes to signal molecules to begin the inflammation process and send white blood cells to the site.

Mutations in *NLRP3* are associated with familial cold autoinflammatory syndrome. Researchers think that the changes cause cryopyrin to be overactive, leading to episodes of cold, fever, and inflammation that exposure to cold triggers. *NLRP3* is inherited in an autosomal recessive pattern and is located on the long arm (q) of chromosome 1 at position 44.

NLRP12

Mutations in the *NLRP12* gene, officially called the "NLR family, pyrin domain containing 12" gene, cause familial cold autoinflammatory syndrome. Normally, *NLRP12* provides instructions for making the protein monarch-1. Monarch-1 is a member of a group of proteins called the nucleotide-binding domain and leucine-rich repeat-containing (NLR) proteins. These proteins are found in the cytoplasm of the cell and especially in white blood cells. Like the proteins made by *NLRP3*, monarch-1 is involved in the immune process, especially in the process of inflammation.

Two mutations in *NLRP12* have been found in families with the condition that live in the area of Guadeloupe in Mexico. The mutations appear to disrupt the work of the monarch-1 protein, leading to the episodes of fever, cold, and inflammation. The *NLRP12* gene is inherited in an autosomal recessive pattern and is located on the long arm (q) of chromosome 19 at position 13.42.

What Is the Treatment for Familial Cold Autoinflammatory Syndrome?

The symptoms of this disorder can be quite debilitating. Patients may try to compensate for the cold and fever with over-the-counter treatments or even by moving to a warmer climate. Some medications have been shown to be effective. Injections of certain immunosuppressants or monoclonal antibodies may also be used.

Further Reading

"Familial Cold Autoinflammatory Syndrome." 2010. Genetics and Rare Diseases Information Center. National Institutes of Health (U.S.). http://rarediseases.info.nih.gov/GARD/Condition/9535/Familial_cold_autoinflammatory_syndrome.aspx. Accessed 1/10/12.

"Familial Cold Autoinflammatory Syndrome." 2010. Genetics Home Reference. National Library of Medicine (U.S.). http://ghr.nlm.nih.gov/condition/familial-cold-autoinflammatory-syndrome. Accessed 1/10/12.

Familial Dysautonomia (FD)

Prevalence Occurs primarily in people of Ashkenazi Jewish heritage; affects about 1 in 3,700 of this population; extremely rare in general population

Other Names FD; HSAN3; HSAN Type III; HSN-III; Riley-day syndrome

Because of the particular lifestyle of the Ashkenazi Jews of central and eastern Europe, many genetic disorders are prevalent in the populations. These diseases may have been in the original settlers from Israel and then have developed through genetic drift as the population intermarried. In 1949, Conrad Milton Riley and Richard Lawrence Day wrote about patients who had central autonomic nervous system problems and defects in tear production in the journal *Pediatrics*. The condition was known for some time by the name Riley-Day syndrome. Today, it is more commonly called familial dysautonomia, or FD.

What Is Familial Dysautonomia (FD)?

Familial dysautonomia is a condition of the autonomic nervous system that affects involuntary processes as well as the sensory system. The name of the disorder "familial dysautonomia" combines the term "familial," which means the condition has been noted in the family or ancestors, with dysautonomia. The word "dysautonomia" is from two Greek words: *dys*, meaning "with difficulty," and *autonomia*, relating to the "establishment of one's own laws or rules." In this sense, people with dysautonomia have difficulty in the system that maintains certain automatic activities of the body.

The autonomic nervous system controls involuntary actions such as breathing, digestion, production of tears, regulation of body temperature, and blood pressure. These actions are all things that one's body does without thinking about or choosing to act. This condition also affects the five senses. FD is a member of a group of disorders known as hereditary sensory and autonomic neuropathies, or HSAN. All HSAN have widespread sensory dysfunction and variable autonomic system dysfunction.

The problems of FD appear early in infancy and may show the following signs or symptoms:

- Insensitivity to pain
- Inability to produce tears
- Poor muscle tone or hypotonia
- Very poor growth and failure to thrive
- Feeding difficulties
- Frequent lung infections
- Difficulty maintaining body temperature
- Unstable blood pressure; at time blood pressure may be high and at other times low
- Frequent vomiting
- Problems with the gastrointestinal tract

As the children become older, additional symptoms may occur such as bedwetting, poor balance, abnormal curvature of the spine, increased risk of bone fractures, and kidney and heart problems. A major crisis may occur when a condition called aspiration pneumonia emerges. The individual regurgitates food, which is then breathed into the lungs causing infections. Another condition, called "dysautonomia crisis," may occur when a constellation of symptoms appear all at once due to physical or emotional stress. Almost all the children have learning disabilities and attention disorders. Adults with FD experience repeated infections, impaired kidney function, and worsening of vision due to shrinking of the optic nerve.

What Is the Genetic Cause of Familial Dysautonomia (FD)?

Mutations in the *IKBKAP* gene cause familial dysautonomia. The official name of the gene is "inhibitor of kappa light polypeptide gene enhancer in B-cells, kinase complex-associated protein." Normally, the gene *IKBKAP* instructs for making a protein call IKK complex-associated protein (IKAP). Found especially in the brain cells, IKAP is part of a six-protein complex called the elongator complex, which is important in transferring information in genes to the areas that make proteins. The elongator complex is especially active in the transcription of proteins that affect the framework of the cells called the cytoskeleton and of cell movement. The cytoskeleton is critical for the growth of nerve cells, especially the axons and dendrites. Cell movement is critical to get nerve cells to their proper place in the brain.

Mutations of *IKBKAP* reduce the amount of the important protein IKAP. The brain cells may have little of this protein. The reduced amount of IKAP and reduced placement of brain cells due to improper motility lead to the signs and symptoms of dysautonomia. *IKBKAP* is inherited in an autosomal recessive pattern and is located on the long arm (q) of chromosome 9 at position 31.

What Is the Treatment for Familial Dysautonomia (FD)?

Currently no cure exists for FD, and affected individuals usually die by age 30. Treatment of the various dangerous symptoms may be in order, and constant attention to prevention of dysautonomic crisis is essential. Some new therapies that elevate the level of IKAP are being investigated.

Further Reading

"Familial Dysautonomia." 2010. Genetics Home Reference. National Library of Medicine (U.S.). http://ghr.nlm.nih.gov/condition/familial-dysautonomia. Accessed 5/12/12.

Familial Dysautotomia Society. http://www.familialdysautonomia.org. Accessed 5/12/12.

"*IKBKAP*." 2010 Genetics Home Reference. National Library of Medicine (U.S.). http://ghr.nlm.nih.gov/gene/IKBKAP. Accessed 5/12/12.

Fanconi Anemia

Prevalence About 1 in every 350,000 births; higher frequency in two ethnic groups, Afrikaaners (Dutch settlers) of South Africa and Ashkenazi Jews

Other Names Anemia, Fanconi's; Fanconi's pancytopenia syndrome; Fanconi's panmyelopathy, Fanconi's refractory anemia

Do not confuse this condition with Fanconi syndrome, a serious disease of the kidneys.

In 1927, Guido Fanconi, a Swiss pediatrician, described a family of three boys who had serious malformations at birth and serious malfunctions of all the various blood cells, called pancytopenia. The boys were very small for their age and had underdeveloped sex organs. They also possessed unusual skin pigmentations and defects of the thumb and radial bone of the arm. Since Fanconi's description of the rare disorder, more than 2,000 cases have been described in the medical literature. The condition Fanconi anemia was named for this physician.

What Is Fanconi Anemia?

Fanconi anemia (FA) is a condition that affects the bone marrow. The bone marrow is responsible for the production of all types of blood cells. While most anemias involve only the red blood cells, FA is a type of aplastic anemia in which the bone marrow stops or does not make enough of all three types of cells. Thus, the white blood cells, red blood cells, and platelets, the cells that aid in clotting of the blood, are decreased.

The problems with the lack of various blood cells lead to serious disorders. When one does not have enough white blood cells, the staple of the immune system, the person may have many infections. If one does not have enough red blood cells, the person may experience fatigue, dizziness, and other problems. If the person does not have a normal number of platelets, the person may experience excessive bleeding.

The following other symptoms may occur:

- Short stature and small head
- Small sex organs such as the testicles and other genitalia
- Problems with missing or extra fingers; problems with bones in the hands or lower arms
- Abnormal heart, lungs, and digestive tract
- Bone problems in hips, spine, ribs; scoliosis or curvature of the spine may occur
- Skin pigmentations with darkened areas called café-au-lait spots and vitiligo
- Deafness due to abnormal ears
- Eye or eyelid problems
- Mental retardation; learning disabilities

The condition may be more common in males. In 2000, the median age of death was about 30 years of age.

What Are the Genetic Causes of Fanconi Anemia?

FA is caused by a genetic defect in a number of genes that are responsible for proteins that repair DNA. At least 13 genes are related to the condition: *FA*, *FANCA*, *FANCC*, *FANCD1*, *FANCD2*, *FANCE*, *FANCF*, *FANCG*, *FANCI*, *FANCJ*, *FANCL*, *FANCM*, and *FANCN*. All these genes are inherited in an autosomal recessive pattern. One other gene, *FANCB*, is on the X chromosome. The mutations in these genes keep the cells from normally repairing the DNA.

Because of the DNA-repair disorder, 20% or more of FA patients often get myelogeous leukemia or other cancers. Because of the DNA repair problems, the person may not respond to drugs that treat cancer by DNA cross-linking. Likewise, people with FA who survive to adulthood may develop a variety of cancerous tumors, including mouth, tongue, throat, or esophagus.

What Is the Treatment for Fanconi Anemia?

The symptoms of Fanconi anemia may be treated by several means. The first line of treatment may be administering growth factors. Newer medical procedures include blood and marrow stem cell transplants.

Further Reading

"Fanconi Anemia." 2010. Medscape. http://emedicine.medscape.com/article/960401-overview. Accessed 5/13/12.

"Fanconi's Anemia." 2010. MedlinePlus. National Library of Medicine (U.S.). http://www.nlm.nih.gov/medlineplus/ency/article/000334.htm. Accessed 5/13/12.

"Fanconi's Anemia." 2003. Orphanet. http://www.orpha.net/data/patho/Pro/en/FanconiAnemia-FRenPro634.pdf. Accessed 5/13/12.

Fanconi Syndrome

Prevalence Rare; attributed to various causes, some acquired and some hereditary

Other Names Fanconi's syndrome; nephrotic-glucosuric dwarfism with hypophophateric rickets; renal tubular acidosis

Do not confuse this disease with Fanconi anemia, an inherited blood disorder.

Beginning in 1903, several investigators began to report on a condition that combined dwarfism, severe rickets, and kidney or liver disorders. Many names were given to the syndrome. However, it was Guido Fanconi, the Swiss pediatrician, who in 1935 brought all the studies together and named the disorder nephrotic-glucosuric dwarfism with hypophophateric rickets. Later researchers

Scientists examine the genealogical tree of a family with Fanconi syndrome. (Richard Howard / TIME & LIFE Images / Getty Images)

confirmed Fanconi's finding and used the name Fanconi syndrome to describe this complex disease with multiple causes.

What Is Fanconi Syndrome?

Fanconi syndrome is a disease of the kidney. The proximal tubules of the kidney are the first part of the kidney to process fluids from the body after it is filtered through the glomerulus. The glomerulus is a group of capillaries at the beginning of the renal tube that filters out waste products and begins the process of urine production. The proximal end of the kidney is that area that is near the midline. Fanconi syndrome may be inherited or caused by drugs or exposure to heavy metals. Different forms of Fanconi syndrome affect different functions of the proximal tubules and result in the some of the diverse symptoms of the disorder.

The abnormal tubule function results in excess amounts of glucose, bicarbonate, phosphates, uric acid, potassium, and other amino acids being lost through the urine. The loss of these valuable substances causes the symptoms of Fanconi syndrome, which are:

- Excretion of a large amount of urine called polyuria: Excessive amounts of mineral and important materials are lost through the loss of water.
- Dehydration: This loss of water may also be related to fever, especially in infants.
- Rickets in children or soft bones in adults, a condition called osteomalacia: Loss of calcium and phosphate in the urine may cause these conditions.
- Growth failure: Loss of calcium and phosphorous and multiple metabolic abnormalities disrupt the normal growth process.
- High blood acid: This causes a condition known as acidosis; bicarbonate and potassium levels are lower.

Fanconi syndrome affects the processing of many substances and reflects a defect in the general function of the proximal tubules and not just one specific pathway.

What Are the Genetic Causes of Fanconi Syndrome?

Although Fanconi is not specifically an inherited disease, the underlying conditions are. The inherited diseases are cystinosis, Wilson disease, Dent disease, Lowe disease, tyrosinemia, Type I, galactosemia, glycogen storage disease, and fructose intolerase.

- Cystinosis: This condition occurs when the amino acid cystine builds up in the cells. This is the most common cause of Fanconi syndrome in children. This disorder is caused by a mutation in the *CTNS* gene located on the short arm (p) of chromosome 17 at position 13.

- Wilson disease: This condition is an inherited condition of copper metabolism. (*See also* Wilson disease.)
- Lowe syndrome: This disease, inherited in an X-linked pattern, affects also the eyes. (*See also* Lowe syndrome.)
- Dent disease: This hereditary disease can lead to Fanconi syndrome and is also X-linked.

Fanconi syndrome can be acquired later in life. Ingesting certain dugs such as expired tetracyclines may cause the disorder. Also, Fanconi syndrome has been seen in the HIV population when certain antiretroviral medications were used. Heavy metals such as lead may lead to the condition.

What Is the Treatment of Fanconi Syndrome?

Treatment involves replacement the minerals or materials lost through urination.

Further Reading

"Fanconi Syndrome." 2010. Medscape. http://emedicine.medscape.com/article/981774-overview. Accessed 5/13/12.

"Fanconi Syndrome." 2010. Merck Manuals. http://www.merckmanuals.com/home/sec11/ch146/ch146f.html. Accessed 5/13/12.

Feingold Syndrome

Prevalence Rare; until 2003, 79 patients reported worldwide

Other Names microcephaly-mesobrachyphalangy-trachesophageal fistula (MMT); microcephaly-oculo-digito-esophageal-duodenal (MODED) syndrome; oculo-digito-esophagoduodenal (ODED) syndrome

From the long descriptions in the other names, this syndrome affects many parts of the body. Murray Feingold, an American physician, first put all these symptoms together and called it a syndrome in 1975. The name Feingold syndrome is much simpler than the long descriptive names that are sometimes given.

What Is Feingold Syndrome?

Although Feingold syndrome affects many body parts, the first noticeable symptoms shared by all individuals with the disorder are the characteristic hand and toe abnormalities. The hand abnormality is called "brachymesophalangy." The terms come from three Greek words: *brachy*, meaning "short"; *meso*, meaning "middle"; and *phalang*, meaning the "digits of the toes and fingers." Thus, brachymesophalangy refers to a very short second and fifth finger. Also the fifth finger may curve inward, a condition called clinodactyly, and underdeveloped thumbs called thumb hypoplasia are evident. The second and third toes or the fourth and fifth toes may be fused, a condition called syndactyly.

Several other conditions may affect the person with Feingold syndrome. The individual may have all or combinations of the following symptoms:

- Face and head: The individual may have a very small head with an extremely small jaw and a narrow opening of the eyelids. Anteverted (turned up) nostrils and ear anomalies may be present.
- Digestive tract: The individual is born with a blockage of the digestive tract, a condition known as gastrointestinal atresia. Sometimes the blockage is in the esophagus or upper part of the small intestine called the duodenum.
- Other areas: The person may have impaired hearing, slow growth, kidney disorders, and heart defects.
- Cognitive impairment and learning disabilities.

The signs and symptoms of the disorder may vary among individuals and even within the same family.

What Is the Genetic Cause of Feingold Syndrome?

Changes in the "neuroblastoma-derived-V-myc avian myelocytomatosis viral-related oncogene," or *MYCN* gene, causes Feingold syndrome. Normally, the *MYCN* gene instructs for making a protein that is essential for the development of bones and tissues during the embryonic stages. It is related to several important systems including formation of the lungs, heart, kidneys, nervous system, digestive system, and respiratory system. In addition, *MYCN* regulates other genes as it binds to specific areas of the DNA, making it an important transcription fact.

MCYN is a member of an important class of genes known as *MYC* oncogenes. The oncogenes are related to cancer formation. When a mutation occurs in oncogenes, the normal cells can grow abnormally and become cancerous.

The 29 mutations in *MYCN* cause Feingold syndrome. In some cases the *MYCN* gene is completely missing. However, most of the changes in the gene stop the production of the protein important in embryonic development. It appears that the absence or reduced amount of the protein is responsible for the varied symptoms of Feingold syndrome. *MYCN* is inherited in an autosomal dominant pattern and is located on the short arm (p) of chromosome 2 at position 24.1.

What Is the Treatment for Feingold Syndrome?

Because Feingold syndrome is a genetic disorder with a variety of symptoms, no treatment or reversal exists. Thus, only the symptoms can be treated with surgery.

For example, blockage of the esophagus or intestines must be corrected. Other defects must be handled as they occur.

Further Reading

"Feingold Syndrome." Humpath.com. http://www.humpath.com/?Feingold-syndrome. Accessed 5/13/12.

"Feingold Syndrome." RightDiagnosis.com. http://www.rightdiagnosis.com/medical/feingold_syndrome.htm. Accessed 5/13/12.

"*MYCN*." 2010. Genetics Home Reference. National Library of Medicine (U.S.). http://ghr.nlm.nih.gov/gene/MYCN. Accessed 5/13/12.

Fetal Alcohol Spectrum Disorders (FASD)

What Are Fetal Alcohol Spectrum Disorders (FASD)?

The umbrella term "fetal alcohol spectrum disorders" (FASD) describes a group of defects caused by the mother's drinking of alcohol during pregnancy. These effects can include physical, mental, behavioral, and learning disabilities that will be present throughout life.

To assess the central nervous system (CNS) damage, Janet Lang and a team at the Fetal Alcohol Diagnosis Program in Duluth, Minnesota, proposed 10 Brain Domains that cover mental, physical, and behavioral concepts.. They include: achievement, adaptive behavior, attention, cognition, executive functioning, language, memory, motor skills, sensory integration, and social communication.

FASD is not a diagnostic term but covers a variety of conditions. In all of the following conditions, prenatal use of alcohol by the mother was the paramount reason for the disorder:

- Fetal alcohol syndrome (FAS): The child has distinctive facial features such as a smooth groove between the nose and upper lip, a thin upper lip, small eye openings, growth deficiencies, and central nervous system (CNS) defects.
- Partial fetal alcohol syndrome (PFAS): Two of the three facial anomalies are present. There many be some damage to the central nervous system with 3 or more of the 10 brain domains.
- Alcohol-related Neurodevelopment Disorder (ARND): Growth or height may range from normal to slightly under normal. There may be minimal or no FAS facial features. There will be a distinct deficit in 3 of the 10 Brain Domains.
- Alcohol-related birth defects (ARBD): Damage to organs such as brain, bones, or muscles is evident.

The term "Fetal Alcohol Effect" (FAE) is an older description that has fallen out of favor among clinicians. People used the term to mean a lesser effect and thought it was not serious because the child did not have the facial features. However, the effects were just as detrimental. Now FAE has been recategorized into the above disorders.

About 1% of children are estimated to suffer from the one of the conditions of fetal alcohol spectrum disorder.

What Is the Genetic Cause of Fetal Alcohol Spectrum Disorders (FASD)?

When a pregnant woman consumes alcohol, she may interfere with the development of the fetus so that the infant may have several symptoms, including neurological disorders and learning disabilities. Even moderate drinking during pregnancy can decrease the level of maternal thyroid hormones as well as fetal thyroid hormones. Use of alcohol may impair an enzyme in the fetal brain called iodophyronine deodinase type III, which appears to affect normal development and leads to later learning and behavioral difficulties.

Actually, only a trained professional can make a diagnosis. Teachers and others may suspect FASD but cannot make the diagnosis. Following are some of the signs of the need for assessment:

- Maternal alcohol use
- Sleeping, breathing, or feeding problems
- Small head or facial or dental abnormalities
- Heart defects or other organ dysfunction
- Deformities of joints, limbs, and fingers
- Vision or hearing problems
- Mental retardation or delayed development
- Behavioral problems—may be diagnosed as ADHD, but behavior is more erratic and disruptive

Why is the Diagnosis Important?

Most children, even those whose mothers are suspected of alcohol use, are dismissed from the hospital without a diagnosis. When children get to school, the spectrum may be thought to be the result of poor parenting or other disorders.. However, proper diagnosis is essential to help the person receive appropriate services.

What Is the Treatment for Fetal Alcohol Spectrum Disorders (FASD)?

Treatment for child with an FASD depends on the nature and severity of exposure and is determined case by case. Prevention is the key to this problem. To avoid

FASDs, a woman should not drink alcohol while she is pregnant, or even when she might get pregnant. She may not know of the pregnancy for several weeks or more. The Centers for Disease Control and Prevention recommends: "If you are pregnant, don't drink. If you drink, don't get pregnant."

See also Embryopathies

Further Reading

"Children with Fetal Alcohol Spectrum Disorders (FASD) Have More Severe Behavioral Problems Than Those with ADHD." 2009. *Science Daily*. http://www.sciencedaily.com/releases/2009/07/090716164335.htm. Accessed 12/24/11.

"Fetal Alcohol Spectrum Disorders (FASDs)." 2011. Centers for Disease Control. http://www.cdc.gov/ncbddd/fasd/facts.html. Accessed 12/24/11.

Lang, Jeannette. 2006. "Ten Brain Domains: A Proposal for Functional Central Nervous System Parameters for Fetal Alcohol Spectrum Disorder Diagnosis and Follow-Up." http://www.motherisk.org/JFAS_documents/JFAS_5012_Final_e12_6.28.6.pdf. Accessed 12/24/11.

"Specific Genetic Cause of Fetal Alcohol-Related Developmental Disorders Found." 2009. The Endocrine Society. *Science Daily*. June 10. http://www.sciencedaily.com/releases/2009/06/090610124426.htm. Accessed 2/29/12.

"Understanding Fetal Alcohol Spectrum Disorders: Getting a Diagnosis." 2005. Substance Abuse and Mental Health Service Administration. http://fasdcenter.samhsa.gov/documents/WYNKDiagnosis_5_colorJA_new.pdf. Accessed 12/24/11.

Fetal Isotretinoin Syndrome

Prevalence Between 1982 and 2003 in the United States, more than 2,000 women have become pregnant while taking the drug; many pregnancies ended in abortion or miscarriage. About 160 babies with birth defects were born.

Other Names Tretinoin Embryopathy Syndrome (TES), Isotretinoin dysmorphic syndrome; Isotretinoin Teratogen/dysmorphic syndrome; retinoic acid embryopathy

Many teenagers and even adults know the psychological impact of severe acne. To find a drug that would treat the aggravating condition would be a miracle to these sufferers. As early as the 1930s, it was known that strong doses of vitamin A could treat troublesome skin conditions. After World War II, the age of the "miracle drugs" developed, and to find a drug that would cure acne would be a "miracle" for the individuals with acne. Isotretinoin appeared to be that wonder. In 1984, Hoffman-LaRoche derived a product from vitamin A called the tretinoin—one of the retinoids—that gave hope to people with severe skin problems. Their drug was called Accutane (the pharmaceutical name), and it immediately became popular. However, hanging over this hope was a dark cloud. Children with serious birth defects were born to women who had used tretinoin. The miracle retinoids turned out to be a teratogen, a term arising from the Greek words *teratos*, meaning "monster," and *gen*, meaning "giving birth to." The terms used to describe the condition all mean the same thing: accutane embryopathy, retinoic acid embryopathy, and tretinoin embryopathy syndrome.

What Is Isotretinoin?

Isotretinoin, a human-made derivative of vitamin A, is the generic name for the pharmaceutical Accutane. These vitamin A derivatives are also called retinoids. Vitamin A, found in many fruits and vegetables and also as a synthetic vitamin, is a nutrient essential for good growth. Vitamin A has been known for a long time to help with growth of skin cells. After treating with isotretinoin or Accutane, the vitamin A derivative, the lesions usually clear for a year or more. However, some people do have side effects. Depression, including thoughts of suicide, has been reported.

The U.S. Food and Drug Administration (FDA) approved the use of Accutane in September 1982. When laboratory animals were tested, it was shown to cause birth defects in the offspring. The drug was granted approval only if the label would carry the information about birth defects. The warning advised women taking the medication to avoid pregnancy.

What Are the Genetic Effects of Fetal Isotretinoin Syndrome?

It only took a year on the market for the first case of an infant with Accutane-related birth defects to be reported. Then 10 more cases surfaced, and by 1985, 154 cases were reported. Each had used the drug during pregnancy, and the offspring exhibited a pattern of defects that included the ears, face, and heart; some of the babies were stillborn. Sometimes the term retinoic acid embryopathy is used. Another drug, Tegison, also a tretinoin derivative, is used to treat psoriasis; this drug may cause birth defects even after the person stopped using it years before.

Isotretinoin embryopathy is not inherited but is caused as the embryo is developing in the uterus. The exact nature of the deformity depends upon the development of the pregnancy and the length of exposure. Range and severity vary from cases to case. The set of malformations usually involve the central nervous system, head, and face as well as other malformations of the heart. The retinoids have been known to interfere with the HOX signaling pathways that are used to pattern the branchial arches during the fourth week of pregnancy. The branchial arches are the formations that lead to the nervous system, head and facial appearance, and heart.

The danger period is from 15 days after conception to about four weeks after the mother's last monthly period. The dose of isotretinoin used in Accutane appears insignificant, and the drug does not appear to damage the fetus when stopped before pregnancy.

What Are the Signs and Symptoms of Fetal Isotretinoin Syndrome?

Multiple abnormalities as well as a host of associated conditions may be seen. Following are signs of isotretinoin embryopathy syndrome: face and head abnormalities, face not symmetrical, receding chin, uvula (the flap hanging down at the back of the throat) anomaly, limbs deformed, one side of the body different from the other, and double vision. Some other diseases and complications may result: adrenal problems, aortic arch abnormalities, blind external ear canals, cataracts, cleft lip or palate, congenital clubbed foot, neural tube defect, facial paralysis, hydrocephalus, inguinal hernias, spina bifida, vertebral, and hand anomalies.

Isotretinoin is classified as FDA Pregnancy Category X, and participants are warned about use in pregnancy. The FDA introduced a program called iPLEDGE on August 12, 2005, in an attempt to ensure that female patients receiving isotretinoin do not become pregnant. As of March 1, 2006, only physicians registered in iPLEDGE are able to prescribe isotretinoin, and only pharmacists registered and qualified in iPLEDGE will be able to dispense isotretinoin. The iPLEDGE program also applies to males, even though there has been no evidence of isotretinoin excretion through seminal fluids.

What Is the Treatment for Fetal Isotretinoin Syndrome?

Treatment of the child affected by isotretinoin depends upon the severity of the case and is symptomatic. For example, if the heart is affected, surgery may be performed. Likewise, deformities of the face can be treated by surgery. The child qualifies for special services under the Individuals with Disabilities Education Act.

See also Embryopathies

Further Reading

"Accutane Embryopathy." 2005. http://www.healthline.com/galecontent/accutane-embryopathy-1. Accessed 5/13/12.

Dolan, Siobhan. 2004. "Isotretinoin and Pregnancy: A Continued Risk for Birth Defects." Medscape News Today. http://www.medscape.com/viewarticle/492119. Accessed 2/29/12.

"Retinoic Acid Embryopathy." http://www.aarpmedicareplans.com/galecontent/accutane-embryopathy?print=true. Accessed 5/13/12.

FG Syndrome

Prevalence 1 to 1,000 in the Utah Valley; several hundred cases reported worldwide; may be overdiagnosed because of the similarity to other disorders

Other Names FGS; FGS1; Keller syndrome; mental retardation, large head, imperforate anus, congenital hypotonia, and partial ageneisis of the corpus callosum; OKS; Opitz-Kaveggia syndrome

J. M. Opitz and E. G. Kaveggia published the first study of a syndrome with multiple anomalies and mental retardation. They observed two sisters who had five sons with the syndrome. The researchers took the letters of the surnames of the two sisters who had the sons—F and G. Opitz concluded in a 2008 study that Kim Peek, who was the basis for the character played by Dustin Hoffman in the movie *Rain Man*, probably had FG syndrome.

What Is FG Syndrome?

FG syndrome affects many body parts and is found mostly in males, indicating a relationship with the X chromosome. Almost all of the individuals with FG have cognitive disability and a distinct behavior pattern. When compared with people with other intellectual disorders, the individuals are friendly, sociable, and inquisitive. They appear to have strong social and daily living skills, although their language skills may be comparatively weak. The behavior and intellectual disabilities may arise from abnormalities in corpus callosum, the tough, connective tissue between the two hemispheres of the brain. This abnormality is called agenesis, which means the body part did not develop during embryonic development.

Often the mother has had a difficult pregnancy and frequent delivery by Cesarean section. The child has communication difficulties and special sensory problems. He may overreact to certain stimuli of touch, sound, light, crowds, temperature changes, and emotional pressure. One unique behavior may be the aversion to certain food textures. For example, he may not eat anything that is round such as peas and may "bolt" food into the lungs and develop pneumonia. Temper tantrums and withdrawal may be triggered by small, insignificant things. He may appear fascinated by mechanical toys and objects and can sit for hours "tinkering" with the object. He is often diagnosed as having autism, Asperger syndrome, or pervasive development disorder.

The physical symptoms are as follows:

- Head and face: The individual has a very large head compared to the size of the body, small, simple ears, and wide-set eyes with the outside corners of the eyes pointing downward. The forehead is tall and prominent. He may have an "open-mouthed" expression and a thin upper lip but a thick lower lip, which gives a pouting expression.
- Body: The person has short stature and may have joints that are fixed and tend to stay in one place.
- Gastrointestinal problems: Individuals have constipation, reflux, and/or a possible obstruction in the anal opening called imperforate anus.

- Medical problems: Heart defects, seizures, undescended testicles, and a hernia in the lower abdomen have been reported in affected individuals.

What Is the Genetic Cause of FG Syndrome?

FG syndrome is caused by mutations on the X chromosome. The responsible gene is *MED12*, whose official name is the "mediator complex subunit 12" gene. Normally, *MED12* instructs for making the mediator complex subunit 12 protein, one of about 25 proteins that regulate gene activity. The mediator complex acts as a transcription factor to turn genes off and on. Researchers also note the MED12 protein is involved in early brain development of nerve cells in the brain.

Mutations in the *MED12* gene disrupts the structure of the MED12 protein and leads to the behavioral and physical symptoms of FG. *MED12* is located on the long arm (q) of the X chromosome at position 12-Xq21 and is inherited in a recessive pattern. There are a few other areas of the X chromosome that are questionable gene sites. The other locations are Xq28, Xp22.3, Xp11-p11.3, and Xq22.3. When a set of symptoms are caused by gene mutations at different locations, the disorder is called heterogeneous. FG appears to be one of those conditions.

What Is the Treatment for FG Syndrome?

Correct diagnosis is mandatory. Treating the symptoms of constipation, reflux, hyperactivity, or seizures is the rule. Physical interventions such as surgery may be needed. Patients should be placed in a proper school setting to learn communication skills using the computer as early as possible.

Further Reading

"FG Syndrome." 2003. Gale Encyclopedia of Public Health. http://www.healthline.com/galecontent/fg-syndrome. Accessed 5/13/12.

FG Syndrome Family Alliance, Inc. http://www.fgsyndrome.org. Accessed 5/13/12.

Opitz, John M. 2003. "FG Syndrome." Orphanet. http://www.orpha.net/data/patho/Pro/en/FGSyndrome-FRenPro1053.pdf. Accessed 5/13/12.

Fibrodysplasia Ossificans Progressiva (FOP)

Prevalence Very rare condition; occurs in about 1 in 2 million people worldwide; only several hundred cases reported

Other Names fibroplasia ossificans progressiva; FOP; myositis ossificans; myositis ossificans progressive; progressive myositic ossificans; progressive ossifying myocitis; Stoneman's disease

People Who Turn to Bone

In 1692, a French physician probably wrote the first description of FOP: "I saw a woman today who finally became hard as wood all over." This description of fibrodysplasia ossificans progressiva used the metaphor of wood; others have used the stone metaphor referring to the disorder as "Stoneman's disease."

However, a group of courageous doctors have been interested in unlocking the secrets of this disorder. In 1977, physician Michael Zosloff's first patient with FOP was an eight-year-old girl, who fortunately had been identified with the disease. Misdiagnosis happens in 80% of the cases, and probing test treatment exacerbates the condition.

When Zasloff went to work at the NIH, he was ridiculed for wanting to concentrate on this rare, unimportant disease. He contended that if a person has that disease, the condition is very important. One patient was a woman in her late 20s, named Jeannie Peeper. Jeannie had FOP diagnosed at the age of four; however, soon after graduating from college, she fell and broke her hip. Zasloff had to tell her that not a lot could be done, but that he would like to introduce her to other people with FOP and gave her a list of 11 more people. Now that group has gown to the International FOP Association (IFOPA), a research and support foundation, incorporated in June 1988, which has a website at http://www.ifopa.org.

In October 1995, the skeleton of Harry Eastlack arrived for a two-day symposium in Philadelphia. Over 43 families attended to hear orthopedic surgeons, geneticists, and other doctors discuss FOP. The contribution of Eastlack's skeleton to science has enabled scientists to know and understand the mysteries of this disease. Most skeletons are a pile of bones that must be wired together. Harry Eastlack's skeleton is like a standing man. Viewing this skeleton at the Mutter Museum in Philadelphia gives the viewer an astounding picture of what FOP does to a body.

Throughout medical history, there have been references of men who turned to stone. Beginning about 1800, some written references became more specific. But in the 1900s, Harry Eastlack (1933–1973) became the best known case. At the age of 10, Harry began to develop bony lesions in his back and from then until he died at the age of 40; his body had slowly turned to bone, with only his lips being able to move.

Before his death, Harry requested that his body be donated to science and his skeleton be preserved for study. Today, Harry's skeleton can be seen in the Mutter Museum in Philadelphia. Harry's skeleton has enabled scientists to discover some of the details of this rare disorder that turns people to stone. The condition is known as fibrodysplasia ossificans progressiva, or FOP.

What Is Fibrodysplasia Ossificans Progressiva (FOP)?

Fibrodysplasia ossificans progressiva is a condition in which the body's fibrous tissue turns to bone. The words in the long title indicate what happens: fibrodysplasia comes from the root words *fibro*, meaning "thread or fiber," and "plasia," which comes from *plas* meaning "form or mold"; ossificans refers to the making of the bone, and progressiva means that the transformation is gradual. Hence, fibrodysplasia ossificans progressiva is a disorder in which the connective tissue, tendons, and ligaments are gradually replaced with bone, forming a skeleton outside the skeleton.

The first clue of FOP is present at birth. The big toe is deformed, possibly with a missing joint or a large lump at the joint. The appearance of a bony lump is called a flare-up. The first flare-up usually appears before age 10 and looks like a bump on the shoulder or back area. The bumps may then flare up on the arms, chest, and finally the feet. The extra-skeletal growth, which is called heterotropic bone, is independent of the normal skeleton and appears to follow the pattern of bone development in the embryo. It is unknown why some fibrous tissue, such as the diaphragm, tongue, eye muscles, and cardiac and smooth muscle, appear to be spared from ossification. However, because the joints are affected, the person with FOP will eventually not be able to eat or speak. Growth around the lungs may cause breathing difficulties.

Any trauma to the muscles such as a fall or even invasive surgeries may inflame the muscles and cause the bone to be deposited in the injured area. Because the condition is so rare, the condition is often misdiagnosed as cancer or neurofibromatosis, leading the doctors to order invasive biopsies that increase the deposit of bone. In fact, researchers from the Children's Hospital at the University of California, San Francisco, report the misdiagnosis rate at about 80%.

FOP is a progressive disease and typically gets worse as one ages. However, the rate of progression is individual and the progression is quite unpredictable.

What Is the Genetic Cause of Fibrodysplasia Ossificans Progressiva (FOP)?

Mutations in the *ACVR1* gene, officially known as the "activin A receptor, type 1" gene, is the cause of FOP. Normally, *ACVR1* instructs for production of activin receptor type 1 protein, a member of a large protein family called bone morphogenetic protein (BMP) type 1 receptors. BMP receptors are on the cell membranes, with one end inside the cell and the other projecting outside the cell. The receptor can then receive messages from outside the cell and send them inside the cell. These receptors are found in many body tissues, controlling the growth and repair of those tissues. They are especially related to the gradual replacement of cartilage into bone as the person grows into maturity.

The mutation in the *ACVR1* gene occurs when the amino acid arginine is substituted for histidine. This substitution of only one building block may disrupt the process of bone formation by changing the shape of the ACVR1 protein. As a result, the receptor is constantly activated, causing the overgrowth of bone. The *ACVR1* gene is inherited in an autosomal dominant pattern and is located on the long arm (q) of chromosome 2 at position 23-q24.

What Is the Treatment for Fibrodysplasia Ossificans Progressiva (FOP)?

Because this is a genetic condition, no treatment currently exists. Surgery is not an option because removal of bone may cause more bone to grow. Researchers are investigating new treatments for FOP. Medication may help the symptoms of FOP. Also, a new drug is being developed to control bone growth.

Further Reading

"*ACVR1*." 2010. Genetics Home Reference. National Library of Medicine (U.S.). http://ghr.nlm.nih.gov/gene/ACVR1. Accessed 5/13/12.

"Fibroplasia Ossificans." 2010. Medscape. http://emedicine.medscape.com/article/1112501-overview. Accessed 5/13/12.

"Fibrodysplasia Ossificans Progressiva." 2010. University of California, San Francisco, Benioff Children's Hospital. http://www.ucsfbenioffchildrens.org/conditions/fibrodysplasia_ossificans_progressiva. Accessed 5/13/12.

IFOPA: International Fibrodysplasia Ossificans Progressiva Association. A foundation that supports research, information, and communication among those who have the disorder of FOP and that ultimately hopes to find a cure. http://www.ifopa.org.

46,XX Testicular Disorder of Sex Development

Prevalence About 1 in 20,000 individuals with male appearance

Other Names XX male syndrome; XX sex reversal

What Is 46,XX Testicular Disorder of Sex Development?

46,XX testicular disorder is a condition in which the fetus develops as a male but has two X chromosomes. People with the condition have underdeveloped external male genitalia, such as a small penis and undescended testes. The urethra is on the underside of the penis. Some of the individuals have ambiguous sex organs that are neither male nor female. However, the children are generally raised as males. Adults with the disorder are shorter than average and unable to father children.

What Is the Genetic Cause of 46,XX Testicular Disorder?

In a normal individual, females have two X chromosomes, and males have one X and one Y chromosome. An area on the Y chromosome determines the male sex and sex characteristics. In this area is a gene called *SRY*.

This disorder is not inherited. Mutations in the *SRY* gene, officially known as the "sex determining region Y" gene, are related to 46,XX testicular disorder. Normally, *SRY* provides instructions for the transcription factor that binds to areas of DNA and determines whether the fetus develops as a male.

About 80% of the cases of this disorder are the result of an abnormal exchange of genetic material between chromosomes. This is a random event occurring during spermatogenesis, when the sperm is formed. The exchange of material called translocation places the *SRY* gene onto an X chromosome. If the offspring gets this X chromosome with the *SRY* gene, it will be a male even though it does not have a Y chromosome. The condition is called *SRY*-positive 46,XX testicular disorder. The remaining cases are called *SRY*-negative 46,XX testicular disorder. They are more likely to have both male and female genitalia. *SRY* is located on the short arm (p) of the Y chromosome at position 11.3.

What Is the Treatment for 46,XX Testicular Disorder?

Administering the male hormone testosterone at puberty can aid in the development of secondary sex characteristics such as growing facial and pubic hair, deepening of the voice, and preventing breast development.

Further Reading

"46,XX Testicular Disorder of Sex Development." 2011. Genetics Home Reference. National Library of Medicine (U.S.). http://ghr.nlm.nih.gov/condition/46xx-testicular-disorder-of-sex-development. Accessed 12/13/11.

"46,XX Testicular Disorder of Sex Development." 2011. Orphanet. http://www.orpha.net/consor/cgi-bin/OC_Exp.php?lng=en&Expert=393. Accessed 12/13/11.

Vilain, Eric J. 2009. "46,XX Testicular Disorder of Sex Development." *GeneReviews*. http://www.ncbi.nlm.nih.gov/books/NBK1416. Accessed 12/13/11.

47,XX,+21. *See* Down Syndrome (DS)

47,XXX. *See* Triple X Syndrome

47,XXY. *See* Klinefelter Syndrome

47,XY,+21. *See* Down Syndrome (DS)

47,XYY Syndrome

Prevalence About 1 in 1,000 newborn boys; 5 to 10 boys born with syndrome in the United States each day

Other Names Jacobs syndrome, XYY syndrome; XYY karyotype; YY syndrome

Not often does a genetic finding cause controversy, but some studies involving the XYY chromosome type certainly caused a furor when they were published. In December 1965, British geneticist Patricia Jacobs and colleagues in Edinburgh released a chromosome study of 315 males at the state hospital for the developmentally disabled in Lanarkshire, Scotland, and found that nine patients, ages 17 to 36 and six feet tall, had 47XYY karyotypes. Jacobs characterized the men with the XYY pattern as aggressive and violent criminals.

Over the next decade, all published XYY studies were with height-selected institutionalized males and tightly clung to the myth that men with XYY chromosomes were genetically bound to be criminals. Some women during the 1970's even had abortions if they knew the fetus carried that extra Y chromosome. However, Beckwith worked hard to disprove this idea as a myth, and in 1974 showed that the studies linking XYY and criminal behavior were flawed.

What Is 47,XYY Syndrome?

Jacobs syndrome is now generally referred to as 47,XYY syndrome to separate it from the stigma that was created when Jacobs released the report of prisoners in 1966. The syndrome that has an abnormal number of sex chromosomes is called an aneuploidy. The male normally has an X chromosome from the mother and a Y chromosome from the father, making the total 46. The number 47 indicates an extra chromosome, in this instance the Y chromosome. The condition causes no unusual physical features and is not detected unless there is a genetic test for some other reason.

47,XYY have a normal sex development and no problem with fertility when they mature. Some researchers think that the most prevalent symptoms are delayed emotional development and learning problems in school. They claim that up to

50% of boys with the extra Y chromosome have speech development problems compared to only 10% of men without the extra Y.

Some of the following characteristics may be present:

- Immaturity
- Increased height
- Swollen joints
- Arthritis
- Joint stiffness
- Impaired joint mobility
- Chest pain
- Delay of motor skills such as sitting or walking
- Hand tremors
- Autism spectrum disorder in a small percentage of cases.

These features vary greatly among boys and may not necessarily be related to the 47,XYY.

What Is the Genetic Cause of 47,XYY?

The genetic cause of 47,XYY is an extra Y chromosome present in the cells. Normally, individuals have 46 chromosomes. Females have two X chromosomes (XX) and males have an X and Y. Males with 47,XYY have an extra chromosome in each one of their cells, and it is not clear whether the extra Y is connected with tallness and learning problems. Some males have an extra Y only in some of their cells. This condition is called 46,XY/47,YY and is referred to as a mosaic condition.

Neither 47,XYY nor 46XY/47,XYY appears to be inherited. The XYY phenomenon probably occurred during spermatogenesis or the formation of sperm. An error in cell division called nondisjunction can result in sperm cells resulting in an extra Y chromosome. The mosaic condition probably occurred as a random event in early embryonic development.

What Is the Treatment for 47,XYY Syndrome?

Because the condition itself is basically asymptomatic, there is no treatment. Tallness is not considered a problem and may be an asset. The learning problems or other possible problems can be handled with basic intervention techniques. The male will never know that he has XYY unless he is tested for some other condition.

Further Reading

"47XYY Syndrome." 2010. Genetics Home Reference. National Library of Medicine (U.S.). http://ghr.nlm.nih.gov/condition/47xyy-syndrome. Accessed 5/1/12.

"Symptoms of Jacobs Syndrome." 2010. Right Diagnosis. http://www.rightdiagnosis.com/j/jacobs_syndrome/symptoms.htm. Accessed 5/1/12.

48,XXYY Syndrome

Prevalence Affects about 1 in 18,000 to 50,000 males

Other Names XXYY syndrome

What Is 48,XXYY Syndrome?

48,XXYY syndrome is an abnormality of chromosome formation that affects normal development in males. In addition to sexual development, the boy will have medical and behavioral disorders. Instead of the normal 46 chromosomes, people with this pattern have 48 chromosomes—two X chromosomes and two Y chromosomes.

The following symptoms are characteristic of this disorder:

- Abnormal male sexual development: Small testes produce only small amounts of testosterone leading to reduced facial and body hair, increased breast development, and infertility. These men will not be able to father children.
- Poor muscle development
- Low energy levels
- Height: The boys are taller than their peers.
- Delayed motor development: As infants, the child may be slow in sitting, standing, and walking.
- Tremor that starts in teenage years and worsens over time.
- Dental issues: Both primary (baby) and secondary teeth are slow to erupt. The teeth may have thin enamel, misalignment, and multiple cavities.
- Vascular disorders: Blood vessels in the legs may narrow, causing clots that occur in the legs, called deep vein thrombosis.
- Skin ulcers
- Flat feet
- Elbow disorders
- Many medical problems: The person may develop allergies, asthma, type 2 diabetes, heart defects, and seizures.
- Speech and language disorders
- Learning disabilities: The boys especially have difficulty reading but may be able to function well in math, visual-spatial skills, such as puzzles.

- Behavioral disorder: The boys appear to be inclined for attention deficit hyperactivity disorder (ADHD), mood disorder, and autism spectrum disorders.

What Is the Genetic Cause of 48,XXYY Syndrome?

This condition is not inherited. It is the result of a random event during the formation of sperm and egg. The random error is called nondisjunction. In 48,XXYY, the extra sex chromosome arises in the sperm cell, which has gained two extra sex chromosomes in a sperm cell. The sperm has one X and two Y chromosomes. When that sperm fertilizes a normal egg with one X chromosome, the child will have two X chromosomes and two Y chromosomes in each of the body cells. In a small percentage of cases, the nondisjunction may occur after fertilization.

What Is the Treatment for 48,XXYY?

Because this is a serious chromosomal anomaly, no cure for 48,XXYY exists. The long list of problems must be treated when the symptoms occur. The boy will qualify for special education in school.

Further Reading

"48,XXYY Syndrome." 2011. Genetics Home Reference. National Library of Medicine (U.S.). http://ghr.nlm.nih.gov/condition/48xxyy-syndrome. Accessed 12/13/11.

Lolak, Sermsak, MD; Elisa Dannemiller, BS; and Francis Andres, MD. 2005. "48,XXYY Syndrome, Mood Disorder, and Aggression." *American Journal of Psychiatry*. http://ajp.psychiatryonline.org/article.aspx?Volume=162&page=1384&journalID=13. Accessed 12/13/11.

The XXYY Project. "What Is XXYY Syndrome." 2011. http://www.xxyysyndrome.org/whatisxxyysyndrome.html. Accessed 12/13/11.

Fragile X Syndrome

Prevalence Occurs in 1 in 4,000 males and 1 in 8,000 females

Other Names Escalante's syndrome (in Brazil and South America); fra(x) syndrome; FRAXA syndrome; FXS; Marker X syndrome; Martin Bell syndrome; X-linked mental retardation and macroorchisism

When a condition appears in a family, researchers begin creating a pedigree of the family tree marking the incidence of a disorder in the family. In 1943, Martin and Bell traced a pedigree of people with long, narrow faces and mental disability to

Although intellectually challenged, as may be people with Fragile X syndrome, this man is still able to navigate his way through the Internet. (CDC/Amanda Mills)

the females of families. In 1969, Herbert Lubs, studying the genetic patterns, found an unusual "marker X chromosome" related to mental disability. Later, in 1970, another scientist, Frederick Hecht coined the term "fragile site." The condition that these three scientists were researching is now called Fragile X syndrome.

What Is Fragile X Syndrome (FXS)?

Fragile X syndrome is an X-linked genetic condition, which results in a spectrum of intellectual and behavioral actions ranging from mild to severe. Because of the relationship to the X chromosome, males are mostly affected; however, if the female gets one gene from the father and one from the mother, she may also have fragile X syndrome.

In addition to learning disorders and cognitive disability, the person with FXS may have the following characteristics:

- Facial appearance: Long face with large or protruding ears, protruding jaw, and high forehead. The characteristics may become more pronounced and abnormal appearing with age.
- Behavior: The person displays extreme hyperactivity and attention disorders. They may display fidgeting, hand flapping, impulsivity, and problems focusing on tasks.

- Communication: About one-third of the males have autistic-like behaviors that affect social interaction and communication.
- Speech: Speech may appear cluttered and nervous.
- Seizures: These occur in about 15% of males and 5% of females.
- Body features: Fingers may be unusually flexible with double-jointed thumbs; after puberty, testicles may be enlarged, a condition known as macroorchidism.
- Flat feet.
- Low muscle tone.

Fragile X is the most common single gene cause of autism and the most common inherited cause of intellectual disability.

What Is the Genetic Cause of Fragile X Syndrome (FXS)?

Changes in the *FMR1* gene, officially known as the "fragile X mental retardation 1" gene, cause Fragile X syndrome. Normally, *FMR1* instructs for a protein called fragile X mental retardation 1, or FRMP. This protein plays an important role in the communication between the nerve cells, in an area called the synapse. Important for a condition called synaptic plasticity, the protein aids in creating the conditions for memory and learning. The protein acts as a shuttle within cells carrying mRNA from the nucleus to the area where proteins are assembled. Many of the proteins are related to nerve action.

One area of the gene has the trinucleotide sequence CGG on the X chromosome. If the repeats are between 29 and 31, the person will be normal for this gene and not affected by the syndrome. The full mutations may have more than 200 CGG repeats and the person will be affected with FXS. The expanded CGG repeat turns off *FMR1*, disrupting the creation of the fragile X mental retardation 1 protein. This disruption leads to the signs and symptoms of Fragile X. People with 55 to 200 repeats may have a permutation and be intellectually normal but have mild versions of the physical and behavioral attributes such as prominent ears or hyperactivity. About 20% of women with a permutation stop their menstrual cycles before the age of 40. One other situation may occur with permutation; the person may develop fragile X–associated tremor/ataxia (FXTAS), resulting in a progressive condition movement, tremor, memory loss, and mental and behavioral changes.

FRMX1 is inherited in an X-linked dominant pattern and is located on the long arm (q) of the X chromosome at position 27.3.

What Is the Treatment for Fragile X Syndrome (FXS)?

Because Fragile X is a genetic condition, there is no cure, although several medications have been proposed. Currently, behavior therapy, special education, and physical therapy may help improve the quality of life for individuals with Fragile X.

See also X Chromosome: A Special Topic

Further Reading

"*FMR1*." 2010. Genetics Home Reference. National Library of Medicine (U.S.). http://ghr.nlm.nih.gov/gene/FMR1. Accessed 12/16/11.

"Fragile X Syndrome." 2010. MedlinePlus. National Library of Medicine (U.S.) http://www.nlm.nih.gov/medlineplus/fragilexsyndrome.html. Accessed 12/16/11.

"Fragile X Syndrome (FXS)." 2010. Center for Disease Control and Prevention (U.S.). http://www.cdc.gov/ncbddd/single_gene/fragilex.htm. Accessed 12/16/11.

National Fragile X Foundation. 2010. http://www.fragilex.org. Accessed 5/13/12.

Freeman-Sheldon Syndrome

Prevalence Rare; exact prevalence unknown; up to 1990, only 65 cases reported in literature

Other Names arthrogryposis distal type 2A; craniocarpotarsal dysplasia; craniocarpotarsal dystrophy; DA2A; FSS; whistling face syndrome; whistling face–windmill vane hand syndrome

In 1938, E. A. Freeman and J. H. Sheldon wrote about a condition in which the children had severe hand malformations and unusual faces. The faces appeared like masks with a small, puckering mouth that appeared like the person was constantly whistling. The patients also had a bent-over hand that looked like the shape of a windmill. Some people dubbed the disorder the whistling face–windmill hand syndrome. It is now called Freeman-Sheldon syndrome after the two doctors who studied and first wrote about it.

What Is Freeman-Sheldon Syndrome?

Freeman-Sheldon syndrome (FSS) affects the face, hands, and feet. FSS is one of a group of disorders known as distal arthrogyrposis. The word "arthrogyrposis" comes from three Greek roots: *arthro*, meaning "joint"; *gyrpos*, meaning "curved"; and *osis*, meaning "condition." Thus, the condition describes a joint in a flexed or contracted position. The word "distal" refers to the body parts that are far from the body core, such as the hands or feet. Freeman-Sheldon syndrome is the most serious of the distal arthrogyrposis.

There are three main symptoms of this disorder involving the face, hands, and feet.

- Distinctive facial appearance: The first thing that one notices is the appearance of the face. The individual has droopy strabismus (crossed or wandering eyes) and a small, puckered mouth with pursed lips that look like the person is attempting to whistle. The mouth, tongue, and jaw are abnormally small. The person eats with difficulty and can possibly choke on food. There is a prominent forehead. Eyes appear to be a bit sunken and eyelids droop; thus, vision is impaired. The philtrum, the distance between the nose and mouth, is extended. The chin has an "H"- or "Y"-shaped dimpling. The palate may be high, affecting speech and making a nasal sound.
- Hands: The hands point outward and fists can be clinched. Hand function is impaired.
- Feet: The feet are clubbed, turning inward or outward. The child has difficulty walking.

In addition, other symptoms may include:

- Scoliosis: The person may have a curved spine that appears as bowed or in an "S" shape. If left untreated, this deformity can compress the lungs, making it difficult to breathe.
- Joint deformities that restrict movement.
- Malignant hyperthermia: If the person has surgery, which may be recommended for treating the many symptoms, the muscles of the person may freeze up and temperature may spike. Certain drugs may also spark the same reaction. Malignant hyperthermia is life-threatening.
- Respiratory problems due to the constrictions of the chest.
- Walking difficulties.

Intelligence is not generally affected in most people, but about one-third have some degree of learning disabilities.

What Is the Genetic Cause of Freeman-Sheldon Syndrome?

Changes or mutations in the *MYH3* gene cause Freeman-Sheldon syndrome. Normally, the *MYH3* gene instructs for a protein called "embryonic skeletal muscle myosin heavy chan 3" and belongs to a large group of proteins known as myosins. Myosins are involved in cell movements and transport of materials in and out of cells. Myosin combines with actin to make muscle fibers and is responsible for the contraction of muscles. As indicated in the name, the protein is formed during embryonic development and is very important for the development of muscles.

Mutations in this gene disrupt many of the function of the normal development of the embryo. Fetal muscles do not contract properly causing many of the symptoms of Freeman-Sheldon syndrome.

FSS is inherited in different patterns. Some cases involve an autosomal dominant pattern and others have an autosomal recessive pattern. In other cases, the inheritance pattern is unknown. *MYH3* is located on the short arm (p) of chromosome 17 at position 13.1

What Is the Treatment for Freeman-Sheldon Syndrome?

No standard protocol of treatment exists. The disorder is so rare, and the deformities and very variable. The hand and foot disorders require conservative orthopedic and plastic surgery. Correction of the deformities of the face is important not only because of aesthetics, but also for vital life functions such as eating and breathing. The deformities of the spine are evaluated but may be difficult to correct because of the thickening of the various tissues along the spine. An overall concern is the patient's reaction to drugs and anesthesia, which may cause malignant hyperthermia and pulmonary complications. Speech and hearing treatment is also advised.

Further Reading

Freeman-Sheldon Research Group Support Page. 2012. http://www.fsrgroup.org. Accessed 5/13/12.

"Freeman Sheldon Syndrome." 2010. Genetics and Rare Diseases Information Center (GARD). National Institutes of Health. (U.S.). http://www.rarediseases.org/search/rdbdetail_abstract.html?disname=Freeman%20Sheldon%20Syndrome. Accessed 5/13/12.

Friedreich Ataxia (FRDA)

Prevalence 1 in 40,000 people; found individuals with European, Middle Eastern, or North African ancestry; rare in other ethnic groups

Other Names FA; FRDA; Friedreich spinocerebellar ataxia

In 1863 German physician Nicolaus Friedreich described five patients in Heidelberg who had a condition characterized by poor balance, loss of muscle control, and slurred speech. The disease that began in childhood affected the voluntary muscles the child got older. A 1984 study found 40 cases of Friedreich ataxia (FRDA) in both men and women in 14 French Canadian families. The cases were traced to one couple, Jean Guyon and Mathurine Robin, who arrived in the eastern part of Canada in 1634, indicating an autosomal recessive condition. Friedreich's name was given to the condition because of his concentrated research and publication on the disease.

What Is Friedreich Ataxia (FRDA)?

Friedreich ataxia is a disorder that affects the nervous and muscular systems. The child will first begin to have poor coordination, an unusual walking pattern or gait, and slurred speech. From the difficulty walking, ataxia spreads to the arms and trunk. The symptoms typically begin during the ages of 5 and 15 and progressively worsen over time. About 10 years after first symptoms, the person is usually confined to a wheelchair.

However, about 25% of the people with Friedreich ataxia may develop an atypical form of the condition between the ages of 26 and 39, called late-onset Friedreich ataxia (LOFA); after the age of 40, it is referred to as very late-onset Friedreich ataxia (VLOFA). The condition can begin as late as age 75. LOFA and VLOFA progress more slowly than the classical form.

In FRDA, the brain and spinal cord degenerate. Especially affected is the cerebellum, the part of the brain that controls balance and movement. Peripheral nerves that go to the arms and legs slowly become thinner. The following symptoms may develop as FRDA progresses:

- Loss of sensation in the extremities: This loss may progress to other parts of the body.
- Slurring of speech: This condition is known as dysarthia.
- Heart disorders: The person may experience a form of heart disease called hypertrophic cardiomyopathy, in which the heart is enlarged and the heart muscle is weakened. Chest pains and heart palpitations may occur.
- Diabetes: People with the condition have a tendency to develop this diabetes. About 20% of people with FRDA have carbohydrate intolerance, and 10% develop diabetes mellitus.
- Loss of tendon reflexes, especially in the knees and ankles.
- Scoliosis or abnormal curving of the spine.
- Vision and hearing impairment.
- Very high plantar arches.

The disorder does not appear to affect cognitive facilities or learning ability.

What Is the Genetic Cause of Friedreich Ataxia (FRDA)?

Mutation or change in the gene "frataxin" or *FXN* causes Friedreich ataxia. Normally, *FXN* instructs for making a protein called frataxin that is found in all the cells of the body. Frataxin is related to energy production and is found in abundance in the mitochondria of heart, spinal cord, liver, pancreas, and voluntary muscles. The mitochondria are the bean-shaped organs in the cells known as the powerhouses of the cell. The protein fratoxin works within the mitochrondria to collect iron and sulfur molecules, which are needed for producing energy.

A section of *FXN* is made of a series of building blocks called GAA repeats. This repeat or trinucleotide of guanine-adenine-adenine is normally 12 to 33 times. However, in Friedrich ataxia, this section of the gene can have as many as 66 to more than 1,000 times GAA repeats. The appearance of the disorder seems to be connected to the number of repeats. Those with fewer than 300 repeats tend to have the late-onset or very-late-onset types. Disruption in the *FXN* gene appears to cause the shortage of the enzyme and decrease the energy-producing cells in specific areas thus affecting their proper function. *FXN* is inherited in an autosomal recessive pattern and is located on the long arm (q) of chromosome 9 at position 21.11.

What Is the Treatment for Friedreich Ataxia (FRDA)?

Because FRDA is a genetic disorder, no cure presently exists. However, many types of interventions for the symptoms are available. Surgery may help alleviate the conditions of the spine and heart. A metal rod may be inserted to slow the progression of scoliosis. Also, assistive devices such as a walker, cane, or wheelchair may assist mobility. Some experimental procedures are being tried. In Canada a drug called idebenone has been approved, but the U.S. Food and Drug Administration (FDA) has not approved this drug for the United States and has pulled all trials of the drug.

Further Reading

"Friedreich Ataxia." 2010. *GeneReviews*. National Center for Biotechnology Information (U.S.). http://www.ncbi.nlm.nih.gov/books/NBK1281. Accessed 5/13/12.

"NINDS Friedreich's Ataxia Information Page." 2010. National Institute of Neurological Disorders and Stroke (U.S.) http://www.ninds.nih.gov/disorders/friedreichs_ataxia/friedreichs_ataxia.htm. Accessed 5/13/12.

Fryns Syndrome

Prevalence Not exactly known; present in large cohorts of individuals with congenital diaphragmatic hernia (CDH)—about 4% to 10% of persons with CDH

Other Names diaphragmatic hernia, abnormal face, and distal limb anomalies

Do not confuse with Lujan-Fryns, caused by a gene located on the X chromosome with symptoms similar to Marfan syndrome.

In 1979, Jean-Pierre Fryns described two stillborn siblings born with coarse facial features, cleft palates, hole in the diaphragm, and a clouded cornea. During the mother's pregnancy, the doctor noted that there was an unusual amount of amniotic fluid, a condition known as polyhydamnionos. Later in 1989, the syndrome was found in seven cases per 100,000 live births in a French population. The complex syndrome was named for Dr. Fryns.

What Is Fryns Syndrome?

Fryns syndrome affects many areas of the body and is diagnosed by excluding other conditions that are similar. The features overlap with the other conditions, and a diagnosis of Fryns is exceedingly difficult to make. It was first thought that children with Fryns syndrome died at birth, but later reports showed that some live a longer time.

The first signal of Fryns syndrome is the defect in the diaphragm. The diaphragm is a large, flat muscle that separates the abdominal area from the chest cavity. When the diaphragm descends, the person brings air into the lungs or inhales; when it relaxes, the individual exhales. In congenital diaphragmatic hernia (CDH), a hole develops in the diaphragm before birth. This hole or opening allows the stomach and intestines to move into the chest area crowding the heart and lungs. The infant cannot breathe properly, and a life-threatening condition occurs.

Several other symptoms may be present:

- Face and head: Some unusual coarse facial features include widely spaced and very small eyes, a broad and flat nasal bridge, and a long philtrum (the space between the upper lip and nose). The cornea of the eyes may be clouded. The upper part of the mouth may have an opening called a cleft palate, and lips may have split, called a cleft lip. Ears may be low set.
- Body: The person may have short and stubby fingers with small or no fingernails. In addition to lung and breathing disorders, the person may have problems with the heart, gastrointestinal system, kidneys, and genitals.
- Severe developmental delay and mental retardation for those who live into childhood.

What Is the Genetic Cause of Fryns Syndrome?

Actually, the specific genetic cause is unknown. It is believed to be inherited in an autosomal recessive pattern. It is the most common autosomal recessive syndrome associated with CDH and is diagnosed after excluding the following genetic disorders that have specific symptoms:

- Simpson-Golabi-Behmel syndrome, a X-linked disorder
- Cornelia de Lange syndrome
- Donnai-Barrow syndrome
- Matthew-Wood Syndrome
- Several other chromosomal abnormalities, such as isochromosome 12p or monosomy 15q26.

What Is the Treatment for Fryns Syndrome?

Treating the physical manifestation of Fryns syndrome by surgery or other supportive measures is important. Treatment of the hernia of the diaphragm is most important. The other anomalies may dictate consultations by a pediatric neurologist, cardiologist, and a plastic surgeon team to correct certain deformities.

Further Reading

"Fryns Syndrome." 2010. Genetics Home Reference. National Library of Medicine (U.S.). http://ghr.nlm.nih.gov/condition/fryns-syndrome. Accessed 5/18/12.

"Fryns Syndrome." 2010. National Center for Biotechnology Information (U.S.) http://www.ncbi.nlm.nih.gov/books/NBK1459. Accessed 5/18/12.

Fucosidosis

Prevalence Rare; 100 cases reported worldwide; seen mostly in Italy, Cuba, and the southwestern United States

Other Names alpha-fucosidase deficiency; fucosidase deficiency; fucosidase deficiency disease

In 1865, Archibald Garrod coined the term "inborn error of metabolism." He noted that a family of children had black urine and surmised that the problem was in the way the body broke down certain substances. Since that time, many conditions have been identified which are inborn errors of metabolism. One such error of metabolism occurs when the body cannot break down certain sugars, fat, and proteins, allowing the broken molecules to build up in the body and damage certain parts. The condition fucosidosis is one of these rare metabolic disorders.

What Is Fucosidosis?

Fucosidosis is a condition that affects many body parts especially the brain. This condition is one of nine lysosomal storage diseases. In fucosidosis, an enzyme called alpha-fucosidase is missing, allowing dangerous molecules to build up in the cells and cause the following symptoms:

- Intellectual disability that worsens with age and may lead to dementia in later life
- Coarse facial features
- Abnormal bone development and impaired growth
- Delayed motor skills; as they grow older, they may lose the skills they have acquired
- Seizures
- Clusters of small blood-red growths on the skin called angiokeratomas
- Recurring respiratory infections
- Very large abdominal organs

Severe cases may appear in early infancy and the child lives for only a few years; Milder conditions may begin around ages one to two, and the individual may survive into adulthood. At one time, physicians thought there were two types of the disorder. Now most researchers believe that it is a single disorder with a range of symptoms.

What Is the Genetic Cause of Fucosidosis?

Changes in the "fucosidase, alpha-L1 tissue" gene, or *FUCA1* gene, cause fucosidosis. Normally, *FUCA1* instructs for making an enzyme alpha-L-fucosidase. This enzyme is found in the lysosomes, the small organelles in the cells that are responsible for digesting and recycling materials. Within the lysosome this enzyme plays a role in breaking down three large molecules: long sugar chains called oligosaccharisde, fats called glycolipids, and certain proteins called glycoproteins. At the end of the breakdown process, alpha-L-fucodase cleaves off a sugar molecule called fucose.

When any one of the 26 mutations occur in the *FUCA1* gene, fucosidosis can occur. The abnormal gene causes the enzyme to be abnormally short and nonfunctional. Without the enzyme, fats and sugars are not broken down, allowing the partially broken compounds to accumulate in the lysosome and cause the cell to malfunction and eventually die. When brain cells are lost, the symptoms of dementia and intellectual disability occur. The errant cells can also accumulate in other organs such as the liver, heart, and kidney.

FUCA1 is inherited in an autosomal recessive pattern and is located on the short arm (p) of chromosome 1 at position 34.

What Is the Treatment for Fucosidosis?

If a physician suspects the disorder, he or she will order a special urine test to detect the presence of the partially broken down oligosaccharides. No cure or reversal of the condition exists at present. Treatment of the symptoms such as controlling seizures may help the patient have a longer life. Some experimental trials have included bone marrow transplants, but the results are not conclusive.

Further Reading

"Fucosidosis." 2010. Genetics Home Reference. National Library of Medicine (U.S.). http://ghr.nlm.nih.gov/condition/fucosidosis. Accessed 5/18/12.

"Fucosidosis." 2010. The Medical Biochemistry Page. http://themedicalbiochemistrypage.org/fucosidosis.html. Accessed 5/18/12.

Kugler, M. 2008. "Fucosidosis." About.com Rare Diseases. http://rarediseases.about.com/od/rarediseasesf/a/fucosidosis.htm. Accessed 5/18/12.

Fumerase Deficiency

Prevalence Rare; about 100 affected people mostly located in a religious community in the southwestern United States

Other Names fumarase deficiency; fumaric aciduria; fumerate hydratase deficiency

In June 2007, a story broke in the *Digital Journal* of a religious sect in West Texas that had most of the world's fumerase deficiency cases. The sect broke away from the church of the Church of Jesus Christ of Latter-day Saints over 70 years ago when the main church banned polygamy or multiple marriages. The group settled in enclaves in Texas, Arizona, and Colorado.

Eighty-five percent of the descendants are blood relative of John Barlow and Joseph Smith Jessup. It is believed that Jessup or his wife had the mutation for fumerase deficiency. In the compound the people marry young and often marry cousins or relatives. Those with the disorder have mental retardation, unusual facial features, brain malformation, and epileptic seizures. According to the report, the people believe the condition is in the water.

What Is Fumerase Deficiency?

Fumerase deficiency is a metabolic disorder that occurs when an enzyme called fumerase does not work properly. The condition affects the nervous system, especially the brain. Following are the symptoms of fumerase deficiency:

- Very small head, a condition known as microcephaly
- Very low IQ, around 25
- Weak muscle tone, a condition called hypotonia
- Failure to thrive and grow at expected rate
- Abnormal brain structure
- Unusual facial feature with a prominent forehead, low-set ears, small jaw, wide-spaced eyes, and a sunken nasal bridge
- An enlarged liver and spleen
- Excess of red blood cells, called polycythemia
- Deficiency of white blood cells, making them open to infections

Most of the children die in infancy, but a few have lived into early adulthood. The condition is sometimes dubbed "polygamists' down syndrome."

What Is the Genetic Cause of Fumerase Deficiency?

A mutation in the fumerase hydratase gene (*FH*) causes fumerase deficiency. Normally the *FH* gene is an important player in creating the enzyme fumerase that acts in the citric acid or Krebs cycle. This cycle describes how cells used oxygen to produce energy in the mitochondria. In the cycle, fumerase, an enzyme, converts a molecule called fumerate to malate.

Changes in the *FH* gene interfere with the enzyme's ability to convert fumerate to malate and thus disrupt the citric acid cycle. The deficiency in the process that allows energy to be produced in the cells is very harmful, especially to the cells in the developing brain. This disruption causes the signs and symptoms of fumerase deficiency.

The *FH* gene is inherited in an autosomal recessive pattern and is located on the long arm (q) of Chromosome 1 at position 42.1.

What Is the Treatment for Fumerase Deficiency?

Prevention here is probably worth a pound of cure. Tackling such beliefs as the cause being in the water is very difficult, especially among groups that live in isolated communities.

Further Reading

"FH." 2010. Genetics Home Reference. National Library of Medicine (U.S.). http://ghr.nlm.nih.gov/gene/FH. Accessed 5/18/12.

"Fumarase Deficiency." 2010. Genetics Home Reference. National Library of Medicine (U.S.). http://ghr.nlm.nih.gov/condition/fumarase-deficiency. Accessed 5/18/12.

Mullins, K. J. 2007. "Mormon Sect's Polygamy Cause Most Called of Fumerase Deficiency." In *Religion*. http://www.digitaljournal.com/article/195535. Accessed 5/18/12.

G

Galactosemia

Prevalence Classic galactosemia, about 1 per 30,000 to 60,000 births; common in the Irish Traveler population because of inbreeding; types II and III, less than 1 in 100,000 newborns

Other Names classic galactosemia; epimerase deficiency galactosemia; Galactose-1-phsopahte uridyl transferase deficiency; galactokinase deficiency; galactose-6-phosphate epimerase deficiency; GALT deficiency

The Irish Travelers are groups of nomadic people living in England, Ireland, and the United States. Sometimes they are referred to as "gypsies" or "tinkers." They have their own language, distinct customs, and unique health problems. For example, over half of the travelers do not live past age 39; women especially have a high mortality rate. They also have a genetic disorder common in the population. That genetic disorder has resulted from marriages within and from the possible descendants of an original Irish carrier.

When Archibald Garrod first used the term "inborn error of metabolism," interest in other possible conditions developed. In 1917, a German physician, F. Goppert, discovered a disease present at birth as a defect in the inborn mechanism of metabolism. In 1956, Herman Kalckar traced the condition to the deficiency of the enzyme that blocked the breakdown of the sugar galactose. Although several terms have been used to name the disorder, the most popular is galactosemia.

What Is Galactosemia?

Galactosemia is a condition in which the body is unable to use the sugar galactose. Along with lactose, galactose is one of the sugars present in milk, cheese, butter, and many other foods in small amounts. The word "galactose" comes from two Greek words: *gala*, meaning "milk," and *heme*, meaning "blood." Thus, the person with galactosemia has failed to break down the sugar galactose into glucose and has an abundance of galactose remaining in the bloodstream. In these people, the enzyme used to metabolize galactose in glucose, a simple sugar that the body can use for energy, is missing or diminished, and the person builds up toxic levels of galactose that can damage body organs.

Three types of galactosemia exist. These types are caused by mutations in three specific genes. The types are classical galactosemia, type II known as galactokinase deficiency, and type III known as galactose epimerase deficiency. All three types affect the enzyme involved in breaking down galactose.

What Are the Types and Genetic Causes of Galactosemia?

Classic galactosemia is the most serious and life-threatening type. The complications appear within a few days after birth, when the infant first gets milk or breast milk. About 75% of infants with the disorder die. Affected individuals may have the following symptoms due the accumulation of galactose:

- Serious feeding difficulties
- Lethargy and lack of energy
- Failure to thrive
- Yellowing of eyes due to liver damage
- Kidney failure
- Cataracts
- Overwhelming bacterial infections, especially by *E. coli.*
- Speech difficulties
- Brain damage

Diagnosis of galactosemia is usually made during the first week of life as part of standard newborn screening. Exclusion of any product with lactose such as milk, cheese, or butter is absolutely essential. Even though the person may be on the severely restricted diet, many of the above complications still occur. A high incidence of speech and learning disabilities exists among those with classic galactosemia.

Mutations in the *GALT* gene, officially known as the "galactose-1-phosphate uridylytransferase" gene, cause classic galactosemia. Normally, *GALT* instructs for making enzymes to help the body process the simple sugar galactose. The enzyme converts a modified form of galactose called galactose-1-phosphate into another simple sugar, glucose, which is the main energy source for cells. Most of the mutated forms of *GALT* completely eliminate the action of the enzyme, and galactose builds up in the body causing the dangerous symptoms of the disorder.

Another *GALT* mutation=s causes a milder form of the disorder called Duarte galactosemia. Many of the complications of the classic disease are not present here. In Duarte galactosemia, the child inherits a classic gene (G) from one parent and the Duarte variant gene (D) from the other. This genetic makeup is referred to as D/G galactosemia. The same diagnostic test for newborns is used to determine this disorder, and individual may be placed on the restricted diet for a trial period. The *GALT* gene is inherited in an autosomal recessive pattern and is located on the short arm (p) of chromosome 9 at position 13.

Type II galactosemia causes fewer medical problems than the classic type. The individual may develop cataracts but generally has fewer medical complications. Mutations in the *GALK1* or "galactokinase 1" gene cause this type of galactosemia. Normally, *GALK1* instructs for an enzyme, galactokinase 1, a path in the process of converting galactose into the simple molecules that the body can use. The mutations in *GALK1* change a single protein building block or delete a portion of the gene disrupting the making of the enzyme. Thus, galactose and another sugar called galactitol build ups and damage the lens of the eyes causing cataracts. The presence of cataracts is the primary symptom of type II galactosemia. *GALK1* is inherited in an autosomal recessive pattern and is located on the long arm (q) of chromosome 17 at position 24.

Type III galactosemia has many of the signs and symptoms of the classic form but varies from mild to severe. The conditions include cataracts, liver disease, kidney disease, delayed growth, and intellectual disability. Mutations in the *GALE* or "UDP-galactose-4-epimerase" gene cause type III galactosemia. Normally, *GALE* instructs for making the enzyme UPD-galactose-4-epimerase that helps process galactose into glucose. Most of the more than 20 mutations involve a change in one of the protein building blocks. Disruption of the enzyme makes it unstable and unable to properly process galactose, causing the damaging symptoms of galactosemia. *GALE* is inherited in an autosomal recessive pattern and is located on the short arm (p) of chromosome 1 at position 36-p35.

What Is the Treatment for Galactosemia?

The treatment for classic galactosemia is to eliminate foods with lactose and galactose from the diet. Even then, the person may develop some of the symptoms, especially learning disabilities.

Galactosemia must not be confused with lactose intolerance. These individuals have an acquired or inherited shortage of the enzyme lactase and experience abdominal pains after ingesting such dairy products.

Further Reading

"Galactosemia." 2010. MedlinePlus. National Library of Medicine (U.S.). http://www.nlm.nih.gov/medlineplus/ency/article/000366.htm. Accessed 5/18/12.

"Galactosemia." 2010. National Center for Biotechnology Information (U.S.). http://www.ncbi.nlm.nih.gov/books/NBK1518. Accessed 5/18/12.

Galactosemia Foundation. http://www.galactosemia.org. Accessed 5/18/12.

Galactosialidosis

Prevalence Unknown; more than 100 cases reported, with 60% of those being the adult form; more common in people of Japanese descent

Other Names deficiency of cathepsin A; Goldberg syndrome; lysosomal protective protein deficiency; neuraminidase deficiency with beta-galactosidase deficiency; PPCA deficiency

Some conditions appear to be concentrated in people of a certain area. This rare but complex condition is one of these. Galactosialidosis appears to be mostly in the Japanese population and has been studied by Japanese researchers as well as Western physicians because of its genetic causes, which are part of a large family of genes called the cathepsins.

What Is Galactosialidosis?

This disorder disturbs the function of the lysosomes and affects many body systems. It has three forms or subtypes. The following forms manifest themselves according to age:

- Early infantile form: This subtype is the most serious and is related to extensive accumulation of fluid before birth, a condition called fetal hydrops (hydrops fetalis). The early infantile form is associated with several systemic problems: an enlarged liver and spleen, a hernia in the lower abdomen called an inguinal hernia, skeletal weakness, large intestinal organs, kidney failure, general overall swelling, and early death. The facial features are described as "coarse," and cherry-red spots are characteristic. Diagnosis is made between birth and three months.
- Late infantile form: The late type has some of the same symptoms as the early infantile type but appears less severe and begins later in infancy. Developing the symptoms within the first year of life, these children may have short stature, liver and heart problems, and the same coarse facial features. Life expectancy varies according to the severity of the symptoms.
- Juvenile/adult form: The symptoms of this form of galactosialidosis are different from the first two types. Around the age of 16, the individual may begin to have coordination problems, muscle twitches, and seizures. They may have the typical dark red spots on the skin and abnormalities in the bones of spine. They also develop coarse facial appearances. However, in spite of the problems, the individuals have a normal life expectancy.

What Is the Genetic Cause of Galactosialidosis?

Changes in the *CTSA* gene, officially known as the "Cathepsin A" gene, cause galactosialidosis. Normally, *CTSA* instructs for making the protein cathepsin A, an enzyme that is active in the lysosomes. The lysosomes are small organelles in the cells that act as garbage disposals, breaking down old parts of cells and recycling others. Cathepsin A combines with two other proteins—neuraminidase 1 and beta-galactosidase—to form a large protein complex that breaks down sugar molecules (oligosaccharides) and fats (glycolipids). Also, on the surface of the cell, cathepsin A forms a large complex with neuraminidase 1 to form elastin. Elastin is a protein that forms the connective tissues of the body.

Galactosialidosis is part of a large family of disorders related to the lysosomal storage. When the enzymes do not function properly, the function of the lysosome breaks down and does not properly dispose of the discarded cell parts. Unwanted substances then build up in the lysosomes, causing the symptoms of galactosialidosis. *CTSA* is inherited in an autosomal recessive pattern and is found on the long arm (q) of chromosome 20 at position 13.1

What Is the Treatment for Galactosialidosis?

Because this condition is genetic, there is no prescribed treatment at this time. Treating the symptoms of seizures or other abnormalities is according to individual cases.

Further Reading

"*CTSA*." 2010. Genetics Home Reference. National Library of Medicine (U.S.). http://ghr.nlm.nih.gov/gene/CTSA. Accessed 5/18/12.

D'Azzo, A., et al. 2001. "Galactosialidosis." In *The Metabolic and Molecular Bases of Inherited Disease*, chap. 152. http://books.mcgraw-hill.com/getommbid.php?isbn=0071459960&template=ommbid&c=152. Accessed 5/18/12.

"Galactosialidosis." 2010. Genetics Home Reference. National Library of Medicine (U.S.). http://ghr.nlm.nih.gov/condition/galactosialidosis. Accessed 5/18/12.

Gastroschisis

Prevalence About 1 in 2,000 births in the United States

Other Names paraomphacele

Some defects occur during embryonic development and are evident at birth. The word itself gives a clue to the nature of this condition. The word "gastroschisis" comes from two Greek roots: *gastro*, meaning "stomach" or "intestines," and *schsis*, meaning "cut or split." The latter word has a similar meaning to the familiar object "scissors." Thus, the condition gastroschisis involves the intestine's cutting through the abdominal wall.

What Is Gastroschisis?

Gastroschisis is among the most commonly occurring physical abnormalities in which the body contents protrude through the abdominal wall. The term "herniation" is used to describe a condition when an organ protrudes through an area that normally contains it. The defect is usually about four centimeters wide and is located to the right of the junction of the umbilicus (belly button) and normal skin. With the gastroschisis, no sac encloses the organs.

A similar birth defect, called an omphalocele, involves the umbilical cord. In this condition, the organs remains enclosed in a sac, and the defect is much larger.

What Is the Cause of Gastroschisis?

In order to understand what happens in this condition, it is essential to understand what happens as the embryo develops. Most of the genetic mutations show up in early formative days. In addition to genetic disorders, a group of environmental conditions may occur that affect the development of the embryo and fetus.

The term "embryopathies" refer to congenital abnormalities that occur during the embryonic period. The development of the offspring can be thought of in several stages:

- Weeks 1–2 or the first 14 days: The first 14 days after fertilization, an insult or injury generally does not result in an abnormality. The problem generally repairs itself or the fertilized egg dies, a condition known as spontaneous abortion.
- Third week to eighth week: The embryo is very vulnerable. During these weeks, numerous divisions occur, and organs begin to form. The embryo grows at a relatively enormous rate. The term "noxae" is used to describe anything that is harmful to health. The word has the same root and meaning as the word noxious, meaning harmful or not healthful.
- Fetal period, beginning at nine weeks, to birth. The term "fetopathy" is used to describe abnormalities that happen during this period. The sensitivity to injury now is greatly reduced. Most of the organs are formed. The one exception is the cerebral cortex, the forward and thinking part of the brain. This organ continues to be susceptible to noxae between the 8th and 15th weeks of development.

The pre-embryonic period entails the first two weeks of human development beginning at fertilization. At this point the fertilized egg is called a zygote, which then undergoes division to become a blastocyst. Some of these cells undergo division to form three layers that become the cells of the embryo. These cells are called primary germ layers or stem cells and will become the following body tissues:

- Ectoderm: The outer layer that gives rise to the outer epithelium such as hair, nails, and skin, the sense organs, the brain and spinal cord. The epithelium forms the epidermis and other structures.
- Mesoderm: The middle layer that gives rise to the bones, muscle, connective tissue, circulatory system, and most of the excretory and reproductive systems.
- Endoderm: The inner layer that gives rise to the epithelial linings, the cells that from the linings of the body cavity such as the digestive tract, most of the respiratory tract, bladder, liver, pancreas, and some of the endocrine glands.

In gastroschisis and omphalocele, most of the focus is on the mesodermal phase of development. Several hypotheses have evolved for the cause of the disorder:

- Mesoderm failure: The mesoderm is the middle layer that forms several systems. For some reason the processes necessary for proper development are not at play. For example, cell death or apoptosis may occur.
- Rupture of the amnion around the umbilical ring: The embryo becomes surrounded by the amnion, which is a bag of protective fluid. This sac may break and surround the umbilical ring. The umbilicus is the area where the umbilical cord attaches from the mother's placenta to the developing baby.
- Problems within the umbilical vein: In the umbilical cord, veins and arteries are the connection for nutrients and oxygen getting to the embryo and for waste products to be removed. If this vein is weakened, the body wall may weaken and the intestine will protrude.
- Abnormal folding of the body wall: During the fourth week of development, the lateral body folds in a specific way to become the body wall. If the fusion is not complete, the bowel may herniate and protrude to the right of the umbilicus.

Generally, it is accepted that the failure to close the umbilical ring during the critical mesodermal period causes the incomplete closure and herniation.

What Is the Genetic Cause of Gastroschisis?

If the gastroschisis occurs alone without other hernias or abnormalities, researchers think it may be inherited in an autosomal recessive pattern. Also, it may be traced to a sporadic mutation.

Epidemiological data over the last 40 years suggest that the condition is associated with advanced maternal age. It is also associated with trisomy 13, 18, and 21 (accounting for 25% to 50% of cases) and Beckwith-Wiederman syndrome. More recent research has shown that gastroschisis has increased in recent years with young maternal age and low gravidity. Infection, smoking, drug abuse, or anything that contributes to low birth weight may increase risk. One study indicates that a change in paternity or childbearing with different fathers may cause the mother's immune system to play a role in the gastroschisis.

What Is the Treatment for Gastroschisis?

Surgery is essential, and techniques have improved over the years. In 1960, the survival rate was only 60%; today, more than 90% survive. Prenatal tests can reveal gastroschisis before birth and amniocentesis can discover if there are chromosomal disorders.

Further Reading

Gastroschisis Support Center. http://www.gastroschisis.co.uk. Accessed 5/18/12.

Sorrels, W. 2008. "Pediatrics: Gastroschisis Complications in Later Life." AllExperts. http://en.allexperts.com/q/Pediatrics-1429/2008/7/Gastroschisis-complications-later-life.htm. Accessed 5/18/12.

Gaucher Disease

Prevalence Occurs in 1 in 50,000 to 100,000 in the general population; Type I frequent in the Ashkenazi Jewish population, affecting about 1 in 500 to 1,000 people; other forms not frequent in the Jewish population

Other Names Cerbroside lipidosis syndrome; Gaucher's disease; Gaucher splenomegaly; Gaucher syndrome; GD; glucocerbrosidosis; glucocerebrosidase deficiency; glucosylceramidase deficiency; glucosylceramide beta-glucosidase deficiency; glucosylceramide lipidosis; glucosyl cerebroside lipidosis; kerasin lipoidosis; kerasin thesaurismosis; lipoid histiocytosis

Because of their isolation and intermarriage, many disorders are prevalent in the Ashkenazi Jewish population. It is believed that when the Jews spread from the area of Israel to eastern Europe and other areas, the gene was present in the founding group. Years of intermarriage presented the recessive gene as endemic in the population. In 1882, Phillippe Gaucher, a French physician, first described the disease that was so prevalent in the Jewish population and lent his name to the disorder. However, it was not until 1965 that the biochemical basis for the disorder was found.

What Is Gaucher Disease?

Gaucher disease is a metabolic disorder that occurs when a fatty substance called glucocerebroside accumulates in the organs, causing serious damage to many body parts. Gaucher is one of several diseases that affect lipid or fat storage. The following three types occur:

- Type 1: Type 1 Gaucher disease is the most common and may begin early in life or in adulthood. Also the symptoms may range from mild to severe and include enlargement of the liver or spleen, a low number of red blood cells, skeletal disorders, and sometimes lung and kidney impairment. People with this type may bruise easily and experience fatigue because of anemia. Type one does not affect the brain and spinal cord. This type is prevalence in people of Ashkenazi Jewish descent.

- Type 2: Type 2 Gaucher is a very serious form. Evidence of liver and spleen enlargement is evident at about three months. This type is known as neuropathic because the brain and central nervous system are affected. This brain damage is progressive and the child usually dies before the age of two.
- Type 3: Type 3 Gaucher disease has mild to severe enlargement of liver and spleen depending on the individual. Brain damage may occur in the form of seizures and become progressively worse over the years. Type 3 also has eye movement disorders, skeletal deficits, and blood disorders.
- Perinatal lethal disease: This most severe type of disorder starts before birth with extensive swelling, distinctive facial features, scaly skin, and serious neurological problems. These children live only a few days after birth.
- Cardiovascular Gaucher disease: This type is rare and primarily affects the heart valves, causing them to calcify.

All of the disorders have the same thing in common: a deficiency of the enzyme that breaks down or recycles glucocerebroside.

What Is the Genetic Cause of Gaucher Disease?

Changes in the *GBA* gene, officially known as the "glucosidase, beta, acid" gene, cause Gaucher disease. Normally, *GBA* instructs for the making of the enzyme beta-glucocerebrosidase, an enzyme active in the lysosomes. The lysosomes are small organelles within the cell that act as the garbage disposal or housekeeper of the cell. In the lysosome, enzymes act to break down old cell parts or package them for recycling. Beta-glucocerebrosidase is a housekeeping enzyme that breaks down the huge molecule called glucocerebroside into glucose or sugar and a simpler fat molecule called ceramide.

More than 200 mutations in *GBA* are related to Gaucher disease. A change in a single building block or amino acid disrupts the structure of the enzyme beta-glucocerbrosidase and keeps it from working normally. Thus, the molecule glucocerebroside and other substances build up in the white blood cells in the spleen, liver, bone marrow, and other organs. The accumulation of these substances shuts down the cells in the organs and tissues down and keeps them from working, causing the symptoms of Gaucher disease.

GBA is also associated with Parkinson disease and a brain disorder called dementia with Lewy bodies. *GBA* is inherited in a recessive pattern and is located on the long arm (q) of chromosome 1 at position 21.

What Is the Treatment for Gaucher Disease?

Many genetic disorders do not have treatments; however, because of its prevalence in the Jewish population, several studies have been successful in finding a treatment for types 1 and 3. In 1995, the U.S. Food and Drug Administration approved an effective treatment called Ceredase. Later, in 2001, an improved drug, Cerezyme, was found to be more effective. This drug is a highly effective enzyme replacement that reduces the size of the liver and spleen and reverses other symptoms of the disorder. However, no treatment exists for the brain damage in types 2 and 3.

Further Reading

"Gaucher Disease." 2010. Genetics Home Reference. National Library of Medicine (U.S.). http://ghr.nlm.nih.gov/condition/gaucher-disease. Accessed 5/18/12.

National Gaucher Foundation. 2010. http://www.gaucherdisease.org. Accessed 5/18/12.

"NINDS Gaucher Disease Information Page." 2010. National Institute of Neurological Diseases and Stroke (U.S.). http://www.ninds.nih.gov/disorders/gauchers/gauchers.htm. Accessed 5/18/12.

Gene Therapy: A Special Topic

What Is Gene Therapy?

Gene therapy is a set of approaches designed to correct defective genes responsible for disease development. The technique is based on the transfer of a "normal" gene into an individual's cells and tissues to treat an "abnormal" hereditary disease-causing gene. Several questions are considered before undertaking gene therapy:

- Does the condition result from mutations in one of more genes?
- Do researchers know where the gene is located?
- Can copies of the gene be made in the laboratory?
- What is known about the disorder?
- What tissues does the disorder affect?
- Is there a protein known to be related to the disorder?
- Will adding a normal copy of the gene fix the problem?
- Do scientists have the ability to effectively deliver functioning genes into cells where the gene defect exists?

Gene therapy is actually a sophisticated extension of conventional medicine. Rather than treat a patient with drugs or surgery, the patient receives DNA. Several approaches may be used for correcting the defect:

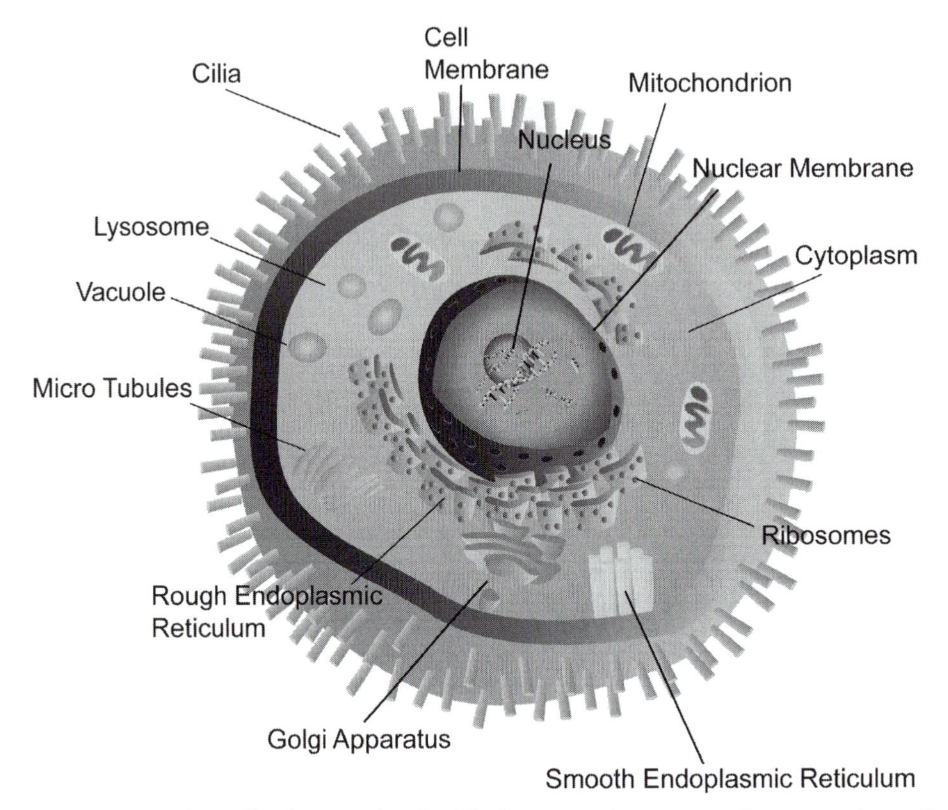

Composition of a cell. This graphic highlights some important elements of a cell, including the nucleus, cytoplasm, mitochondrion, lysosome, and ribosomes. (Mark Rasmussen/Dreamstime.com)

- Gene insertion: A normal gene is inserted into a nonspecific location within the genome to replace a nonfunctional gene. This approach is the most common.
- Gene modification: A normal gene may be swapped for an abnormal gene using recombination.
- Gene surgery: An abnormal gene may be repaired by reverse mutation or by changing the defect to return the gene to its normal function.
- Gene regulation: The degree to which the gene is turned off or on is altered. For example, certain genes may regulate the process of producing proteins; targets to change the regulation of these genes are a possibility.

Types of Gene Therapy

Gene therapy seeks to supplant genes that are not doing their job. Theoretically, it is possible to transfer at two levels: somatic and germ-line cells:

1. Somatic gene therapy: This type introduces therapeutic genes at the tissue or cellular level to treat a specific disorder. Only the person receiving the treatment is affected. This procedure can be done in two ways. First, there is *ex vivo* (or *in vitro*) therapy, in which cells are modified outside the body for later transplantation back into the body. A second strategy involves *in vivo* therapy, in which cells are changed directly in the body.
2. Germ-line therapy: The germ cells are the egg and sperm. In this form of therapy, genes are inserted into reproductive cells or possibly into embryos to treat diseases that could be passed on to future generations. This form is the most controversial. Some people fear that germ-line therapy could be used to control human development in ways not connected with disease (e.g., to control the development of intelligence or appearance).

Scientific Developments Essential for Gene Therapy: The Human Genome and Research Involving Human Proteins

The Human Genome Project has identified the layout of genes on chromosomes, but how these genes work normally and abnormally are found in proteins. Control of the functions of the human body is not simple, and lack of scientific data limits efforts at gene therapy. Some of the pathways are very complex and difficult to determine. Although the idea of gene therapy is quite simple, obstacles and challenges have proven to be quite demanding.

Making Copies of the Gene

For gene therapy to take place, copies of the normal gene are made. This process is called cloning. Then the copies must be multiplied. Kary Mullis, a biochemist working in a California biotechnology company, invented an ingenious way of making multitudes of DNA copies in a process called polymerase chair reaction or PCR. PCR enables repetitive DNA replications over a limited region in the DNA where the known gene is located.

The PCR machine looks like a simple unimpressive box with lots of buttons and knobs. PCR is valuable to researchers because it enables them to replicate unique DNA sequences in a short period of time. This first step of making copies of the gene is very important in gene therapy.

Getting the Genes into Cells

How do the genes get into the cells? Scientists must use molecular delivery trucks called vectors. Many kinds of vectors have been proposed, all with the purpose of getting the genes to cells. However, that is where the similarity ends. Following are some of the vectors used or proposed for use in gene therapy:

- Viruses as vectors: Viruses are sneaky little microbes that are associated with the great scourges of history such as smallpox and influenza, but actually they are really simple little organisms. They have only a few genes and are usually

Work done by scientists like American biochemist Edward L. Tatum, who won a share of the 1958 Nobel Prize for medicine or physiology for his study of genetics, led to advances in gene therapy. (Hulton Archive/Getty Images)

only single-stranded ribonucleic acid (RNA). Viruses cannot replicate without other cells, and they find a way to get into the cells for replication. Scientists must find a way to keep the viruses from replicating its own genome but get into the cell of the other animal efficiently. Using viruses as vectors, the viral gene is taken out and the normal gene is inserted into the cell. Following are the common vectors under investigation:

- Retroviruses: Genetic material in retroviruses is in the form of RNA molecules, which insert their genetic material into a host that they invade, in the form of DNA. The host cell is the cell that is present in an organism. The retrovirus cannot live on its own. So when it moves into a host cell, it takes the RNA, combines it with enzymes in the cells, and literally takes over the cell. Retroviruses can create double-stranded DNA copies of their RNA in a procedure called reverse transcriptase. They can make RNA into DNA that is integrated into the host cell. Scientists easily cloned these viruses, removing critical retroviral genes so the virus cannot produce after it delivers its cargo. Retroviruses are used primarily *ex vivo*—outside the body.
- Adenoviruses: These viruses have a double-stranded DNA and cause respiratory infections and eye disorder in humans. They enter a cell efficiently and have high levels of replication and expression. Researchers remove large amounts of DNA of this virus to keep the immune system from rejecting it.
- Adeno-associated viruses: These viruses are small, single-stranded DNA viruses that can insert their genetic material at a specific site. They cause no known disease in humans and have long-term expression. They have the ability to target nondividing cells located in the muscle, brain, liver, and lungs and can insert their genome into the genome of the recipient. They show great promise as vectors.
- Lentiviruses: These viruses are derived from a special group of viruses, of which HIV is a member. It may become a very efficient vector. Removing just six genes from HIV can make it less virulent. So scientists must find ways to make it less dangerous. Lentiviruses have great potential in drug discovery. Two other lentiviruses are herpes viruses and poxviruses.

- Nonviruses as vectors
  - Naked DNA: This nonviral method is simple to prepare and has good safety levels in tests. The amount of naked DNA or pure DNA is unlimited but it may not be efficient compared to viruses. A newer method called a gene gun shoots DNA-coated gold particles into cells using high-pressure gas.
  - Facilitated DNA or liposomes: This technique is in its infancy. An artificial lipid ball with an aqueous core is treated so that it can carry therapeutic DNA. This method is not as efficient as viruses. However, some new molecules called lipoplexes and polyplexes have been created to protect DNA from degrading. Lipoplexes have been used to target cancer cells.
  - Human artificial chromosomes: This idea builds a chromosome from the ground up using a telomere, centrosome, and therapeutic material. A large

number of genes could be inserted. The downside would be that it is such a large molecule that it would not be efficient.

- Infectious mammalian chromosomes: This technique synthesizes viral and nonviral methods. Researchers produced a component of the Epstein-Barr virus (EBV) in the form of a large circular molecule that shows stable expression for one year. In early mouse studies, the technique was 25% more efficient that nonviral vectors.
- Starburst dendrite and new polymers: These polymers are shaped like a star and can release DNA from the ends of the points.
- Endothelial cells: Modifying endothelial cells derived from subcutaneous fat could provide a specific delivery system.

The 1990s: Decade of Hope and Despair for Gene Therapy

During the 1980s, the attitude was developing that science could conquer all diseases. Powerful antibiotics and new surgical strategies were healing terrible diseases. In 1987, Harris and Associates conducted a poll revealing that 87% of U.S. adults were willing to use gene transfer technology to cure a fatal disease. So the decade that saw the beginning of the Human Genome Project, discovery of genes that caused various disorders, and a first successful human project ended in great despair for gene therapy research.

W. French Anderson performed the first successful gene therapy on four-year-old Ashanthi DeSilva in the spring of 1991. Ashi had a condition known as adenosine deaminase deficiency (ADA). Anderson became instantly famous and known as "The Father of Gene Therapy." But Anderson did not hesitate to let people know that there was a risk for gene therapy. He feared the worst if the public expected so much. His prescient words did not lessen the impact of what happened in 1999.

Jesse Gelsinger's Story

The gene therapy community was riding high during the 1990s, and failures were few. In 1995, James Wilson and Mark Batshaw became interested in a genetic deficiency of a liver enzyme, ornithine transcarbamylase (OTC). OTC causes ammonia to build up in the blood that can lead to convulsions, vomiting, coma, and death. Batshaw convinced Wilson that this condition should be the subject of their first gene therapy trial. Rather than using a retrovirus as a vector, they decided on an adenovirus. They submitted results of experiments with animals or the pre-clinical phase and gathered data on the details and the protocol. There were granted a phase 1 gene therapy trial for adult patients with OTC deficiency.

Jesse Gelsinger was an 18-year-old Tucson, Arizona, teenager with OTC deficiency. He could control this condition with diet and 32 pills a day. The protocol that he volunteered for had no chance of providing him with any benefit. The written protocol was to test only the safety of a treatment for babies with OTC

Phases Leading to FDA Approval

Research Phase	Goals of the Trial
Preclinical research	This is the basic research phase during which the idea for treatment is tested in many trials. This is done generally with small animals, usually mice. With the animal model of the disease developed, the experiments are replicated many times, After mice, experiments may extend to larger animals such as dogs, pigs, or monkeys.[1]
Phase I	After preclinical research, applications are made to the FDA, the National Institutes of Health (NIH), and the Recombinant DNA Advisory Committee (RAC) that address trials for gene therapy. Phase I trials are considered safety trials and use only a small number—from 2 to 20—adult subjects who are fully informed about the nature of the test. The FDA carefully reviews the data in the investigational new drug (IND) application, looking for adverse effects.
Phase II	If the drug appears safe in humans, investigators recruit a large number of subjects—from 100 to 300—to continue safety studies and evaluate how well the drug works. Researchers carefully evaluate all data; the FDA closely monitors the study again looking for adverse events and whether the drug is doing what it is supposed to do. This procedure is very time-consuming and costly. Many studies are discontinued at this stage.
Phase III	In this phase investigators recruit thousands of people from a variety of population centers. A new entity emerging worldwide in research is the contract research organization (CRO), which recruits subjects to participate in trials and conducts the day-to-day administration of research. Massive amounts of data are collected before presenting a new drug application (NDA) to the FDA. If the trials are accepted by the FDA, the drug is approved for marketing.
Phase IV	After approval, the drug's performance is monitored for long-term effects in a follow-up that may take from 10 to 20 months. The FDA may pull the drug from the market if problems arise.

[1]Most gene therapy trials in the United States are currently in preclinical stage; a few have advanced to Phase I.

Source: Developed by Evelyn Kelly in *Gene Therapy* (Westport, CT: Greenwood, 2007).

deficiency. Jesse was deemed eligible and assigned to the group who received the highest dose of AD virus. He died on the second day of the trial on September 17, 1999. The FDA immediately stopped all gene therapy experiments.

Jesse's death sounded a death knell for gene therapy. It had the same effect on the gene therapy community that the *Columbia* space shuttle disaster had on NASA. Both events brought attention to the need to prevent future accidents.

In their eagerness, the researchers had ignored warnings that could have prevented the death of Gelsinger. His family sued the university and researchers in civil court. The court's suit cited the following:

- Wrongful death: The defendants were careless and reckless in their conduct and failed to properly assess the suitability of admitting Jesse to the trial.
- Survival: Defendants kept Jesse from earning money he would have made during his lifetime.
- Product liability: One of the researchers was founder and owner of the biotech company that manufactured the AD virus. The product was poorly tested.
- Lack of informed consent: Jesse had not been warned of all the risks.
- Misleading information: Jesse had not received information about how the tests would be performed.
- Fraud on the FDA: The team deliberately made false representations to the FDA.

The university settled out of court in 2001 for an undisclosed amount.

Ensuring That the Patient Will Not Be Harmed

Although the regulations are complex and confusing, the U.S. government seeks to assure that it approves only ethical and responsible research. Human gene therapy must be seriously and cautiously evaluated. The sidebar "Phases Leading to FDA Approval" reflects the rigid standards for both drug approval and gene therapy.

Gene Therapy Science Moves Forward

During the decade 2001 to 2010, gene therapy had to slowly recover. The proof of the principle had been shown in the success of Ashanthi DeSilva, who in 2001 was a healthy teenager. To give up would be denying great potential for healing many disorders. By 2006, 1,192 trials were being conducted. Most of them are in preclinical or phase 1. Gene therapy is on the cutting edge of medical revolution, and in spite of setbacks, trials are progressing.

See also Adenosine Deaminase Deficiency

Further Reading

Kelly, Evelyn. 2007. *Gene Therapy*. Westport, CT: Greenwood Press.

"Vaccines, Blood, and Biologics." 2012. Food and Drug Administration (U.S.). http://www.fda.gov/cber. Accessed 5/19/12.

"What Is Gene Therapy?" 2012. Genetics Home Reference. National Library of Medicine (U.S.). http://ghr.nlm.nih.gov/handbook/therapy/genetherapy. Accessed 1/27/12.

Genetic Counseling: A Special Topic

Beth and John have just been informed that the baby they have waited so long for may have Down syndrome. They had heard of children with "mongoloid appearances" and even knew a girl with the condition who used to yell out in church. But that was the extent of their knowledge. They had science in high school many years ago but had little understanding of genetics with its genes and chromosomes. Who was going to answer all their questions, give them resources, and tell them what to expect? Their doctor suggested that they see a genetic counselor. Their first question: What is a genetic counselor?

What Is a Genetic Counselor?

Genetic counselors are health care professionals with specialized graduate training in both genetics and counseling Only 100 years ago, people did not know about the role of chromosomes and genes in passing traits from one generation to another. With the Human Genome Project's (HGP) unraveling the secrets of genes, human genetics is playing an increasingly important role in the diagnosis, monitoring, and treatment of disease. The information available is often overwhelming. When

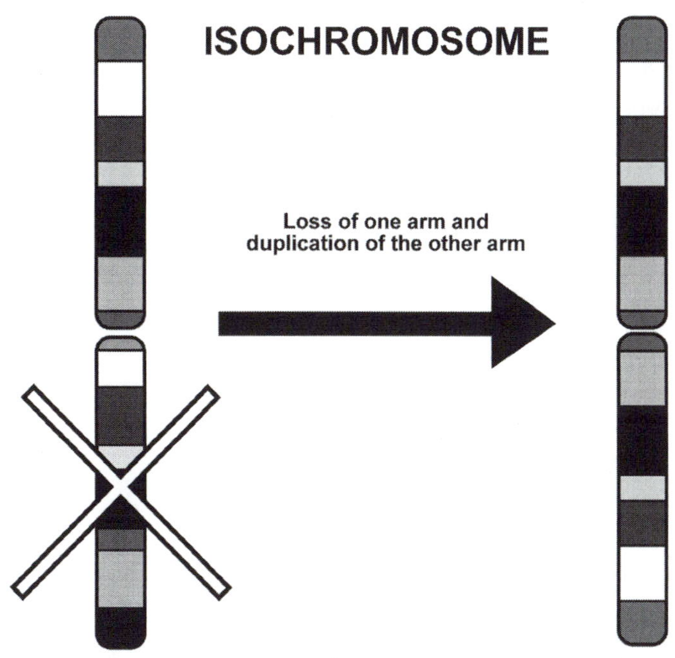

An isochromosome is an abnormal chromosome, occurring when one arm of the chromosome is deleted, and the other arm is duplicated. Such abnormalities can cause problems in human growth and development, which a genetic counselor can explain so that patients can make informed choices. (ABC-CLIO)

people have the information, they are not sure what it means, what to do with it, or how to act on it. Genetic counseling helps people cope with these dilemmas.

Genetic counselors help people to understand genetic conditions and to adapt to the medical, psychological, and family implications of the role of genetics in the disease. They interpret information about complex conditions to people with potential problems. They add a human touch to impersonal information. A physician usually recommends counseling, as part of a team of caregivers. Although it would be ideal to do this before pregnancy when indicated, more than likely it is done when a problem is suspected.

Who May Be Referred to a Genetic Counselor?

Any person may seek out genetic counseling for a condition that he or she suspects they may have inherited or may pass on to children. Many individuals with a family member with Alzheimer disease, Parkinson disease, Huntington disease, or breast cancer seek out answers for themselves or for their offspring and other relatives. Following are some of the situations in which individuals may seek genetic testing:

- Family history of chromosome abnormality such as trisomy 21 (Down syndrome) or XXX syndrome.
- Prenatal screening or diagnosis: Abnormal serum screening or ultrasound findings have indicated a problem. The client has been advised that the screen showed a high risk for chromosomal abnormalities or other disorders such as spina bifida.
- Molecular test for a single gene disorder such as cystic fibrosis.
- Increased maternal age: Women over the age of 35 have a tendency to have children with certain chromosomal disorders as Down syndrome.
- After the birth of a child with a condition.

The counselor can order tests for many genetic diseases. However, following are the most common genetic diseases:

- Down syndrome
- Cystic fibrosis
- Sickle cell disease
- Tay-Sachs disease, a fatal disease affecting the nervous system
- Spina bifida, an embryonic condition where part of the spinal column does not close

The job of the genetic counselor is to comprehend the medical facts of a case and to explain how heredity is involved. The counselor discusses available tests and facilitates arrangements. Options relating to a course of action must take into consideration risk, family goals, and ethical and religious standards. Counselors also serve as patient advocates in the maze of the medical world and also educate the public and other health professionals about genetic diseases.

What Is the Procedure for Genetic Counseling?

The procedure of counseling proceeds through several phases: intake, initial contact encounter, summary, and follow-up. In the first encounter, the counselor will seek to establish rapport with the person or family and determine why they are seeking assistance. While usually only one or two family members are present, occasionally cousins, in-laws, siblings, or grandparents may come to the sessions. The counselor sees not just an affected person, but a whole family as potentially touched by a genetic condition.

The personal and family medical history is taken. The family is advised to ask all related family members about conditions that were present in their families. For example, sometimes women asking about breast cancer will tell only her mother's history; however, mutations in the *BRCA* gene can be passed down through the father's family.

Using information about as many relatives as possible, the counselor will attempt to create a pedigree, a way of visualizing genetic disease in the family. Such probing into one's background may reveal information that is secret and not known to all family members. Counselors are extremely cautious about maintaining confidentiality. Insurance companies have used tests from pedigrees to deny life or health insurance to a person at risk to develop a genetic disease.

Once the pedigree is complete, family members may be asked to have tests, such as X-ray, ultrasound, urine analysis, skin biopsy, or general physical examinations. Other tests may include a karyotype or chromosome study, DNA analysis, amniocentesis, chorionic villus sampling, and fetal cell sorting. Costs can range from hundreds to thousands of dollars, depending on the sizes of the genes and the number of mutations tested for. Scientists have developed several hundred DNA-based tests for different conditions, but mostly these are for basic research, not for use in tests that can be useful in genetic counseling. Of the 100 gene tests available commercially, most are related only to a single gene.

With over 3,000 genetic diseases and so many cultures represented in the United States, counselors have a demanding job balancing personal, moral, ethical, religious, and cultural values for their clients. When options are laid out, individuals then make decisions as to the best course for their personal lives. The genetic counselor lays out the framework for decision making. For many of the conditions, no therapy or treatment is available. So, for example, parents must decide how they want to proceed after hearing that their child has a potential deadly condition.

The National Society of Genetic Counselors defined the following four tasks for the genetic counselor:

- Interpret the family medical histories and assess the chances of disease occurrence or reoccurrence

- Educate the client about testing, inheritance, prevention, management, and resources
- Counsel to promote informed choices
- Help parents adapt to the risk of the condition

Making Genetic Rounds: A Personal Story of a Genetics Counselor

Most of these genetic conditions in this encyclopedia have no cure and some symptoms that may be treated only with great expense. However, the personal story and travail of having a life-threatening genetic disorder or being a parent of one with such a disease is not revealed in the definition and symptoms in an encyclopedia.

Dr. Robert Marion in his book, *Genetic Rounds*, gives readers a snapshot of what it is like to personally encounter such diseases. In 17 chapters, he reveals the emotional side of stories, which occurred over 30 years of his practice. As of this writing, Dr. Marion is a professor of pediatrics and obstetrics and gynecology at the Albert Einstein College of Medicine. He is presently director of Montefiore Medical Center in the Bronx and Blythedale Children's Hospital in Valhalla, New York. He recounts stories of triumph and heartbreak as he deals with some of the world's most horrific diseases. He is frustrated over ethical issues such as whether he should tell friends and family that their loved one will die from an uncommon disorder. He shares his frustration of being able to diagnose a disease knowing there is no cure for it. However, he is gratified when parents of such children cope and manage chronic conditions that without care would have been an early death sentence.

One story that he tells is of an infant referred to him by a pediatrician who thought the boy had a neurodegenerative disease. The doctor suspected Tay-Sachs disease or Niemann-Picks disease, both serious diseases that damage the brain and spinal cord. The mother was from Trinidad and had moved to New York to escape the heat. The mother was in complete denial of medical problems for her son and was convinced that it was the heat that was affecting the boy and that prayer would heal him. The father thought that he would never be any good. Running the tests for neurodegenerative diseases did not show anything conclusive. The mother brought the boy back after several months convinced that prayer had healed him.

Dr. Marion saw a boy sitting up and smiling and definitely not what one would see in a neurological case. But something did not look right. He had no teeth and no hair. Dr. Marion knew immediately what it was. Putting all the pieces of the puzzle together, he recognized it as hypohydrotic ectodermal dysplasia (HED), an X-linked disorder. The mother's hair was sparse, the heat bothered her because she could not sweat, and she still had baby teeth. The relatively mild condition that was present in the mother became severe in her son.

Dr. Marion followed the boy as a teenager. He did well in school but of course had to have an air-conditioned classroom. His appearance, however, was still a problem and he developed serious emotional problems, making few friends and staying in his room most of the time. Dr. Marion realized his depression and tried to get him help, but as of the book's publication date, the teenager still did not realize that he had a lot to offer the world. Such stories bring home the personal and emotional trauma of not only parents but the people who experience genetic disorders. Dr. Marion in his book brings genetic disorders to life.

Further Reading

American Board of Genetic Counseling. http://www.abgc.net/ABGC/AmericanBoardofGeneticCounselors.asp. Accessed 5/19/12.

"Genetic Counseling." 2008. Human Genome Project. http://www.ornl.gov/sci/techresources/Human_Genome/medicine/genecounseling.shtml. Accessed 5/19/12.

"Genetic Counseling." MedlinePlus. National Library of Medicine (U.S.). http://www.nlm.nih.gov/medlineplus/geneticcounseling.html. Accessed 5/19/12.

Marion, Robert. 2009. *Genetic Rounds: A Doctor's Encounters in the Field that Revolutionized Medicine*. New York: Kaplan Publishing.

Genomic Testing: A Special Topic

When the human genome was sequenced, the next logical question was how to use this information to improve lives and health. This soon led to the testing of individuals for many purposes relating to genetics. For example, if someone has symptoms of a disorder that is of a possible genetic cause, testing can help with the diagnosis. A person may know of a family condition of a specific disorder and wonder if that individual is a carrier of the gene, although he or she may not have the disorder. Prenatal genetic testing can tell parents whether their unborn child has a genetic condition. An important area of testing is that of newborn screening, which can show if an infant has a disorder before that condition starts. Genome testing can also be used in forensic medicine to determine if a person was present at a crime. A brand new use of genetic testing is in the field of pharmacogenomics, which can match drug use to treat a specific condition.

What Is Genetic Testing?

Genetic testing subjects the cells of a person to various tests to see if that person's genes have changes or mutations that could indicate a particular disorder. The geneticist may also look for abnormal proteins or abnormal chromosome patterns to determine if a disorder is present. If the test shows some aberration, then the test will possibly indicate the presence of a disorder.

Tests include several types, including tests of individual or short lengths of genes, chromosomes, or special biological tests.

Gene Tests

A person who is undergoing a genetic test will give a sample of saliva, blood, or other body tissue for DNA or RNA study. The test will scan the person's DNA in the following ways:

- DNA probes: A probe is a short string of DNA with a base sequence that can bind to the sequence of the mutated gene. The probes have a tag or marker that is fluorescent and can be readily seen. The probe looks for the piece that matches or complements the altered gene within a person's genome and binds to it. The fluorescent marker then enables the geneticist to identify the gene.
- DNA sequencing: The test compares the sample of the person's DNA to a known normal sample. The test analyzes the DNA base-by-base and will note areas of abnormality. RNA sequencing may also be used.

These gene tests will note mutations caused by a single substitution of one amino acid building block for another or if other small changes have been made such as additions, deletions, repeated patterns, or lost entirely. It can also detect genes that are too active or that have been turned off.

Presently, about 2,000 genetic tests are available in present clinical settings, which means highly skilled geneticists perform these tests to advise people about treatment. About 2,200 disorders have genetic tests. Most of these are for rare single-gene disorders, in which only one gene is responsible for one genetic condition. For example, the normal gene for Fragile X syndrome has perhaps 20 of so repeats of the building blocks cytosine-guanine-guanine repeats or CGG repeats. In the genes of a person with fragile X, the geneticist would find over 200 CGG repeats on the X chromosome. Another single gene condition is Duchenne muscular dystrophy. The genetic pattern will reveal either deletions or additions in the *DMD* gene located on the X chromosome at position 21.2.

Chromosome Tests

In addition to gene tests, a geneticist may study the set of a person's chromosomes. Normally, humans have 22 pairs of body chromosomes, which are numbered 1 through 22, and one pair of sex chromosomes named X and Y. The chromosomes are located in the nucleus of each cell. A geneticist will use chromosome tests to look at the features of the individual's chromosomes, which will include the number, the arrangement, and structure. Missing or added pieces or translocations (switched pieces) can be seen in these tests. Following are the types of chromosome tests:

- Karyotype: This technique has been known for some time, even before the full sequencing of the human genome. (*See also* Karyotyping: A Special Topic.) The karyotype is especially helpful in diagnosing disorders that involve multiple chromosomes. For example, people with trisomy 18 have three chromosome 18s; people with trisomy 13 have three chromosome 13s instead of just two. Children with Down syndrome have three copies of chromosome 21. Many other conditions can be identified with a karyotype of chromosomes.

- Fluorescent in situ hybridization or FISH analysis: This test, which is more sophisticated than karyotyping, uses fluorescent DNA probes to find pieces of chromosomes that are missing or have extra copies. FISH analysis can locate missing parts of chromosomes that are characteristic of several conditions
- Biochemical tests: These types of tests study the activities of proteins. Because most genes provide instructions for making proteins, a test may determine proteins that are not working. These tests are especially helpful in determining metabolic conditions. Some biochemical tests are part of the complement given in newborn screening. For example, children with phenylketonuria or PKU do not break down the amino acid phenylalanine, which can cause a host of health problems if not detected. Newborn screening at birth can detect this harmful buildup so that physicians may recommend treatment.

What Are the Reasons for Genetic Testing?

Tests may be given for many reasons. Suppose a person has a family history of a particular disorder, such as early-onset Alzheimer disease. The person wonders if they will be the next in the family to have the dreaded condition, which is genetic and traced to chromosome 14. A genetic test could reveal that the mutation is not present and give that person peace of mind. If the person does have the gene, certain behaviors such as diet, exercise, or brain activities may lessen the impact and help them make informed choices about their future. However, it is important to keep in mind that all persons with the particular mutation may never develop the condition.

Following are some of the reasons for genetic testing.

For Diagnosis

A child is born that has some of the physical symptoms of a disorder that doctors may suspect is Down syndrome. The chromosome test can tell if the person has the extra chromosome 21. This can help parents make some informed decisions. For example, early education and training of children with Down syndrome can help them attain a higher level of functioning.

For Prediction of Future Disease Disorders

A daughter has just been informed that her mother has the severe neurological disease called Huntington disease. For a woman, this dominant disorder does not appear until she is into her childbearing years and may have several children.

There is a 50-50 chance that each offspring will have the condition. The daughter wants to know if she carries the gene. If she does not carry the gene, she can be greatly relieved. However, even with this gene, which has multiple repeats of cytosine-adenine-guanine in cells, she may have some shorter repeats that will not produce full symptoms of the disorder. Here is where a genetic counselor can help interpret.

People may want to know if they have certain inherited genetic cancer such as breast or colon cancer, or diseases such as type 2 diabetes. These conditions involve multiple genes. One condition in which genetic tests determine if a person has mutations in protective genes such as *BRCA1* and *BRCA2*, which may cause some breast cancers and ovarian cancers. Having the gene simply indicates a risk of getting the disorder, and the person may not have the disorder at all. In the case of many of these conditions, changing lifestyles of diet, exercise, or even taking preventive medications may help.

For Preconception and Carrier Testing

These tests can tell if a person has mutations in one of the body chromosomes for a variety of genetic diseases. These disorders are recessive, meaning that the carrier may not have the disease but might pass it on to the children. If both parents are carriers, the chances of an individual child having the disease are one in four. Several recessive genes are found in people of Ashkenazi Jewish descent, whose ancestors came from eastern Europe. One example is Tay-Sachs disease.

For Prenatal Testing During Pregnancy

The following reasons may lead a pregnant woman to seek prenatal testing:

- Older mothers: Women over the age of 35 are at risk for having children with chromosomal abnormalities or birth defects.
- A family history: Many diseases are said to run in families. People with Jewish backgrounds might want to test for a variety of diseases, including thalassemia or Tay-Sachs disease. People of African descent would want to test for sickle-cell disease.
- Some conditions that occur during pregnancy, such as spina bifida, a condition in which the area around the vertebra does not close, can be revealed in prenatal testing.
- Three testing technologies for prenatal testing:
    - Ultrasound: The mother goes in for an ultrasound test, which involves moving an instrument over the abdominal area and reading the picture of the developing fetus on the screen.
    - Amniocentesis: A large needle is inserted into the abdominal wall into the uterus drawing out a sample of the amniotic fluid, the fluid surrounding the embryo. Testing can tell if there are abnormal cells or fluids in the amnion.

- Chorionic villus sampling or CVS: This procedure takes cells from the placenta, which carries fetal cells.

For Newborn Screening

This type of testing is the most familiar and the most widespread. In the United States, this program is controlled by the state, but all states are required to test for 21 disorders and some test for many more. (*See also* Newborn Screening: A Special Topic.) A small sample of blood is taken from the heel of the child and then tested for a variety of disorders. Many of these are metabolic disorders that with prompt attention can avoid life-threatening conditions.

For Pharmacogenomic Testing

This newest field of genetic testing is growing exponentially. The person's genes can reveal how certain drugs or pharmaceuticals can be broken down in the body. The goal is to find drug treatments that are best for the person.

For Forensic Testing

Forensic tests can use DNA sequencing for legal purposes. This testing is not used to determine genetic diseases but to see if a person was involved in a crime. The test is also used to determine paternity.

A Word of Caution

Genetic testing must be timely and be performed by reliable sources, in order to provide health care providers and patients with the needed information to make informed decisions about treatment protocol. Many genetic disorders still have no treatments; however, knowing that one carries the condition can help that person make future plans. A variety of tests are offered directly to consumers, over the Internet, and for a price. Some of these are offered via television advertisements for products that give consumers genetic information without involving a doctor. These kits, usually involving the use of scrapings from the inside of the cheek, arrive by mail, and consumers are notified of the results by mail or over the telephone. Because there is no medical person involved to assist in the interpretation, use of these tests is not advised by the medical community.

Further Reading

"Gene Testing." 2010. Human Genome Project Information. http://www.ornl.gov/sci/techresources/Human_Genome/medicine/genetest.shtml. Accessed 12/24/11.

"Genetic Testing:" 2012. MedlinePlus. National Library of Medicine (U.S.). http://www.nlm.nih.gov/medlineplus/genetictesting.html. Accessed 2/13/12.

"What Is Genetic Testing?" 2012. Genetics Home Reference. National Library of Medicine (U.S.). http://ghr.nlm.nih.gov/handbook/testing/genetictesting. Accessed 2/13/12.

Giant Axonal Neuropathy (GAN)

Prevalence Rare; incidence unknown

Other Names GAN; neuropathy, giant axonal

Some disorders are named from the part of the biological system that they affect. The name "giant axon neuropathy" describes the condition in each part of the name. Of course, "giant" means huge or large. An axon is an important part of the nerve cell, which is made of the dendrite, the cell body, and a long straight process called the axon. Nerve impulses move from the dendrite, which means "little tree," to the cell body and then through the axon across the gap of the synapse to the dendrite of the next neuron. The axon is covered with a protective sheath called the myelin sheath, and the axon and sheath are referred to as a nerve fiber. The word "neuropathy" comes from two Greek words: *neuro*, meaning "nerve," and *path*, meaning "disease." Thus, in the context, the name describes a condition in which neuropathy is caused by giant axons.

What Is Giant Axonal Neuropathy (GAN)?

Giant axonal neuropathy is a progressive condition in which the protein that is essential for the nerve fibers to form is missing. This missing protein causes the axons to swell up and tangle, appearing as large abnormal structures under the microscope. Hence, the term "giant" is used to describe them. Because the nerve fibers or filaments are disorganized, the impulse cannot move through them, and the axons deteriorate.

The peripheral nervous system controls nerve signals to the extremities of the arms, legs, and other body parts. Symptoms of GAN originate in this system. The following symptoms usually begin before the age of five and are progressive as the axons are destroyed:

- Clumsiness
- Muscle weakness in the legs that progresses to a "waddling gait"
- Numbness and lack of feeling in the arms and legs typical in neuropathy
- Seizures
- Hearing and visual problems
- Nystagmus, or a rapid movement back and forth of the eyes
- Mental retardation

A unique characteristic of the disease is dull, tightly curled kinky hair that is very different from the hair of either of the parents.

Doctors diagnose GAN using several tests. One test that diagnoses neuropathy is called a nerve conduction velocity test, which measures the degree of electrical impulses going to the body periphery. A brain MRI may be used. More definitive is a biopsy of a small amount of tissue from a peripheral nerve, which will reveal the giant axons under the microscope.

What Is the Genetic Cause of Giant Axonal Neuropathy (GAN)?

Changes in the gigaxonin gene (*GAN* gene) cause giant axonal neuropathy. Normally, *GAN* instructs for making the protein gigaxonin used in destroying damaged or excess proteins. A complex process then occurs in a system known as the ubiquitin-proteasome system. Using many steps, this system attaches to an unwanted protein and then a host of enzymes destroys that protein. Scientists believe that gigaxonin is part of a family of ubiquitin ligase complex that targets certain proteins for destruction.

In people with GAN, mutations either change or remove just one of the building-block proteins altering the shape of gigaxonin. This change keeps gigaxonin from binding to and destroying targeted proteins. The absence of the properly functioning protein allows the accumulation of unwanted proteins and results in disorganization of the cell. The axons of the nerves swell and become large, keeping them from transmitting nerve signals properly. Although damage begins in the peripheral nervous system, over time the damage can affect the central nervous system. *GAN* is inherited in an autosomal recessive pattern and is located on the long arm (q) of chromosome 16 at position24.1.

What Is the Treatment for Giant Axonal Neuropathy (GAN)?

Treatment is symptomatic. The physicians and therapists of a medical team work to minimize physical and mental deterioration. Most children become wheelchair-dependent in the second decade of life. Some children with GAN have survived to early adulthood.

Further Reading

"*GAN*." 2010. Genetics Home Reference. National Library of Medicine (U.S.). http://ghr.nlm.nih.gov/gene/GAN. Accessed 5/18/12.

"Giant Axonal Neuropathy." 2010. Genetics Home Reference. National Library of Medicine (U.S.). http://ghr.nlm.nih.gov/condition/giant-axonal-neuropathy. Accessed 5/18/12.

"Giant Axonal Neuropathy." Medic8. http://www.medic8.com/genetics/giant-axonal-neuropathy.htm. Accessed 5/18/12.

"NINDS Giant Axonal Neuropathy Information Page." 2010. National Institute for Neurological Diseases and Stroke (U.S.). http://www.ninds.nih.gov/disorders/gan/GiantAxonalNeuropathy.htm. Accessed 5/18/12.

Gitelman Syndrome

Prevalence Affects about 1 in 40,000 worldwide

Other Names familial hypokalemia-hypomagnesemia; Gitelman's syndrome; GS; hypokalemia-hypomagnesemia, primary renotubular, with hypocalciuria; tubular hypomagnesemia-hypokalemia with hypocalcuria

In 1966, Dr. Hillel Gitelman and colleagues wrote about a rather mild kidney disorder that appeared to run in families. They found that people with the condition had abnormally low amounts of calcium, magnesium, and potassium in the bloodstream. The disorder was named for this doctor—Gitelman disease.

What Is Gitelman Syndrome?

Gitelman syndrome is a disorder of the kidneys. Located near the middle of the back just below the rib cage, the kidneys are two bean-shaped organs that are essential to process waste materials from body metabolism. Each kidney is made up of millions of tiny cells called nephrons that filter the blood and remove waste products and water, which becomes urine. The kidney is a major chemical factory that must maintain a balance of basic atoms needed by the body, such as potassium, magnesium, and calcium. Gitelman syndrome causes an imbalance of these chemicals.

Following are the symptoms of Gitelman syndrome, which usually appear in late childhood or adolescence:

- Painful muscle spasms
- Muscle weakness or cramping
- Tingling sensation in the skin, usually in the face
- Excessive fatigue
- Low blood pressure
- A painful joint condition, called chondrocalcinosis.
- Heart disorders

The symptoms of the disorder can vary even among members of the same family.

What Is the Genetic Cause of Gitelman Syndrome?

Two genes—*SLC12A3* and *CLCNKB*—are associated with Gitelman syndrome.

SLC12A3 Gene

Mutations in the *SLC12A3* gene, officially known as the "solute carrier family 12 (sodium/chloride transporters), member 3" gene, cause most cases of Gitelman syndrome. Normally, *SLC12A3* provides instructions for making the protein NCC, a sodium chloride cotransporter. A cotransporter moves charged sodium and chlorine particles across the cell membranes. NCC is critical for kidney function. It provides the mechanism that leads kidneys to reabsorb salt from the urine back into the bloodstream. This retention of salt is what affects body fluid levels and helps maintain blood pressure.

Over 140 mutations have been associated with the *SLC12A3* gene. Most of the mutations involve a change in only one amino acid building block in the NCC protein. The single mutation disrupts the protein from reaching the cell membrane or alters the ability to move across the cell membrane. Some mutations either add or take away portions of genetic matter and produce a short, nonworking version of the NCC protein. Thus, the kidneys do not reabsorb salt and allow the excess to accumulate in the urine. Because the salt transport is nonfunctioning, other chemicals, such as potassium, magnesium, and calcium are not reabsorbed, either. *SLC12A3* is inherited in an autosomal recessive pattern and is located on the long arm (q) of chromosome 16 at position 13.

CLCNKB Gene

Mutations in the *CLCNKB* gene, officially known as the "chloride channel Kb" gene, cause Gitelman syndrome. Normally, *CLCNKB* provides instructions for making a chloride channel called C1C-Kb, which is found abundantly in the kidneys. C1C-Kb works with several other proteins to control movement of ions in and out of the kidney cells, called the nephrons. This channel is part of the system that allows the reabsorption of salt—sodium chloride, or NaCl—from the urine back into the bloodstream. Along with the protein instructed by the gene *SLC12A3*, which also aids in the retention of salt, the CLCNKB protein affects the body's fluid levels and maintains blood pressure.

Mutations in the *CLCNKB* gene cause Gitelman syndrome. The mutations are not as common as the mutations in the *SLC12A3* gene. Changes in the gene cause abnormalities in the C1C-Kb channel and disrupt the reabsorption of salt and other ions including potassium, magnesium, and calcium. *CLCNKB* is inherited in an autosomal recessive pattern and is located on the short arm (p) of chromosome 1 at position 36.

What Is the Treatment for Gitelman Syndrome?

Compared to many kidney disorders, Gitelman syndrome is relatively mild. Treatment depends on getting an accurate diagnosis and then keeping the blood potassium, sodium, and chloride at normal levels. The person must eat a diet rich in foods that provide these nutrients and by taking supplements. Additional medicine may supplement electrolyte loss.

Further Reading

"Gitelman Syndrome." 2009. The Bartter Site. http://barttersite.org/gitelman-syndrome. Accessed 1/8/12.

"Gitelman Syndrome." 2010. Genetics Home Reference. National Library of Medicine (U.S.). http://ghr.nlm.nih.gov/condition/gitelman-syndrome. Accessed 1/8/12.

"Gitelman Syndrome." 2010. MedlinePlus. National Library of Medicine (U.S.). http://www.nlm.nih.gov/medlineplus/kidneydiseases.html. Accessed 1/8/12.

Glaucoma, Early Onset. *See* Early-Onset Glaucoma

Glucose-6-Phosphate Dehydrogenase Deficiency (G6PD)

Prevalence Estimated 400 million people worldwide; occurs more frequently in parts of Africa, Asia, and Mediterranean; affects about 1 in 10 African American males in the United States

Other Names deficiency of glucose-6-phosphate dehydrogenase; G6PDD, G6PD deficiency

The discovery of some diseases follows strange paths. In history, this disorder is traced to the broad bean, known as the fava bean. Fava beans were known by the Greeks to cause favism, a type of anemia. Pythagoras had the answer: Avoid eating the beans. More recently, three scientists investigating the cause of hemolytic anemia found that certain people who had taken the drug primiquine developed the condition. In 1956, they experimented, using volunteers from the Illinois State Prison, and concluded that the drug primiquine caused the destruction of red blood cells. They surmised that this disorder was probably inherited. This type of human experiment, even with prisoners who volunteer, cannot be done today. However, the experiments did eventually lead to the discovery of inborn error of metabolism called glucose-6-phosphate dehydrogenase deficiency.

What Is Glucose-6-Phosphate Dehydrogenase Deficiency (G6PD)?

Glucose-6-phosphate dehydrogenase deficiency (G6PD) is a condition that affects the red blood cells and is found mostly in males. The red blood cells carry oxygen from the lungs to the body cells. People with G6PD deficiency lack an enzyme called glucose-6-phosphate dehydrogenase. The lack of this enzyme causes red blood cells to break down or be destroyed prematurely, a condition known as hemolysis. This destruction of the red blood cells or erythrocytes leads to hemolytic anemia. The following symptoms of G6PD deficiency result from the destruction of the red blood cells:

- Paleness
- Fatigue
- Enlarged spleen
- Yellowing of the whites of the eyes
- Jaundice or yellowing of the skin
- Dark urine

- Shortness of breath because oxygen is not getting to the lungs
- Rapid heart rate
- Rarely, kidney failure and death

G6PD deficiency may also cause mild to severe jaundice in newborns.

The people most likely to have G6PD deficiency are males with an African American or Middle Eastern background, especially those of Kurdish or Sephardic Jewish descent. Following are items that may trigger the red blood cell destruction:

- Antimalarial drugs such as quinidine or quinine
- Aspirin
- Nonsteroidal anti-inflammatory drugs (NSAIDs)
- Sulfa drugs
- Bacterial or viral infections
- Broad beans, also known as fava beans
- A family history of G6PD deficiency

What Is the Genetic Cause of Glucose-6-Phosphate Dehydrogenase Deficiency (G6PD)?

Changes in the "glucose-6-phosphate dehydrogenase" gene, or *G6PD* gene, cause glucose-6 phosphate dehydrogenase deficiency. Normally, *G6PD* provides the instructions for making the enzyme of the same name, glucose-6 phosphate dehydrogenase. This enzyme is critical for protecting red blood cells from damage and premature destruction. Found in all cells, the enzyme is also important because it has a role in the normal processing of carbohydrates. The enzyme begins the conversion of glucose to ribose-5-phosphate, a compound important for building DNA and RNA. One of the chemicals in the pathway is a molecule called NADPH, which helps protect cells. If anything disrupts the pathway, toxic chemicals can build up in the cells. The loss and destruction of the red blood cells causes the symptoms of hemolytic anemia.

The *G6PD* gene provides instructions for making an enzyme called glucose-6-phosphate dehydrogenase. This enzyme, which is active in virtually all types of cells, is involved in the normal processing of carbohydrates. It plays a critical role in red blood cells, which carry oxygen from the lungs to tissues throughout the body. This enzyme helps protect red blood cells from damage and premature destruction.

G6PD has over 400 variations. (Some gene variants provide advantages—G6PD variants protect against malaria. The term "mutant" suggests dysfunction. "Polymorphism" is another term that is more neutral.) The mutant gene may vary from person to person and also may differ in certain populations. For example, in Egypt only one variant, called the "Mediterranean allele," is in the population; in

Japan the one variant is called "Japan allele." In the United States, the disorder common among black males may show several of the different types of alleles. *G6PD* is inherited in an X-linked pattern and is located on the long arm (q) of the X chromosome at position 28. All X-linked genes are likely to affect more males than females. As long as there is one good copy of the gene in the female, a normal enzyme will be produced. In males, because there is only one X chromosome, one defective *G6PD* gene passed on by the mother can cause the deficiency.

What Is the Treatment for Glucose-6-Phosphate Dehydrogenase Deficiency (G6PD)?

Basically, the best prevention for the flare-ups or episodes is to avoid the things that trigger the disorder, such as foods or drugs. Some medications may be used to fight infection, and in some cases, a transfusion may be necessary. Most patients recover from a hemolytic episode over a period of time.

Further Reading

"*G6PD*." 2010. Genetics Home Reference. National Library of Medicine (U.S.). http://ghr.nlm.nih.gov/gene/G6PD. Accessed 5/18/12.

"Genetics of G6PD Deficiency." 2009. http://rialto.com/g6pd/genetics.htm. Accessed 5/18/12.

"Glucose-6-Phosphate Dehydrogenase Deficiency." 2010. Genetics Home Reference. National Library of Medicine (U.S.). http://ghr.nlm.nih.gov/condition/glucose-6-phosphate-dehydrogenase-deficiency. Accessed 5/18/12.

"Glucose-6-Phosphate Dehydrogenase Deficiency." 2010. MedlinePlus. National Library of Medicine (U.S.). http://www.nlm.nih.gov/medlineplus/ency/article/000528.htm. Accessed 1/25/11.

Glucose-Galactose Malabsorption (GGM)

Prevalence Rare; only a few hundred cases reported; may be more prevalent with mild cases not reported

Other Names carbohydrate intolerance; complex carbohydrate intolerance; GGM; monosaccharide malabsorption

When the average person generally thinks of sugar, he means the white crystals that sweeten cakes and beverages. However, for scientists and physicians, sugar is much more complex. The white, crystal table sugar is called sucrose. Sucrose is a complex sugar called a disaccharide. Another complex sugar, lactose, is found in milk. These two sugars cannot be used and absorbed in the body; thus they must be broken down into simpler sugars called monosaccharides.

In order for the body to use sugars, sucrose is broken down into glucose and another simple sugar, fructose. Lactose is broken down into the simple sugars fructose and galactose. Sucrose, lactose, and other molecules made from sugars are called carbohydrates. When a person's body cannot break down these molecules properly, he or she may be carbohydrate intolerant because of glucose-lactose malabsorption.

What Is Glucose-Galactose Malabsorption (GGM)?

Glucose-galactose malabsorption is a rare condition described well by its name. The term "malabsorption" comes from two Latin words: *mal*, meaning "bad"; and the base word *absorb*, meaning "to take in." GGM occurs when the body cannot move glucose and galactose across the intestinal lining and the substances cannot be digested.

GGM becomes evident on the first day or in the first few weeks of the baby's life when they experience the following symptoms:

- Severe diarrhea
- Dehydration because of the loss of fluid
- Increased acid in the blood and tissues
- Weight loss when they take in milk or breast milk
- Rapid death if glucose/galactose are not removed from the diet.

The children are able to tolerate fructose or soy-based products that do not contain glucose or galactose. As the children get older, they may gain a little tolerance. However, they may pass small amounts of glucose in the urine or develop kidney stones or deposits of calcium in the kidneys.

What Is the Genetic Cause of Glucose-Galactose Malabsorption (GGM)?

Changes in the *SLC5A1* gene, officially known as the "solute carrier family 5 (sodium/glucose cotransporter), member 1" gene, cause glucose-galactose malabsorption. In the space enclosed by the small intestine called the lumen, the enzyme lactase breaks down lactose into glucose and galactose; the enzyme sucrase breaks down sucrose and fructose. Normally, *SLC5A1* then instructs for making the sodium/glucose cotransporter protein that moves the simple sugars across the intestinal lining. This protein is critical in the workings or the epithelial cells, which line the wall of the intestines. These cells have small fingerlike projections called villi, where nutrients are absorbed in the intestine. The cluster of villi is referred to as the brush border. It is the sodium/glucose cotransporter protein that moves the simple sugars across the brush border for absorption in a process called active transport. Sodium and water are also transported with the sugars.

More than 40 mutations have been identified in the *SLC5A1* gene. These mutations may cause a shortened protein or be the result of improper folding of the protein. A disruption in the process prevents glucose and galactose from being absorbed, causing the buildup of glucose and galactose in the intestinal tract. Nothing goes across the brush border, including the water that is necessary for body cells. The individual then develops life-threatening diarrhea and dehydration. *SLC5A1* is inherited in an autosomal recessive pattern and is located on the long arm (q) of chromosome 22 at position 13.1.

Half of the cases of GGM results from intermarriage within families. It is surmised that at least 10% of the general population has a mild form of glucose intolerance.

What Is the Treatment for Glucose-Galactose Malabsorption (GGM)?

The treatment depends on rapid diagnosis to determine glucose/galactose intolerance. No milk can be given to this person. The diet must be controlled throughout life to avoid all substances with sugar and milk products.

Further Reading

"Glucose-Galactose Malabsorption." 2010. Genetics Home Reference. National Library of Medicine (U.S.). http://ghr.nlm.nih.gov/condition/glucose-galactose-malabsorption. Accessed 5/18/12.

"Glucose Galactose Malabsorption." 2010. National Center for Biotechnology Information (U.S.). http://www.ncbi.nlm.nih.gov/books/NBK22210. Accessed 5/27/12.

"*SLC5A1*." 2010. Genetics Home Reference. National Library of Medicine (U.S.). http://ghr.nlm.nih.gov/gene/SLC5A1. Accessed 5/18/12.

GLUT1 Deficiency Syndrome

Prevalence Very rare; fewer than 100 cases reported since its identification in 1991; however some researchers believe the condition in increasing

Other Names DeVivo disease; encephalopathy due to GLUT1 deficiency; glucose transport defect, blood-brain barrier; glucose transporter protein syndrome; glucose transporter type 1 deficiency syndrome; GLUT1 DS; GTPS

Many genetic disorders are not treatable, leaving physicians to focus only on treating the symptoms until the person dies. However, GLUT1 deficiency is treatable. Animal models have been created that have brought considerable insight into the disorder. Along with diet and care, this disorder can be brought under control.

What Is GLUT1 Deficiency Syndrome?

GLUT1 deficiency syndrome primarily affects the brain. The condition is caused by impaired glucose transport into the brain and a low concentration of glucose in the cerebrospinal fluid. Infants with GLUT1 deficiency may start life normally but within a few months may have seizures and show slow growth in brain and head. These children with early-onset epilepsy appear to be resistant to anticonvulsant medications.

If not treated, some of the following symptoms may develop:

- Developmental delay
- Intellectual disability
- Stiffness and spasticity of the muscles
- Poor coordination of movements
- Speech difficulties
- Lack of energy
- Headaches
- Twitching muscles
- Erratic eye movements

Electroencephalography (EEG) of the brain waves shows improvement after food intake of a ketogenic diet. A ketogenic diet is high in fat and has lots of protein, but is low in carbohydrates. Doctors use this diet to treat children with difficult-to-control epilepsy. The diet works like many of the high-protein weight loss plans. It forces the body to burn fats rather than carbohydrates. Normally carbohydrates in food are converted to glucose, which is important in proper brain function. Very little carbohydrates in the diet cause the liver to convert fat into fatty acids and ketone bodies, which then pass into the brain and replace glucose as an energy source. This state, called ketosis, reduces the frequency of epileptic seizures. Some children with a milder form do not have seizures but may exhibit some of the features because of the glucose deficiency.

What Is the Genetic Cause of GLUT1 Deficiency Syndrome?

Mutations in the *SLC2A1* gene, officially known as the "solute carrier family 2 (facilitated glucose transporter), member 1" gene, cause GLUT1 deficiency syndrome. Normally, *SLC2A1* instructs for a protein called the glucose transporter protein type 1 (GLUT1). Located in the cell membrane, this protein carries the simple sugar glucose from the blood into the cells that use it to produce energy. Many substances cannot move across this blood-brain barrier, the tiny capillaries that connect arteries and veins. The blood-brain barrier keeps unwanted materials from entering the delicate nerve tissues of the brain. However, the GLUT1 transporter carries glucose across this blood-brain barrier to the very important brain cells.

The more than 50 mutations of the *SLC2A1* gene either reduce or eliminate the action of the GLUT1 transporter, and glucose does not get to the cells of the brain. When glucose is reduced or disrupted, the person may experience the signs and symptoms of GLUT1 deficiency syndrome. *SLC2A1* is inherited in an autosomal dominant pattern and is located on the short arm (p) of chromosome 1 at position 34.2.

What Is the Treatment for GLUT1 Deficiency Syndrome?

Beginning the ketogenic diet as soon as possible is essential. If the diet does not stop the seizures, an anticonvulsant medication may be required. This disease is being studied in many countries due to an increasing number of patients. Research into the use of ketogenic diets, molecular and biochemical analysis, and development of animal models has made GLUT1 deficiency syndrome a treatable condition.

Further Reading

"GLUT1 Deficiency Syndrome." 2010. Genetics Home Reference. National Library of Medicine (U.S.). http://ghr.nlm.nih.gov/condition/glut1-deficiency-syndrome. Accessed 5/18/12.

"GLUT1 Deficiency Syndrome—2007 Update." 2010. National Center for Biotechnology Information (U.S.). http://www.ncbi.nlm.nih.gov/pubmed/17718830. Accessed 5/18/12.

"*SLC2A1*." 2010. Genetics Home Reference. National Library of Medicine (U.S.). http://ghr.nlm.nih.gov/gene/SLC2A1. Accessed 5/18/12.

GM1 Gangliosidosis

Prevalence Occurs in about 1 to 100,000 to 200,000; Type I most frequent; individuals with Type III of Japanese descent

Other Names beta-galactosidase-1 (GLB1) deficiency; beta-galactosidosis

A metabolic condition involves some deficiency in the way proteins are stored and used in the body. In 1902, Archibald Garrod first described a metabolic disorder when a several patients in one family came to him and their urine appeared black. He surmised that there was some inborn error of metabolism that caused the condition. GM1 gangliosidosis is another one of the inborn errors of metabolism.

What Are GM1 Gangliosidoses?

Written in the plural form indicates there is more than one form of the disorder. This metabolic disorder results from a deficiency of beta-galactosidase, an enzyme that mediates storage and breakdown of lipids or fats especially in the brain and spinal cord. When this enzyme is not present, nerve cells in the brain and spinal cord are destroyed.

Depending on the age of onset, three basic types of GM1 exist:

- Early infantile GM1: This type becomes apparent by the age of six months. The child may have liver and spleen enlargement, coarsening of facial features, poor skeletal development, joint stiffness, distended abdomen, problems with gait, and obvious neurodegeneration. As the disease progresses, the child may lose vision, and a cherry-red spot on the retina is characteristic of this disorder. The child may also develop an enlarged heart. Children with this type generally do not survive past age 2.
- Late infantile GM1: Children with this form have normal early development but then begin to regress. They do not have the coarse facial features, cherry spots on eyes, but may have seizures and startling at loud noises, loss of hearing, and dementia. The symptoms begin around 18 months to about 5 years. The person still has a shortened life expectancy but may live into early adulthood. Tay-Sachs disease, a highly investigated disorder that is present in the Jewish community, is one of these disorders.
- Adult GM1: This form is the mildest end of the spectrum. The condition occurs between the ages of 3 and 30. Symptoms are abnormalities of the backbones or vertebrae, unsteady gait, and neurological disorders. Life expectancy varies.

What Is the Genetic Cause of GM1 Gangliosidosis?

Changes in the *GLB1* gene, officially known as the "gangliosidosis, beta 1" gene, cause GM1 gangliosidosis. Normally, *GLB1* instructs for making the enzyme beta-galactosidase. Located in the lysosomes, or the garbage disposal or housekeeper of the cells, this enzyme helps break down a molecule called GM1 ganglioside, a substance that is important for the normal health of the brain cells.

Mutations in the *GLB1* genes disrupt the proper production of the beta- galactoside enzyme. Substances then accumulate in toxic levels, and progressive damage destroys the important brain cells. Those people with a higher amount of the enzyme have milder signs and symptoms; those with lower amounts have the signs and symptoms of GM1.

This disorder is one of the many lysosomal storage disorders, which arise when the housekeeper of the cells cannot clean the house properly, and unwanted junk accumulates. *GLB1* is inherited in an autosomal recessive pattern. It is located on the short arm (p) of chromosome 21 at position 33.

What Is the Treatment for GM1 Gangliosidosis?

No treatment exists for this any of the gangliosidoses. If the person has convulsions, an anticonvulsant may control. Others treatment includes supportive nutrition and hydration. The National Institute of Neurological Disorders and Stroke

(NINDS) is conducting research on lipid storage disorders and issues grants to researchers across the country.

Further Reading

"*AGL*." 2010. Genetics Home Reference. National Library of Medicine (U.S.). http://ghr.nlm.nih.gov/gene/AGL. Accessed 5/18/12.

"*GLB1*." 2010. Genetics Home Reference. National Library of Medicine (U.S.). http://ghr.nlm.nih.gov/gene/GLB1. Accessed 5/18/12.

"NINDS Generalized Gangliosidoses Information Page." 2010. National Institute of Neurological Diseases and Stroke (U.S.). http://www.ninds.nih.gov/disorders/gangliosidoses/Gangliosidoses.htm. Accessed 5/18/12.

Greig Cephalopolysyndactyly Syndrome (GCPS)

Prevalence Rare; real prevalence unknown; estimated 1 to 9/1,000,000

Other Names cephalopolysyndactyly syndrome; Greig cephalopolysyndactyly (GCPS) syndrome; Greig syndrome

In 1926, David Middleton Greig, a Scot, wrote an article in the *Edinburgh Medical Journal* called "Oxycephaly," in which he described a condition that affects the limbs, head, and brain. The condition was characterized by the possession of extra fingers and toes. The condition was named Greig syndrome after this researcher. Sometimes the name Greig is confused with the Norwegian composer Grieg, but the name is pronounced as "Gregg" with a trilled Scottish "r." Usually now the name is seen as Greig cephalopolysyndactyly syndrome. The word "cephalopolysyndactyly" describes the condition well when it is broken down into four Greek root words: *cephalo*, meaning head; *poly*, meaning many; *syn*, meaning "together"; and *dactyl*, meaning digits of the fingers and toes.

What Is Greig Cephalopolysyndactyly Syndrome (GCPS)?

Greig cephalopolysyndactyly syndrome (GCPS) is a condition that affects the head, digits of the fingers, and toes, and in some cases the limbs. Symptoms of the disease are diverse, ranging from mild to severe, and may display the following:

- Large head: Called macrocephaly, the condition is characterized by an individual whose head appears larger than the body. The person may also exhibit a very high forehead.
- Abnormal digits: The person may have one or more extra fingers and toes or an abnormally wide thumb or big toe.
- Fused skin: The skin between the fingers and toes may appear webbed or fused together, a condition called syndactyly.

- Abnormal eye focus: The person may have widely spaced eyes, a condition called ocular hyperteleorism.
- Occasional central nervous system (CNS) anomalies: The person may have developmental delay or intellectual disability.
- Occasional hernias.

What Is the Genetic Cause of Greig Cephalopolysyndactyly Syndrome (GCPS)?

Changes in the *GLI3* gene, officially known as the "GLI family zinc finger 3" gene, cause GCPS. Normally, this gene belongs to a large family of genes that shape the normal pattern of tissues and organs during embryonic development. *GLI3* instructs for proteins that attach to specific regions of DNA and control gene expression. These proteins are called transcription factors. This protein plays a role in cell growth, cell specialization, and patterning of structures in the brain and hand.

GCPS is caused by several mutations in the *GLI3* gene. The mutations usually include exchanges of one amino acid building block or deletions and insertions in certain areas of the gene. The gene is contained on chromosome 7, and in some cases a large area of that chromosome is affected with insertions or deletions. Reduced amounts of the *GLI3* protein cause the signs and symptoms of GCPS. *GLI3* is inherited in an autosomal dominant pattern and is located on the short arm (p) of chromosome 7 at position 13.

What Is the Treatment for Greig Cephalopolysyndactyly Syndrome (GCPS)?

Treatment of the disorder is symptomatic and variable. Plastic or orthopedic surgery may correct severe anomalies. The prognosis for typically affected patients is good. A slight increase in delay or cognitive impairment may occur. Patients with large deletions in the *GLI3* gene may have a worse prognosis.

Further Reading

Biesecker, L. 2008. "The Greig Cephalopolysyndactyly Syndrome." *Orphanet Journal of Rare Diseases*. http://www.ojrd.com/content/3/1/10. Accessed 5/18/12.

"*GLI3*." 2010. Genetics Home Reference. National Library of Medicine (U.S.). http://ghr.nlm.nih.gov/gene/GLI3. Accessed 5/18/12.

"Greig Cephalopolysyndactyly Syndrome." 2010. Genetics Home Reference. National Library of Medicine (U.S.). http://ghr.nlm.nih.gov/condition/greig-cephalopolysyndactyly-syndrome. Accessed 5/18/12.

"Greig Cephalopolysyndactyly Syndrome." 2010. RightDiagnosis.com. http://www.rightdiagnosis.com/g/greig_cephalopolysyndactyly_syndrome/intro.htm. Accessed 5/18/12.

Guanidinoacetate Methyltransferase Deficiency

Prevalence Very rare; only a few dozen cases worldwide; of these, about one-third are of Portuguese descent

Other Names creatine deficiency, cerebral; creatine deficiency syndrome due to GAMT deficiency; GAMT deficiency

In 1902, Archibald Garrod wrote about the first inborn error of metabolism when he looked at the black urine present in a family. The condition was called alkaptonuria, and he was the first to recognize the genetic nature of a metabolic disease. In 1996, S. Stockler and a German team called guanidinoacetate methyltransferase deficiency the first inborn error of creatine metabolism in man. Creatine is a substance that combines with phosphates to serve as a high energy source for muscle contraction.

What Is Guanidinoacetate Methyltransferase Deficiency?

Guanidinoacetate methyltransferase deficiency, also known as GAMT deficiency, is a metabolic disorder that affects the nervous system and muscles. Guanidinoacetate methyltransferase is an enzyme that synthesizes the compound creatine from the amino acids glycine, arginine, and methionine in a two-step process. The compound produced in the first step is called guanidinoacetate. In the second step, the enzyme then produces creatine. Creatine then combines with phosphate to form phosphocreatine, which serves as a source of high-energy phosphate released in the anaerobic phase of muscle contraction.

The symptoms of GAMT deficiency vary greatly. Following are some of the signs and symptoms of the disorder:

- Muscle weakness
- Intellectual disability
- Seizures
- Problems with communication and social interaction that resemble autism
- Involuntary movements such as facial tics and tremors

Generally, a wide spectrum of symptoms exist in the disease, making it difficult to diagnose without a blood test for it.

What Is the Genetic Cause of Guanidinoacetate Methyltransferase Deficiency?

Mutations in the *GAMT* gene, officially known as the "guanidinoacetate-N-methyltransferase" gene, cause GAMT. Normally, *GAMT* instructs for the production of the enzyme guanidinoacetate methyltransferase. This enzyme is essential in the two-step production of creatine. Creatine is then used in many tissues to store and use energy, including the all-important contraction of muscles and nervous system functioning.

Over 15 mutations in *GAMT* are known to disrupt the enzyme formation. The most common mutation, found in the Portuguese population, occurs when the amino acid serine replaces tryptophan at position 20. All the mutations keep guanidinoacetate methyltransferase enzyme from proper creatine synthesis. The problems then are most serious in organs and tissues that require large amounts of energy, such as the brain and muscles. *GAMT* is inherited in an autosomal recessive pattern and is located on the short arm (p) of chromosome 19 at position 13.3.

What Is the Treatment for Guanidinoacetate Methyltransferase Deficiency?

Because this is a genetic disorder, there is no cure. Some of the symptoms are partially reverse with treatments of oral creatine monohydrate.

Further Reading

"*GAMT*." 2010. Genetics Home Reference. National Library of Medicine (U.S.). http://ghr.nlm.nih.gov/gene/GAMT. Accessed 5/18/12.

"Guanidinoacetate Methyltransferase (GAMT) Deficiency." 2010. Orphanet. http://www.orpha.net/data/patho/Pro/en/GuanidinoacetateMethyltransferaseDeficiency-FRenPro1726.pdf. Accessed 5/18/12.

Gyrate Atrophy of the Choroid and Retina

Prevalence Very rare; more than 150 identified worldwide; one-third from Finland

Other Names HOGA; hyperornithinemia with gyrate atrophy of choroid and retina; OAT deficiency; OKT deficiency; ornithine aminotransferase deficiency; ornithine-delta-aminotransferase deficiency; ornithine keto acid aminotransferase deficiency; ornithinemia with gyrate atrophy

In 1902, Archibald Garrod coined the term "inborn error of metabolism" to indicate a genetic disease caused by chemical imbalance or deficiency. Gyrate atrophy appears only to affect the eye, and to understand this condition, it is essential to know a little about the structure of the eye. When one looks at an eye, he or she sees the iris or colored part of the eye that lets in light. The hard, white part of the eye is called the sclera. The choroid is a dark vascular coat of the eye between the sclera and the retina. It consists of blood vessels. The retina of the eye is the

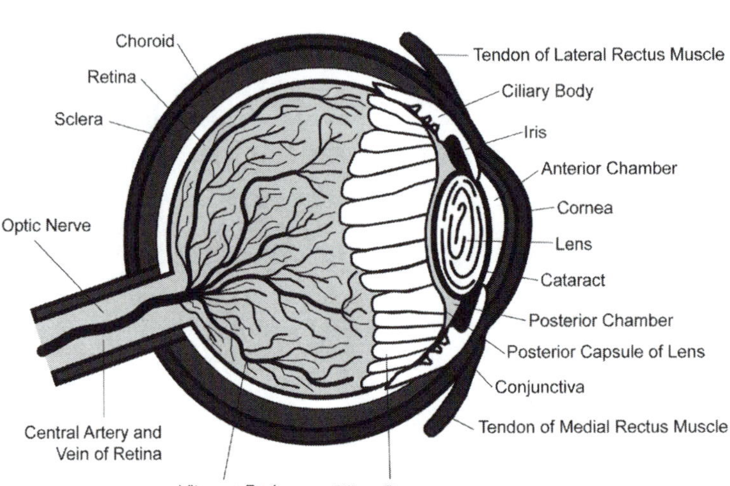

In gyrate atrophy, people with the disorder have continued loss of cells in the retina, resulting in vision loss, usually by the age of 50. (ABC-CLIO)

innermost layer of the eye, which receives the image transferred through the lens and is the intermediate instrument of vision sending the visual impulse through the optic nerve to the brain.

What Is Gyrate Atrophy of the Choroid and Retina?

Sometimes, the disorder, just referred to as gyrate atrophy, is a progressive disease of the eye. The word "gyrate" means ring-shaped or convoluted. Cells in the retina begin to atrophy. As the person goes from childhood to adulthood, the following symptoms may occur:

- Nearsightedness or myopia: The condition begins in childhood and gets progressively worse.
- Night blindness: The person has difficulty seeing anytime lights are dimmed, especially at night.
- Loss of peripheral or side vision.
- Tunnel vision: As the person gets older, the vision begins to narrow.

A form of neonatal gyrate atrophy causes ammonia to collect in the blood and leads to poor feeding, vomiting, seizures, or coma. Usually this condition is observed in the neonatal unit, and if treated promptly, it does not reoccur.

Gyrate atrophy does affect intellectual function, but abnormalities may show on brain imaging or neurological tests. Sometimes the condition will lead to disturbances in the peripheral nervous system, and the person will experience tingling or pain in hands or feet. The vision loss is usually complete between the ages of 40 and 50.

What Is the Genetic Cause of Gyrate Atrophy of the Choroid and Retina?

Mutations in the *OAT* gene, officially known as the "ornithine aminotransferase" gene, cause gyrate atrophy. Normally, the *OAT* gene provides instructions for making the enzyme ornithine aminotransferase that is active in the mitochondria, the powerhouses of the cell. It is in the mitochondria where the enzyme breaks down a substance called ornithine. Ornithine is involved in the urea cycle that processes ammonia, a nitrogen compound generated when protein is broken down. Ornithine also is active in assuring that proper balance of the amino acid building blocks is maintained. In addition, the enzyme helps convert ornithine into a molecule called pyrroline-5-carboxylate (P5C), which is then converted to basic amino acids glutamate and proline.

Gene mutations can change the amount of protein produced. However, if the mutation changes the amino acid sequence of the protein, the amount of protein produced may be normal but the activity/functionality of the protein is reduced or completely lost. The more than 60 *OAT* gene mutations affect the production of ornithine aminotransferase enzyme. When there is a shortage of enzyme activity, ornithine is not properly converted to P5C, resulting in an excess collection of ornithine in the blood. High levels of ornithine result in the progressive vision and other features that cause blindness. *OAT* is inherited in an autosomal recessive pattern and is located on the long arm (q) of chromosome 10 at position 26.

What Is the Treatment for Gyrate Atrophy of the Choroid and Retina?

Although there is no cure, it has been shown that reducing the amino acid arginine in the diet helps most patients. The person is advised to avoid arginine-rich foods, including peanuts, pecans, almonds, Brazil nuts, cashew, filberts, and grains. Currently, work is being done using a mouse model that will lead to further understanding of the disorder and to a cure.

Further Reading

"Gyrate Atrophy of the Choroid and Retina." 2010. Genetics Home Reference. National Library of Medicine (U.S.). http://ghr.nlm.nih.gov/condition/gyrate-atrophy-of-the-choroid-and-retina. Accessed 5/18/12.

"Gyrate Atrophy of the Choroid and Retina." National Center for Biotechnology Information (U.S.). http://www.ncbi.nlm.nih.gov/books/NBK22249. Accessed 5/18/12.

"*OAT*." 2010. Genetics Home Reference. National Library of Medicine (U.S.). http://ghr.nlm.nih.gov/gene/OAT. Accessed 5/18/12.

H

Hair Loss. *See* Androgenetic Alopecia

Hand-Foot-Uterus Syndrome (HFU)

Prevalence Very rare; only a few families reported worldwide; present in all ethnic groups

Other Names hand-foot-genital syndrome; HFG; HFU

Several genetic conditions affect many body parts. This rare disorder is one that causes abnormalities that appear to be disconnected—the hand, foot, urinary tract, and reproductive tract. The gene that regulates these body parts are close together and are expressed during embryonic development. HFU was first described in 1975, based on genital abnormalities in affected males. Authors of this study said that it should be renamed to hand-foot-genital (HGU) syndrome.

What Is Hand-Foot-Uterus Syndrome (HFU)?

Hand-foot-uterus syndrome is a rare syndrome that causes several disorders of the hand, feet, reproductive, and urinary systems. Following are some of the hallmarks of this syndrome:

- Abnormally short thumbs, which are sometimes relocated meaning the thumb will be in a finger position
- Small fifth fingers that curve in, a condition called clindactyly
- Short feet and abnormally short big toes

- Fusion or hardening of bones in the wrist and ankle; other bones in arms and legs are normal
- Defects in the ureters, the tubes that lead from the kidney to the bladder
- Defects in the urethra, the passageway of the urine from the bladder to outside the body
- Urinary tract infections and often with urinary incontinence, which is the inability to control the passage of urine
- Male reproductive problems—half of men have the urethra opening on the underside of the penis, a condition called hypospadias; curved penis
- Female reproductive system—duplications of uterus in nonfunctional form; early development of uterus

The individuals with HFU may still be able to have children, but women are at risk for pregnancy loss, premature labor, and stillbirth. The symptoms of the disorder may vary with different effects and varying severity in the same family.

What Is the Genetic Cause of Hand-Foot-Uterus Syndrome (HFU)?

Mutations in the *HOXA13* gene, officially known as the "homeobox A13" gene, cause hand-foot-uterus syndrome. Normally, *HOXA13* provides instructions for a protein that acts as a transcription factor, meaning it binds to specific region of DNA and regulates the activity of other genes. This gene is part of a large family of homeobox genes that are at work in the embryo that controls how many parts of the body and its functions. *HOXA13* is necessary for forming and developing limbs, the urinary tract, and the reproductive system. *HOXA13* has three areas that make triplets of the amino alanine; these repeats are called polyalanine or poly(A) tracts. *Poly* is the Greek root meaning "many."

About 14 mutations in *HOXA13* cause hand-foot-uterus syndrome. Most of the mutations add additional polyalanine tracts to the gene, which make it very long and unstable. The changed protein is broken down by the cells and is not available to regulate other genes. All these changes affect the systems in the individual. *HOXA13* is inherited in an autosomal recessive pattern and is located on the short arm (p) of chromosome 7 at position 15.2.

What Is the Treatment for Hand-Foot-Uterus Syndrome (HFU)?

No specific treatment exists for HFU syndrome. However, management of many of the symptoms is possible with surgery or drugs. For example, urinary infections or bladder infections can be controlled with drugs targeted for the symptoms.

Further Reading

"Hand-Foot-Genital Syndrome." 2010. Genetics Home Reference. National Library of Medicine (U.S.). http://ghr.nlm.nih.gov/condition/hand-foot-genital-syndrome. Accessed 5/18/12.

"Hand-Foot-Uterus Syndrome." 2002. Gale Encyclopedia of Public Health. http://www.healthline.com/galecontent/hand-foot-uterus-syndrome-1. Accessed 5/18/12.

"*HOXA13*." 2010.Genetics Home Reference. National Library of Medicine (U.S.). http://ghr.nlm.nih.gov/gene/HOXA13. Accessed 5/18/12.

Harlequin Ichthyosis (HI)

Prevalence Very rare; no more than 100 cases reported worldwide

Other Names harlequin baby syndrome; HI; ichthyosis congenital; harlequin fetus type

The word "harlequin" emerged from old French, describing a clown or buffoon who wore a mask and donned an outfit with brightly colored diamond shapes. The clown wore a mask with an evil, smiling grimace and, as early as 1519, was pictured in paintings of Dante's *Inferno* chasing evil souls to go to hell. Harlequin is also used to describe certain animals with the diamond-shaped patterns, especially the Great Dane harlequin. However, the condition harlequin ichthyosis is not innocent or humorous. It is a serious genetic condition of major consequences.

Reverend Oliver Hart first described the condition in 1750 when he visited a child born to Mary Evans in Charleston, South Carolina. He wrote that he saw a "most deplorable object of a child" whose skin was dry and cracked like the scales of a fish. The child had only two holes for eyes and no ears and lived only 48 hours. He was describing the condition harlequin ichthyosis.

What Is Harlequin Ichthyosis (HI)?

Harlequin ichthyosis is a rare skin disorder in which the infant is born with thick diamond-shaped plates of keratin. Keratin is the substance that is present in the hair, nails, and other horny tissue. In HI, keratin is overexpressed in the skin, a condition known as hyperkeratosis. Horny, fishlike scale plates cover the entire body. The plates have a diamond shape like the suit of the harlequin clown. The word "ichthyosis" comes from two Greek words: *ichthus*, meaning "fish," and *osis*, meaning "condition of." Thus, the child with harlequin ichthyosis has the horny plates, like one might see as scales on a fish, separated by raw, cracked areas.

In addition to the diamond-shaped plates, the following abnormalities may be present:

- Eyes: Eyes may appear as a red bloody mass because the eyelids are covered with the horny plates.
- Mouth: Because of the tension of the plates, the mouth is pulled wide open, mimicking the harlequin's staid, smiling mask.

- Ears: Ears may not be present or may be a small stub.
- Nose: The nose may have only two small holes.
- Appendages: Arms, feet, and legs are so deformed with the heavy plates that they cannot bend. Extra fingers or polydactyly may be present.
- Dehydration: The heavy skin plates prevent heat loss, resulting in overheating; the open areas around the plates allow water to evaporate through the skin, causing dehydration.
- Breathing: Because of the tight skin across the chest, breathing is restricted. The surrounding armor-like plates can lead to respiratory failure.

The open areas around the dense plates leave the body open to infection and other metabolic disorders.

What Is the Genetic Cause of Harlequin Ichthyosis (HI)?

Changes in the *ABCA12* gene, officially known as the "adenotriphosphate or ATP-binding cassette, subfamily A, member 12" gene, cause HI. Normally, *ABCA12* instructs for a protein known as ATP-binding cassette (ABC) transporter, which carries molecules across the cell membrane. The ABC family is thought to mediate the lipid or fat transfer system from the cytosol of the comeocyte (skin cells) and then into the lamellar granules. Lamellar granules are highly specialized, lipid-rich organelles (small organs) within the epidermis. The granules then secrete the lipids that maintain the skin layer.

Mutations disrupt the production of the *ABCA12* protein so that even before birth the proteins do not function. In children with HI, these lamellar granules are either absent or abnormal. The skin does not form the protective barrier, and the thick, hard plates cause the symptoms of HI. *ABCA12* is inherited in an autosomal recessive pattern and is located on the long arm (q) of chromosome 2 at position 24.

What Is the Treatment for Harlequin Ichthyosis (HI)?

Until the discovery of isotretinoin, a strong vitamin A product, HI always meant an early death. However, treatment has changed this somewhat. Born in 1984, Nusrit (Nelly) Shaheen has survived with care and treatments. She had eight siblings; four of the children had HI and died as infants. As of this writing, Nelly leads an active lifestyle and is a sports coach.

Further Reading

"*ABCA12*." 2010. Genetics Home Reference. National Library of Medicine (U.S.). http://ghr.nlm.nih.gov/gene/ABCA12. Accessed 5/18/12.

"Harlequin Ichthyosis." 2010. Medscape. http://emedicine.medscape.com/article/1111503-overview. Accessed 5/18/12.

Milner, M., et al. 1992. "Abnormal Lamellar Granules in Harlequin Ichthyosis." *Journal of Investigative Dermatology*. 99: 824–829. http://www.nature.com/jid/journal/v99/n6/abs/5611841a.html. Accessed 5/18/12.

Haw River Syndrome. *See* Dentatorubral-Pallidoluysian Atrophy (DRPLA)

Hemifacial Microsomia (HFM)

Prevalence 1 in 3,500 to 4,500

Other Names craniofacial microsomia; first and second branchial arch syndrome; Goldenhar syndrome; lateral facial dysplasia; oral-mandibular-auricular syndrome; otomandibular dysotosis

The "Little Baby Face Foundation" is an organization that helps children who are born with severe facial deformities. Children with hemifacial microsomia are determined to be one of the conditions. This organization brings the child and family to New York City free of cost to have corrective surgery

The term "hemifacial microsomia" comes from Greek roots: *hemi*, meaning "half"; *facie*, meaning "face"; *micro*, meaning "small"; and *some*, meaning "body." The half of the face with the small, distorted facial structures is characteristic of this condition. Hemifacial microsomia is the second-most common facial birth defect after cleft lip and cleft palate.

What Is Hemifacial Microsomia (HFM)?

Hemifacial microsomia is a condition that affects bone, muscle, fat, and nerves of the lower half of the face. In this condition, that part of the face is underdeveloped and does not grow normally. The condition may affect tissues on one side of the face or on both sides of the face. However, it always involves the ear and the mandible or jaw bone.

The deformity may vary in both severity and in the area of the face that is involved. In milder cases only some of the structures are affected. However, the most severe cases may display the following hallmarks:

- Underdeveloped external and middle ear
- Underdeveloped side of the skull
- Thickness of the cheek tissue

- Underdeveloped lower jaw
- Problems with the teeth
- Disorders in the nerves that affect facial movement

Although there may be problems with the face, some people have cardiac, vertebral, and central nervous system defects.

What Is the Cause of Hemifacial Microsomia (HFM)?

Hemifacial microsomia develops during the embryonic state at approximately four weeks after fertilization. HFM involves the first and second branchial arch, an area of embryonic development. Here, a vascular problem occurs that leads to clotting and a poor supply of blood to the face. A physical trauma or some event may cause this condition that restricts the development of that area of the face.

Although most of the cases are sporadic, there is substantial evidence for genetic involvement in this condition. A genome-wide search in two families with the features of HFM showed that there was a possible linkage to the long arm (q) of chromosome 14 at position 32. Both autosomal dominant and autosomal recessive patterns have been observed. Also, with such a widespread birth defect, there may be many factors of both genetics and environment involved.

What Is the Treatment for Hemifacial Microsomia (HFM)?

Depending on the treatment that is necessary, generally facial reconstruction begins about the age of five. The first operation involves lowering the upper jaw on the affected side to match the opposite side. A piece of rib may be taken from the child to lengthen the lower jaw. This allows the chin to be moved back to the center of the face. For many children, only one operation is needed on the jaws. Work on the ears begins when the child is about six years old. Required are three or four operations to rebuild the child's outer ear. Sometimes, other surgeries may be essential to add bone to build up the cheek. Finally, the patient may need surgery to correct the flattened forehead and to reposition the eye. To treat this complex problem, the child should be under the care of a qualified craniofacial team.

Further Reading

"A Guide to Understanding Hemifacial Microsomia." 2010. Children's Craniofacial Association. http://www.ccakids.com/Syndrome/Microsomia.PDF. Accessed 5/18/12.

"Hemifacial Microsomia." 2010. Children's Hospital of Wisconsin. http://www.chw.org/display/PPF/DocID/21821/router.asp. Accessed 5/18/12.

"Hemifacial Microsomia." 2011. FACES: The National Craniofacial Association. http://www.faces-cranio.org/Disord/Hemi.htm. Accessed 5/18/12.

Hemochromatosis

Prevalence 2 in 1,000 Caucasians; over a million people in the United States carry the gene

Other Names iron overload

Do you have red hair? Did you know that if you have Irish ancestry that you are at a higher risk of hemochromatosis? In fact, hemochromatosis is known as the "Celtic Curse." Sandra Thomas, president of the American Hemochromatosis Society, founded the society in memory of her mother, who had hereditary hemochromatosis. Her mother was of Irish and Scottish descent. One of the goals of this society is for people of Irish descent to get tested and learn how to prevent the condition.

What Is Hemochromatosis?

Hemochromatosis is a condition in which the body accumulates too much iron from foods or other sources such as vitamins containing iron. The extra iron builds up in the body and, over a period of time, damages organs and tissues if not treated. The word "hemochromatosis" comes from the following Greek words: *hemo*, meaning "blood"; *chromat*, meaning "color"; and *osis*, meaning "condition of." Thus, the blood will appear a strong red color because of the overload of iron. Hemochromatosis is generally hereditary or may be the result of a metabolic disorder.

Hemochromatosis is hard to identify early because many of the symptoms resemble those of other conditions. People may have the symptoms at a younger age but do not recognize them until middle age. Following are the early symptoms:

- Fatigue and feeling very tired
- Weakness
- Weight loss
- Abdominal pain
- Joint pain

As iron builds up in the body organs, the following may occur:

- Early menopause of erratic menstrual periods
- Loss of sex drive or impotence
- Loss of body hair
- Shortness of breath
- Arthritis
- Liver problems
- High blood sugar and diabetes

- Heart problems heart failure
- Gray-colored or bronze-colored skin

Most cases in the United States are hereditary. The child individual inherits a gene from both parents, but symptoms rarely appear before adulthood.

What Is the Genetic Cause of Hemochromatosis?

Changes in the *HFE* gene, officially known as the "hemochromatosis" gene, cause hereditary hemochromatosis, Normally, *HFE* provides instruction for a protein that is located on the surface of cells. The most common cells are those in the liver, intestines, and immune system. *HFE* protein acts as a detector and regulator of the amount of iron in the body. The HFE protein regulates the production of another protein called hepcidin, the master iron regulatory hormone. Hepcidin is produced in the liver and calculates how much iron should be absorbed from the diet and how much is released from storage sites in the body. When the proteins are working correctly, iron is tightly regulated. The body absorbs about 10% of iron obtained from the diet.

More than 20 mutations in *HFE* cause hemochromatosis. In one major mutation, the amino acid cysteine replaces tyrosine; in another mutation, aspartic acid replaces histidine. The mutations disrupt the iron regulation mechanism, causing the iron overload that is characteristic of hemochromatosis. *HFE* is inherited in an autosomal recessive pattern and is located on the short arm (q) of chromosome 6 at position 21.3.

What Is the Treatment for Hemochromatosis?

Treatment consists of taking blood from the arm, much like giving blood. The treatment is safe and effective. Patients can expect to live a normal life if treatment is started before organs are damaged.

Further Reading

American Hemochromatosis Society. 2010. http://www.americanhs.org. Accessed 5/18/12.

"Hemochromatosis." 2010. Hemochromatosis Information Center. http://www.hemochromatosis.org. Accessed 5/18/12.

"Hemochromatosis (Iron Storage Disease)." 2010. Centers for Disease Control and Prevention (U.S.). http://cdc.gov/ncbddd/hemochromatosis/facts.html. Accessed 5/18/12.

Hemolytic-Uremic Syndrome

Prevalence About 1 in 500,000 per year in the United States

Other Names atypical hemolytic-uremic syndrome

Two basic types of hemolytic-uremic syndrome exist. The first is called typical. The typical type is caused by a bacterial infection such as *E. coli* that causes chronic diarrhea that can lead to life-threatening end-stage kidney disease. The typical form usually affects children under 10. This article focuses on the atypical form, which appears to involve be hereditary.

What Is Hemolytic-Uremic Syndrome?

Hemolytic-uremic syndrome is a serious, life-threatening disorder that primarily affects the kidneys. In this condition, small clots form in the tiny blood vessels of the kidneys. These clots keep blood from easily flowing to the kidneys, and as the clots grow, they block the blood flow completely. The following three major hallmarks of the disease result from the clots:

- Hemolytic anemia: Anemia is a condition in which the red blood cells do not function properly and fail to carry their load of oxygen to the body cells. When the red blood cells encounter the constricted area around the clot, they must squeeze through. As they try to pass through the small area, some of them are broken apart and destroyed. If they are destroyed faster than the body can make new ones, the person will develop anemia. The individual will appear pale with yellow, jaundiced eyes. He or she will also be very tired, have shortness of breath, and a rapid heart rate.
- Thrombocytopenia: The thrombus or clot is formed when blood platelets combined with some other substances seal an area to keep it from bleeding. However, when the clots form in the little blood vessels of the kidney, the blood platelets are called into action. If another area needs them for an emergency, they are not available. Therefore, the person will bruise easily and experience abnormal bleeding.
- Kidney damage and acute kidney failure: The blockage in the kidney keeps the organ from disposing of wastes and properly filtering the blood. Accumulated poisons damage the organ. All this leads to a life-threatening condition called end-stage renal disease, or ESRD.

What Are the Genetic Causes of Hemolytic-Uremic Syndrome?

Mutations in the *CFH* gene, officially called the "complement factor H" gene, cause hemolytic-uremic syndrome. *CFH* is found in about 30% of cases. Six other suspect genes cause only a small number of cases. Normally, *CFH* and the other genes provide instruction for a protein involved in the body's immune system. When all the factors of the immune system work together, it is known as the complement system. Proteins produced in this group do the work of the immune system: destroying invaders such as bacteria and viruses, producing essential inflammation, and removing unwanted or unneeded material from the body. In order that these proteins will not turn on the body and attack the wrong cells, a highly regulated system is implemented.

Mutations in the genes affect the system of controls and regulation. The complement system becomes overactive and attacks the cells of the kidney, which then lead to the abnormal clotting and inflammation. Damage to the kidney is progressive, leading to kidney failure and eventually end-stage renal disease.

The presence of the genes themselves is usually not enough to cause hemolytic-uremic syndrome. Certain triggers such as medications, especially anticancer drugs, chronic diseases, viral or bacterial infection, cancers, organ transplants, or pregnancy may evoke the condition. Some cases cannot be traced to a trigger or cause; in this case, the disorder is called idiopathic.

What Is the Treatment for Hemolytic-Uremic Syndrome?

Treatment and attention to symptoms is critical. A blood transfusion may be essential for cases of anemia. Kidney failure must be managed by observation and care. Temporary dialysis may be needed while awaiting recovery of the kidneys from the illness.

Further Reading

"Atypical-Hemolytic-Uremic Syndrome." 2010. Genetics Home Reference. National Library of Medicine (U.S.). http://ghr.nlm.nih.gov/condition/atypical-hemolytic-uremic-syndrome. Accessed 5/18/12.

"*CFH*." 2010. Genetics Home Reference. National Library of Medicine (U.S.). http://ghr.nlm.nih.gov/gene/CFH. Accessed 5/18/12.

"Hemolytic-Uremic Syndrome in Children." 2010. National Institute of Diabetes, Digestive Diseases, and Kidney Disorders. http://kidney.niddk.nih.gov/kudiseases/pubs/childkidneydiseases/hemolytic_uremic_syndrome. Accessed 5/18/12.

Hemophilia

Prevalence Most common in males; Hemophilia A, 1 in 400 to 1 in 5,000 males; Hemophilia B, 1 in 20,000 newborn males worldwide

Other Names hemophilia, familial; hemophilia, hereditary; Type B Christmas disease or hemophilia B Leyden.

Hemophilia is an old and important disease affecting the welfare of states and kingdoms. Among the first early references was a statement in the Jewish *Talmud*, written in the second century, that a mother who has had two sons who have died with bleeding disease did not have to have the third one circumcised. Albucasis, an Arab physician, described several families whose males died after minor accidents. Several other allusions to bleeding disease have appeared in historical writings. However, the first account in the nineteenth century was written by Dr. John Otto, a Philadelphia doctor, who described "bleeders" in certain families and then later traced the same condition back to a woman who came to Plymouth in 1720. He implied here a link to women and X-linked genetic conditions.

Sometimes, hemophilia is called "the royal disease" because it was carried in the European royalty. Queen Victoria, who ruled the British Empire throughout most of the nineteenth century, passed the disorder to several of her female descendants. These women became members of the royalty in Spain, Germany, and Russia. Perhaps the most important case was in Russia, where hemophilia played a big part in the Russian Revolution of 1917.

What Is Hemophilia?

Hemophilia is a rare bleeding disorder in which the blood does not clot normally. Following an injury or even a simple cut, the person may bleed profusely. Having surgery or even having a tooth pulled may involve prolonged bleeding. Also, spontaneous bleeding may occur internally without a cut and affect joints, muscles, brain, or other internal organs. A mild form may not involve the profuse and spontaneous bleeding and does not show up unless the person has a tooth pulled or surgery.

A hemophiliac boy in North Carolina, eight years old, prepares his twice-weekly injection of a very expensive medication that his family was able to afford by being enrolled in the Children's Health Insurance Program (CHIP), after being on a waiting list. (Photo by Steve Liss/Time & Life Pictures/Getty Images) (Steve Liss/Contributor / Time & Life Pictures/Getty Images)

Cries throughout the Palace: How Genetics Brought down an Empire

Genetics possibly brought down the powerful rulers of Russia. It began with the birth of little Alexei Nikolaevich, son of the tsar Nicolas II. Alexis had inherited a serious bleeding disease from his mother Alexandra, who was a granddaughter of Queen Victoria of England. According to guides at the beautiful palace in Saint Petersburg, one could hear him screaming and crying throughout the palace.

Getting no assistance from the doctors of the day, his mother looked elsewhere and turned to a strange, Russian mystic named Gregori Rasputin. Rasputin declared that he had supernatural powers and that he could heal the young boy. Using hypnotism combined with herbs, he would bring back the little boy when he was on the brink of death. the tsarina Alexandra and others in the court began to believe that Rasputin had a special relationship with God and trusted him implicitly.

As Rasputin gained the confidence of the Court, he began to get tremendous powers in the affairs of state, and he became known as the "Mad Monk." He was vindictive and erratic. It was rumored that Alexandra had an affair with him. Although genetic analysis did not determine definitely that the young boy had hemophilia until 2009, it is surmised that this genetic disorder helped to discredit the tsarist government. The distrust led to the downfall of the Romanovs in 1917 and to the eventual Bolshevik (Communist) takeover later that year.

The word "hemophilia" comes from two Greek roots: *hemo*, meaning "blood," and *philos*, meaning "love." Hence, the condition reflects the inability of the body to stop bleeding. People with hemophilia have little or no blood clotting factors, essential elements for controlling bleeding. These protein factors work with blood platelets to form the clot. There are two main types of hemophilia, which have similar characteristics. Both have the missing factors that are essential for blood clotting. Internal bleeding may occur with either type. Following are the two main types of hemophilia, which are dependent upon two blood factors:

- Hemophilia A (known as classic hemophilia): People with this most common form of the disorder have a deficiency in clotting factor VIII.
- Hemophilia B (known as hemophilia B Leyden or Christmas disease): People with this rarer form have a deficiency in clotting factor IX.

Acquired hemophilia is a form of the disease that is not inherited and is characterized by abnormal bleeding in the skin and muscles, and usually begins in adulthood.

What Are the Genetic Causes of Hemophilia?

Two mutated X-linked genes are responsible for hemophilia: *F8* and *F9*. Because it is X-linked, the condition is more common in males. Women who are carriers of the defective gene may experience some mild symptoms. In about one-third of the cases, no family history of hemophilia exists; the disorder appears to be a spontaneous mutation.

F8 Gene

Mutations in the *F8* gene, officially called the "coagulation factor VIII, procoagulant component" gene, cause hemophilia A. Normally, *F8* provides instructions for the protein coagulation factor VIII that is essential for the formation of blood clots. When one is injured, the clots seal off the damaged blood vessels to prevent loss of blood. Made by the cells in the liver, factor VIII roams around the bloodstream in an inactive form. It is bound to a molecule called the von Willebrand factor. When a cut occurs, factor VIII rushes to the site and interacts with another factor called factor IX. The two initiate a chain reaction to cause a clot.

Most of the approximately 1,300 mutations in *F8* involve an exchange of only one base pair of proteins. Other mutations may delete or insert certain base pairs. The change in the gene produces a defective or missing protein, which cannot aid in blood clotting. As a result the symptoms of hemophilia occur. *F8* is inherited in an X-linked recessive pattern and is located on the long arm (q) of the X chromosome at position 28.

F9 Gene

Mutations in the F9 gene, officially called the "coagulation factor IX," lead to Hemophilia B. Normally, *F9* instructs for a protein called coagulation factor IX, which is essential for blood clotting. It is also made in the liver and circulated in an inactive form until injury occurs. When an injury happens, the coagulation factor IX interacts with factor VIII and helps form the very important blood clot.

More than 900 mutations in *F9* involve changes in a single DNA building block. The abnormal genes cause a disruption in work of the coagulation factor IX protein to enable clotting and evokes the symptoms of hemophilia. *F9* is inherited in a recessive X-linked pattern and is located on the long arm (q) of the X chromosome at position 27.1-q27.2.

What Are the Treatments for Hemophilia?

Currently, no cure exists for hemophilia. Treatment is available, but it is costly and requires lifelong infusion or clotting factor that is made from human plasma or recombinant technology. In the 1980s, most of the people with severe hemophilia became infected with HIV/AIDS because of contamination of the blood supply. However, since 1986, the regulations and methods of blood transfusion have improved, and the blood supply is now safe. Ryan White was a young American hemophiliac whose story helped to call attention to the problem of the contamination in the blood supply.

Further Reading

"*F8*." 2010. Genetics Home Reference. National Library of Medicine (U.S.). http://ghr.nlm.nih.gov/gene/F8. Accessed 5/18/12.

"*F9*." 2010. Genetics Home Reference. National Library of Medicine (U.S.). http://ghr.nlm.nih.gov/gene/F9. Accessed 5/18/12.

"Hemophilia." 2010. Genetics Home Reference. National Library of Medicine (U.S.). http://ghr.nlm.nih.gov/condition/hemophilia. Accessed 5/18/12.

National Hemophilia Foundation. http://www.hemophilia.org. Accessed 5/18/12.

"What Is Hemophilia?" 2011. National Heart, Lung, and Blood Institute. http://www.nhlbi.nih.gov/health/dci/Diseases/hemophilia/hemophilia_what.html. Accessed 2/2/11.

Hereditary Angioedema

Prevalence	Affects 1 in 50,000 people; type I most common, accounting for 85% of cases; type II, 15% of cases; type III, very rare
Other Names	C1 esterase inhibitor deficiency; C1 inhibitor deficiency; HAE; HANE; Quincke edema; secondary angioneurotic edema

Allergic reactions can occur in many situations. A red itchy rash known as uticaria may appear with no known cause. Hives or large or small wheals or whelps may happen when one comes in contact with certain foods or substances. However, the German physician Heinrich Quincke found a condition in which swelling appeared to run in families. At one time, the condition was named for him, but now is called hereditary angioedema or HAE.

What Is Hereditary Angioedema?

Hereditary angioedema is characterized by local swelling under the skin. The disorder usually appears in the second to fourth decade, but symptoms may appear in childhood and worsen in puberty. Attacks may occur every week or so and last from 3 to 4 days. The condition varies greatly even among members of the same family.

Episodes may involve the following areas: swelling in the limbs and face; swelling in the abdominal tract involving pain, vomiting, and nausea; and swelling in the airways, which can be life-threatening. One-third of people develop a non-itchy rash called erythema marginatum during the attack. Several things, including stress, may trigger an attack, but most attacks are idiopathic or without a known cause.

Three types of hereditary angioedema exist:

- Type I: This type is caused by a deficiency of the C1 inhibitor, a protein that is important in the immune system. A low level of the C1 inhibitor exists in people with type I angiodema.
- Type II: This type is caused when normal levels of the C1 inhibitor are present but are ineffective.
- Type III: This type was once thought to be present only in women, but men may also have the condition. Mutations in a different gene cause similar symptoms to types I and II.

What Are the Genetic Causes of Hereditary Angioedema?

SERPING1 Gene

Mutations in the *SERPING1* gene, officially called the "serpin peptidase inhibitor, clade G (C1 inhibitor), member 1" gene, cause Types I and II of hereditary angiodema. Normally, *SERPING1* instructs for a protein called the C1 inhibitor, a type of serpin or serine protease inhibitors. Serpins block the activity of certain proteins. The C1 inhibitor controls the process of maintaining blood vessels, which includes inflammation that serves as the body's normal response to injury or infection. The action of the C1 inhibitor is to block blood proteins called plasma kallikrein and factor XII, which produce bradykinin, a substance responsible for inflammation. C1 binds these two proteins and prevent them from having adverse effects.

About 250 mutations in *SERPING1* cause HAE types I and II. The mutations cause the C1 inhibitor not to function properly, and therefore it does not block the excessive amounts of bradykinin that are produced. Thus, excess plasma leaks through wall of the blood vessels causing swelling. *SERPING1* is inherited in an autosomal dominant pattern and is located on the long arm (q) of chromosome 11 at position 12-13.1.

F12 Gene

Mutations in the *F12* gene, officially called the "coagulation factor XII, Type III angioedema" gene, causes Type III angioedema. Normally, *F12* provides instructions for the protein coagulation factor XII, a substance that is essential for blood coagulation. When the body is injured, activation of certain factors including Factor XII form a clot. Factor XII has a role in producing inflammation, the body's response to injury. Factor XII is also important in forming bradykinin that has a role in letting fluids go into the injured area, hence cause the familiar swelling.

In hereditary angioedema type III, at least two mutations of *F12* are involved. The mutations are the result of only one change in the amino acid building blocks that result in factor XII. Without factor XII, bradykinin allows fluid to leak through the wall of the blood vessels, leading to swelling and the typical symptoms of hereditary angioedema. *F12* is inherited in an autosomal dominant pattern and is located on the long arm (q) of chromosome 5 at position 33-qter.

What Is the Treatment for Hereditary Angioedema?

To diagnose the condition, clinicians perform a blood test during an episode and analyze the C1 inhibitor levels. The aim of treatment is to stop the progression of edema as soon as possible, especially if it affects the airways. Some treatments in other countries using human blood were not approved in the United States. Newer treatments include several biologics, recombinant drugs such as ecallantide, which inhibits certain plasma proteins. Helping the patient find the cause of the edema attack is important in avoiding contact. Antihistamines or corticosteroids, which help in many allergic reactions, do not appear to help people with HAE.

Further Reading

"Hereditary Angioedema." 2010. Medscape. http://emedicine.medscape.com/article/135604-overview. Accessed 5/18/12.

"*SERPING1*." 2011. Genetics Home Reference. National Library of Medicine (U.S.). http://ghr.nlm.nih.gov/gene/SERPING1. Accessed 5/18/12.

Hereditary Antithrombin Deficiency (AT)

Prevalence 1 in 2,000 to 3,000; for those who have experienced a blood clot, about 1 in 20 to 200 have hereditary antithrombin deficiency

Other Names antithrombin-3 deficiency; congenital antithrombin-3 deficiency; deep vein thrombosis

Travelers on Qantas, the national airline of Australia, are given specific directions for avoiding deep vein thrombosis. Because the patrons must sit for hours on a plane, they are at risk for developing blood clots, which may dislodge and travel through the bloodstream to the brain or lungs. Videos demonstrate how to exercise in one's seats, and passengers are encouraged to move about the cabin every hour or so. In fact, not only air travel, but long jaunts across the country, motionless days in bed in the hospital, and trauma to the legs may put one at risk for DVT, especially if the person has hereditary antithrombin deficiency.

What Is Hereditary Antithrombin Deficiency (AT)?

Antithrombin deficiency is a condition in which the person's blood does not clot properly. These individuals who have the deficiency are a great disk for developing blood clots, usually in the deep veins of the legs. These clots may break away and locate in the lungs, causing a pulmonary embolism.

In 1905, Morovitz conceived the idea that a substance in the blood controlled blood clotting. Thrombin is a blood protein that is extremely important in the process of clot formation. When one is cut, the protein thrombin along with blood platelets forms a web that becomes a clot so that the person's bleeding will stop. However, as Morovitz theorized, the clotting cannot go on unchecked, and a substance in the blood must control the reaction; that substance is antithrombin. When antithrombin does not work or is deficient, the clotting mechanism is disrupted, and the individual is at risk for blood clots, organ damage, and death.

Antithrombin deficiency can be hereditary or acquired under certain circumstances. The following circumstances may be involved:

- Age: AT may show up early in life usually after adolescence when the person has a first blood clot. However, increasing age is a factor in developing antithrombin deficiency.
- Surgery: People who undergo surgery and lie motionless for hours are at risk for blood pools to gather in the legs. Hospitals are usually very proactive in treating patients using leg messaging devices, such as stockings or machines.
- Immobility: People who are immobile for long periods are at risk.
- Pregnancy: Women with AT will be more likely to have a miscarriage or stillborn child.

What Is the Genetic Cause of Antithrombin Deficiency (AT)?

Mutations in the *SERPINC1* gene, officially called the "serpin peptidase inhibitor, clade C (antithrombin), member 1" gene, causes hereditary AT. Normally, *SERPINC1* provides instructions for making the protein antithrombin. This protein blocks clotting in the bloodstream in places where clotting should not take place. Mutations in a single building block of this gene may disrupt the process and cause a clot to form easily under the right conditions. *SERPINC1* is inherited in an autosomal dominant pattern and is located on the long arm (q) of chromosome 1 at position 23.

What Is the Treatment for Antithrombin Deficiency (AT)?

The individual with AT may experience pain or leg swelling or have no symptoms at all. If the thrombus breaks away and lodges in the lung, surgery may be necessary. Preventing the condition is the best practice.

Further Reading

"Hereditary Antithrombin Deficiency." 2010. Genetics Home Reference. National Library of Medicine (U.S.). http://ghr.nlm.nih.gov/condition/hereditary-antithrombin-deficiency. Accessed 5/19/12.

Patnalk, M. M., and S. Moll. 2008."Inherited Antithrombin Deficiency: A Review." National Center for Biotechnology Information (U.S.). *Haemophilia*. November. 14 (6): 1229–1239. http://www.ncbi.nlm.nih.gov/pubmed/19141163?dopt=Abstract. Accessed 5/19/12.

"*SERPINC1*." 2010. Genetics Home Reference. National Library of Medicine (U.S.). http://ghr.nlm.nih.gov/gene/SERPINC1. Accessed 5/19/12.

Hereditary Folate Malabsorption

Prevalence Unknown; 19 families to date identified with the condition

Other Names Congenital folate malabsorption

The study of vitamins and their deficiencies appears to be a never-ending search with many questions still unanswered. The B vitamins are called folates and are the subject of nutritional research and their relationship to the health of an individual. In 1961, Lubhy and colleagues observed a family in which the children were born normal, but in a few months they were very ill and wasting away. The researchers determined the children had a vitamin B deficiency.

What Is Hereditary Folate Malabsorption?

Hereditary folate malabsorption is a condition in which the body does not absorb certain B vitamins from food. The B vitamins or folates are essential for important cell function such as producing DNA and RNA.

Children appear normal at birth because the mother has provided the vitamins before birth, but within a few months, they begin to show the following symptoms:

- Feeding difficulties
- Diarrhea
- Failure to gain weight
- Lack of energy
- Pale skin
- Tingling or numbness in the hands and feet
- Susceptibility to infections
- Bruising and abnormal bleeding
- Developmental delay
- Seizures

The symptoms are caused by disorders in the blood. The individual will develop megaloblastic anemia, a condition with a low number of very large red blood cells. A low number of white blood cells leads to many infections. A low number of blood platelets results in easy bruising and abnormal bleeding. Some of the children will have neurological problems such as intellectual disability and seizures.

What Is the Genetic Cause of Hereditary Folate Malabsorption?

Mutations in the *SLC46A1* gene, officially known as the "solute carrier family 46 (folate transporter), member 1" gene, cause hereditary folate malabsorption. Normally, *SLC46A1* provides instructions for the proton-coupled folate transporter known as PCFT. This transporter is essential for the proper function of the intestinal lining. The lining of the intestine has small fingerlike projections called microvilli where the nutrients are absorbed from food as it passes through the intestine. The microvilli appear as tiny brushes; in fact, the presence of the villi are referred to as the brush border. In a process called active transport, PCTF assists in moving folates across the brush border for digestion. PCTF also is involved in the transport of folates between the brain and spinal fluid.

Mutations in *SLC46A1* disrupt the production of the PCTF protein so that there is little or no activity. When the protein is nonfunctional, the folates are not absorbed from food and the signs of the deficiency occur. *SLC46A1* is inherited in an autosomal recessive pattern and is located on the long arm (q) of chromosome 17 at position 11.2

What Is the Treatment for Hereditary Folate Malabsorption?

Vitamin B must be given to the person with hereditary folate malabsorption. This is usually done with intramuscular injections, although an oral preparation is available. Early treatment before the symptoms appear can prevent the metabolic consequences of the effect on the blood cells. Sometimes, blood transfusions are essential.

Further Reading

Diop-Bove, Ndeye, et al. 2008. "Hereditary Folate Malabsorption." *GeneReviews*. http://www.ncbi.nlm.nih.gov/books/NBK1673. Accessed 5/19/12.

"Hereditary Folate Malabsorption." 2010. Genetics Home Reference. National Library of Medicine (U.S.). http://ghr.nlm.nih.gov/condition/hereditary-folate-malabsorption. Accessed 5/19/12.

"*SLC46A1*." 2010. Genetics Home Reference. National Library of Medicine (U.S.). http://ghr.nlm.nih.gov/gene/SLC46A1. Accessed 5/19/12.

Hereditary Hearing Disorders and Deafness: A Special Topic

Prevalence The most common birth defect; one in 500 newborns has some hearing loss

Loss of hearing is devastating to people. As one of the major senses, hearing is important in daily and professional lives. Because hearing loss has so many causes, it is impossible to mention all of them. Some problems arise from contemporary lifestyles, including listening to loud music. Some experts predict a major epidemic of hearing loss as people who listen to loud music age. Malformations in the inner and outer ear may also occur, as well as many conditions that appear to have no direct cause. Hearing loss can occur at any age.

Hearing loss in children could be described in the following ways:

- Fewer than half of the cases are of environmental or unknown causes.
- About an equal number of cases are genetic that are not related to a syndrome. These genes are called nonsyndromic. This means that hearing loss occurs without other conditions, such as heart defects or other physical problem. Genes that cause deafness can be recessive, dominant, X-linked, or mitochondrial.
- About a quarter of hearing defects are related to syndromes. This means that the condition comes in a package with other physical defects, such as cleft palate, limb problems, or other bodily disorders.

The entry focuses on the genetics of hearing loss. Genes can cause a range of hearing loss from mild to profound. Hearing loss is measured in decibels (dB). Hearing is considered normal if the individuals are thresholds are within 15 dB of normal hearing when measured by a series of tone bursts. The following chart shows the severity of hearing loss in decibels:

Hereditary Hearing Loss

Severity	Hearing Threshold in Decibels
Mild	26–40
Moderate	41–55
Moderately severe	56–70
Severe	71–90
Profound	90

Two types of tests determine the diagnosis.

- Physiologic tests such as auditory brain stem response (ABR), auditory steady-state response testing (ASSR), evoked otoacoustic emissions (EOAEs), and immittance testing. These tests accumulate data and provide quantitative information for analysis.
- Audiometry is more subjective and includes behavioral testing; pure-tone audiometry; air conduction audiometry; bone conduction audiometry; conditioned play audiometry, which is used in children age 2.5 to 5; conventional audiometry, which is used in children five years of age and older; and audioprofile, which is a composite of several audiograms taken over a period of time.

Newborn hearing screening (NBHS) can identify congenital hearing loss. NBHS is required in 33 states plus the District of Columbia. In other states, it is offered but not required.

Nonsyndromic Hearing Loss

Nonsyndromic hearing loss can be inherited in an autosomal recessive, autosomal dominant, X-linked, or mitochondrial DNA pattern.

Autosomal Recessive

More than 50% of deafness in children is caused by genes that are inherited in an autosomal recessive pattern. At present, 40 out of 92 nonsyndromic recessive genes have been identified.

About 50% of these cases are attributed to the disorder DFNB1 (the designation for nonsyndromic hearing loss) that is caused by the *GJB2* gene, officially called the "gap junction protein, beta 2, 2,26kDa" gene. Normally, *GJB2* instructs for a protein known as connexin 26, a member of a family that creates channels called gap junctions that transport nutrients, charged ions, and signaling molecules between cells. Connexin 26 determines the size of the gap functions and is found in all cells in the body, particularly the inner ear and the cochlea.

Hearing happens when sound waves are converted to electrical nerve impulses that then travel through ear structures to the brain, where one perceives sounds. The conversion process requires just the right number of potassium ions. Studies indicate that connexin 26 provides these correct channels and also is required for growth of cells in the cochlea.

More than 90 mutations of *GJB2* cause the protein connexin 26 not to function properly. Gap junctions are not created, and potassium levels build up and affect the survival of the cells. *GJB2* is located on the long arm (q) of chromosome 13 at position 11-q12.

Autosomal Dominant Nonsyndromic Genes

Autosomal dominant genes are found in every generation and can be passed on by a parent with this type of hearing loss. This type of deafness was the first gene located that relates to hearing. A large Costa Rican family provided the information for locating the gene. The family had deafness throughout several generations. There are 24 known autosomal dominant genes.

X-linked

There are two X-linked genes that cause deafness: *PRPS1* and *POU3F4*. For detailed information about the genes that are nonsyndromic, the website for Hereditary Hearing Loss (see below) is a comprehensive site of the latest information on hereditary hearing loss.

Syndromic Hearing Loss

A syndrome is a condition in which several medical conditions may exist. These symptoms may not even appear to be connected. There are over 400 genetic syndromes that include hearing loss. Some of the conditions are very rare, and hearing may be only one symptom that is not as serious as other medical conditions. Below are some of the most common syndromes.

Autosomal Dominant Syndromes

Waardenburg syndrome (WS) This type of autosomal dominant is the most common related to hearing impairment. Mutations affect skin pigment, eyes, and hair as well as the ears.

Branchiootorenal syndrome (BOR) Risk of kidney problems as well as hearing are related to this syndrome.

Stickler syndromeThis condition combines hearing loss, cleft palate, and bone problems resulting in osteoarthritis.

Neurofibromatosis type 2 (NF2) Hearing loss in this syndrome is secondary and possibly treatable. Tumors or growths at the ends of nerve fibers are the primary feature.

Autosomal Recessive Syndromes

Usher syndrome This syndrome is the most common recessive syndrome and is related to loss of sight.

Pendred syndrome This syndrome combines hearing with development of a severe goiter, a growth on the thyroid.

Jervell and lange-nielsen syndrome This type combines congenital deafness with an abnormal heart condition.

Biotinidase deficiency This syndrome is related to a deficiency in biotin, a water-soluble B vitamin.

Refsum disease This progressive type includes deafness and vision disorders.

X-linked Syndromic Hearing Impairment

Alport dyndrome This serious type includes hearing loss and kidney disease, possibly leading to end-stage renal failure.

Mohr-tranebjaerg syndrome This rare syndrome combines progressive hearing loss with eye abnormalities.

Mitochondrial DNA

Hearing loss occurs as additional symptoms in a number of conditions caused by mutations in mitochondrial DNA. For example, mutations in the mitochondrial gene 125 RNA gene increases risk for hearing loss if the person takes certain antibiotic drugs, such as genticide.

Otosclerosis

Some hearing conditions develop later in life. Otosclerosis comes from three Greek root words: *oto*, meaning "ear"; *sclero*, meaning "hardening"; and *osis*, meaning "condition of." This form of genetic disorder may appear in the third decade of life, and by age 50, 90% are affected. The problem appears to be the hardening of the bone in the inner ear, which in turn interferes with conduction. About 10 genes have been located to cause this condition.

Management

The treatment and management of hearing loss depend on early and proper diagnosis. Possibilities include hearing aids, cochlear implants, and certainly education. Children with hearing disability fall under the Individuals with Disabilities Education Act (IDEA) and are eligible for interventions.

See also Alport Syndrome; Jervell and Lange-Nielsen Syndrome; Neurofibromatosis Type 2; Newborn Screening: A Special Topic; Pendred Syndrome; Refsum Disease; Stickler Syndrome; Usher Syndrome; Waardenburg Syndrome (WS)

Further Reading

Berke, Jamie. 2011. "Causes of Hearing Loss." About.com. http://deafness.about.com/od/articlesandnewsletters/a/causes.htm. Accessed 2/5/12.

"Hereditary Hearing Loss Homepage." 2011. Hereditary Hearing Loss.org. http://hereditaryhearingloss.org. Accessed 2/5/12.

Smith, Richard J. H., et al. 2012. "Deafness and Hereditary Hearing Loss Overview." *GeneReviews*. http://www.ncbi.nlm.nih.gov/books/NBK1434. Accessed 2/5/12.

"Understanding the Genetics of Deafness: A Guide for Patients and Families." 2003. Harvard Medical School. http://hearing.harvard.edu/info/GeneticDeafnessBookletV2.pdf. Accessed 5/19/12.

Hereditary Hemorrhagic Telangiectasia (HHT)

Prevalence Difficult to assess; between 1 in 5,000 and 1 in 10,000

Other Names HTT; Osler-Rendu disease; Osler's disease; Osler-Weber-Rendu disease; Osler-Weber-Rendu syndrome; Weber-Osler

People have had nosebleeds throughout history. Nosebleeds do not seem like such a serious condition unless you are the one experiencing a severe one. In the nineteenth century, three English physicians—Henry Sutton, Benjamin Babington, and John Legg—noted that nosebleeds tended to run in families. If the nosebleeds were severe, they were thought to be a form of hemophilia. A French physician, Henri Jules Louis Marie Rendu, noted that skin and other lesions were part of the disease and that it was really not related to hemophilia. The famous Canadian-born Sir William Osler reported in 1901 that the lesions were also in the digestive tract. But none of these pioneers gave their names to the condition. An American physician, Frederic Hanes, in a 1909 article called it "hereditary hemorrhagic telangiectasia," and that name has stuck.

What Is Hereditary Hemorrhagic Telangiectasia (HHT)?

Hereditary hemorrhagic telangiectasia (pronounced tel-AN-jee-eck-TAZE-ee-ya) is a disorder of the blood vessels. As part of the normal circulatory system, the arteries, which have strong, elastic linings, carry blood from the heart to a series of capillaries where the oxygen–carbon dioxide exchange takes place. Then the blood is returned to the heart through the veins, whose walls are not so strong. Walls of the arteries must be strong because the pumping action must take the blood to the farthest parts of the body.

In people with HTT, some of their blood vessels do not function properly. These arteries flow directly into the veins instead of going through the system of capillaries. A blood vessel that is abnormal in this way is called a "telangiectasia" if small blood vessels are involved. It is called an "arteriovenous malformation" or AVM if larger vessels are involved. When the strong pressure from the arteries moves into the weaker veins, bleeding may occur. These places leave obvious red markings called telangiectasia.

Telangiectasia comes from three Greek words: *tele*, meaning "end"; *angeion*, meaning "vessel"; and *ektasis*, meaning "dilation." Thus, it is a vascular lesion caused by a problem in the small blood vessels, in skin, and in other area of weakness. Birthmarks that appear in young children are telangiectasias.

The skin and linings of the body called the mucous membrane are most likely places for the telangiectasia. Especially vulnerable are the mucous membranes of the nose. Other places include the brain, liver, lungs, stomach, intestine, or other organs.

Several forms of HHT exist:

- Type 1: People with type 1 appear to develop symptoms earlier and may have blood vessel malformations in the lungs and brain.
- Type 2: People with type 2 have a higher risk of liver involvement. More women than men are in this group.
- Type 3: Liver involvement is a hallmark of this type.

- Juvenile polyposis/HHT: This involves the HHT condition, arteriovenous malformations, and polyps or growths in the GI tract.

However, individuals with any form of HTT may develop areas of weakness in any part of the body.

What Is the Genetic Cause of Hemorrhagic Telangiectasia (HHT)?

Mutations in three genes cause forms of HTT. The genes are *ENG*, *ACVRLI1*, and *SMAD4*.

ENG

Mutations in the *ENG* gene, officially called the "endoglin" gene, are involved in type 1 HTT. Normally, *ENG* instructs for making a protein called endoglin, which is found on the surface of cells and in developing arteries. The protein combined with other proteins and growth factors to form blood vessels that ultimately become the arteries, veins, and capillaries. *ENG* belongs to a large family of genes called CD.

Several mutations of *ENG* are related to type 1 HHT. Many of the mutations substitute only one amino acid protein building block for another, but the one substitution is enough to disrupt the function of endoglin. The problems with the protein endoglin appear to keep the capillaries from developing properly and enabling blood from the arteries to flow directly into the veins. *ENG* is inherited in an autosomal dominant pattern and is located on the long arm (q) of chromosome 9 at position 33-q34.1.

ACVRL1

Mutations in the *ACVRL1* gene, officially named the "activin A receptor type-like 1" gene, is involved in type 2 HHT. Normally, *ACVRL1* instructs for a protein called activin receptor-like kinase 1, which is found on the surface of the cells and in the lining of developing arteries. This gene is like one piece of puzzle waiting for the correct piece to be inserted. The piece that it is waiting for is called its ligand, and in this case, the ligand is the transforming growth factor beta (TGF-β). The two pieces fit together to play the role of making arteries, veins, and capillaries.

Several mutations in *ACVRL1* are caused when just one building block or amino acid is substituted for another. The exchange of one protein disrupts the production of the ACVR1 protein and causes the symptoms of type 2 HTT. *ACVRL1* is inherited in an autosomal dominant pattern and is located on the long arm (q) of chromosome 12 at position 11-q14.

SMAD4

Mutations in the *SMAD4* gene, officially called the "SMAD family member 4" gene, is related to the juvenile form of HHT, in which polyps are formed.

Normally, *SMAD4* provides instructions for a protein that transmits chemical signals from the cell surface to the nucleus. This signaling pathway is called the transforming growth factor beta (TGF-β) pathway that affects how the cell produces proteins. TGF-β binds to a receptor and activates the SMAD proteins. *SMAD4* is both a transmission factor and a tumor suppressor, which keeps cells from growing and wildly dividing.

At least five mutations cause the juvenile form of HHT. The disorder of the blood vessels that causes hemorrhaging of the blood vessels also increases the risk of developing polyps that can become cancerous. The mutations disrupt the pathway of the TGF-β proteins and result now only in the weakness of the boundaries between arteries and veins, but also in the development of polyps. *SMAD4* is inherited in an autosomal dominant pattern and is located on the short arm (q) of chromosome 18 at position 21.1.

What Is the Treatment for Hereditary Hemorrhagic Telangiectasia (HHT)?

The recommended treatment for a telangiectasia or AVM depends on both its size and location in the body. Nosebleeds may respond to everyday practical treatments such as humidifying the air or keeping the nasal mucous membranes moist. The lung, brain, and gastrointestinal symptoms should be treated by specialists once they occur.

There is a genetic test for HHT, but it should be preceded by clinical determination of the condition.

Further Reading

"Hereditary Hemorrhagic Telangiectasia." 2009. The Merck Manual. http://www.merckmanuals.com/professional/sec11/ch137/ch137d.html. Accessed 5/19/12.

"Hereditary Hemorrhagic Telangiectasia." 2010. Genetics Home Reference. National Library of Medicine (U.S.). http://ghr.nlm.nih.gov/condition/hereditary-hemorrhagic-telangiectasia. Accessed 5/19/12.

Hereditary Hemorrhagic Telangiectasia Foundation International. 2010. http://hht.org. Accessed 5/19/12.

Hereditary Leiomyomatosis and Renal Cell Cancer (HLRCC)

Prevalence Unknown; reported in about 100 families worldwide

Other Names Reed's syndrome or Reed's disease

In 1973, W. B. Reed and colleagues studied two families in which multiple skin and uterine leiomas appeared over three generations. They also made the connection with a 20-year-old woman who had renal cancer. However, the skin-uterus-renal connection was not completely determined until V. Luanonen and team found the condition in two Finnish families in 2001. With that relationship, the condition became officially known as hereditary leiomyomatosis and renal cell cancer.

What Is Hereditary Leiomyomatosis and Renal Cell Cancer (HLRCC)?

Hereditary leiomyomatosis is a condition that affects the smooth muscles in the skin and in the females of the uterus. The word leiomyomatosis comes from four Greek roots: *leio*, meaning "smooth"; *myo*, meaning "muscle"; *oma*, meaning "tumor"; and *osis*, meaning "condition of." Thus the condition leiomyomatosis refers to a condition of tumors in the smooth muscle.

People with HLRCC tend to develop noncancerous tumors in the smooth muscle. When people with the condition are in their 30s, they may begin to notice groups of small growths around the tiny muscles of the hair follicles, which may look like the goose bumps that one may get when very cold. The bumps appear on the body trunk, arms, legs, and sometimes the face. These skin bumps are called cutaneous leiomas. These leiomas may be more sensitive to cold and sometimes be painful.

Women with HLRCC tend to develop uterine fibroids. Although such fibroids are common in the general population, in these women, they may appear earlier and are larger and more numerous. In addition, about 10% to 16% of people with the condition develop renal cell cancer. The initial symptoms will be lower back pain, blood in urine, and a mass that can be felt by the physician. There may be no real signs until the disease is well advanced.

What Is the Genetic Cause of Hereditary Leiomyomatosis and Renal Cell Cancer (HLRCC)?

Mutations in the *FH* gene, officially called the "fumarate hydratase" gene, causes hereditary leiomyomatosis. Normally, the *FH* gene instructs for an enzyme called fumerase that is important in the citric acid cycle or Krebs cycles. This cycle allows cells to use oxygen to generate energy. Fumerase also converts the molecule fumarate to malate.

The approximate 50 mutations in the *FH* gene generally replace one amino acid with another in the fumerase enzyme. The defect in the enzyme disrupts the citric acid cycles, allowing a buildup of fumerase. This excess interferes with the regulation of oxygen in the cells, thereby resulting in a development of the smooth cell tumors and renal cell cancer. *FH* is inherited in an autosomal dominant pattern and is located on the long arm (q) of chromosome 1 at position 42.1.

What Is the Treatment for Hereditary Leiomyomatosis and Renal Cell Cancer (HLRCC)?

Because this is a genetic disease, treatment of the symptoms help the individuals manage. If the lesions are large and painful, a dermatologist may remove them by surgical incision or treat with laser or cryoblation (freezing). Some drugs have been used for treatment. Uterine fibroids are treated in the same manner as other fibroids and will probably require medical or surgical removal.

Further Reading

"Hereditary Leiomyomatosis and Renal Cell Cancer." 2010. Genetics Home Reference. National Library of Medicine (U.S.). http://ghr.nlm.nih.gov/condition/hereditary-leiomyomatosis-and-renal-cell-cancer. Accessed 5/19/12.

Pithukpakom, Manop, and Jorge Toro. 2010. "Hereditary Leiomyomatosis and Renal Cell Cancer." *GeneReviews*. http://www.ncbi.nlm.nih.gov/books/NBK1252. Accessed 5/19/12.

Hereditary Multiple Exostoses

Prevalence One in 50,000 individuals

Other Names Bessek-Hagen disease; diaphyseal aclasis; multiple cartilaginous exostoses, hereditary multiple exostoses;. Includes: Hereditary Multiple Osteochondromatosis, Type I; Hereditary Multiple Osteochondromatosis, Type II

In certain populations, strange bone growths were noted that gave researchers insight into the condition hereditary multiple exostoses, or HME. The condition was found to be prevalent in certain groups, such as the Chamarro people of Guam and the Ojibway Indian population of Manitoba, Canada. Studying these populations and families with several members with HME enabled scientists to locate the genetic source of this painful condition.

What Is Hereditary Multiple Exostoses?

Hereditary multiple exostoses (HME) is a rare condition characterized by many bony spurs or lumps, or exostoses. An exostosis is an abnormal bone growth that protrudes out of the skin and is different from the regular structure of bone. The lumps or tumors are generally benign or noncancerous. However, over a period of years, they may become cancerous and cause a growth called a chondrosarcoma.

Following are some of the characteristics of HME:

- Tumors or bony bumps appear around areas of active bone growth, especially the area called the metaphysis of the long bones.
- Long bones of the legs, arms, fingers, and toes are most affected.
- Hips and shoulder blades may also show the bony growths.
- Face and skull are generally unaffected.
- Bumps may vary in size.
- Boys and girls can both be affected.
- Bumps appear at the joints and may be round or sharp.
- The bumps continue to grow as long as the child is growing.

HME can be very troublesome to the person. Because the tumors are at the growth center, the person may be shorter than normal. The forearm or leg may bow out. The individual may become stiff at the joints and experience severe pain. Young children may develop exostoses inside the knees, which hurt when the knees hit together.

What Are the Genetic Causes of Hereditary Multiple Exostoses?

Two genes cause hereditary multiple exostoses: *EXT1* and *EXT2*.

EXT1

Mutations in the *EXT1* gene, officially known as the "Exotosin 1" gene, cause HME. Normally this gene instructs for a protein called exostosin-1, which is found in the cells in the Golgi bodies. The Golgi bodies are responsible for modifying newly produced enzymes and proteins. Exotsosin-1 attaches to exostosin-2 to process heparin sulfate, a complex of sugar molecules that is added to other proteins called proteoglycans. Heparan works to regulate a variety of processes, including blood clotting, forming blood vessels, and even the spreading or metastasis of cancer cells.

About 200 mutations of *EXT1* have been found to prevent combining with the exostosin-2 protein. This disruption then prevents the adding of heparan sulfate to proteins, possibly leading to the bony growths of HME. *EXT1* is inherited in an autosomal dominant pattern and is located on the long arm (q) of chromosome 8 at position 24.11.

EXT2

Mutations in the *EXT2*, officially called the "exotosin-2" gene, cause HME. Normally, *EXT2* programs for the protein exostosin-2, which combines with exostosin-1 to form heparin sulfate.

More than 90 mutations in *EXT2* may prevent exostosin-2 from working properly and cause the bony outgrowths experienced in HME. *EXT2* is inherited in an autosomal dominant pattern and is located on the short arm of chromosome 11 at position 12-p11.

What Is the Treatment/Management for Hereditary Multiple Exostoses?

Some people may never require any treatment and learn to compensate for the tumors. Others may require surgery, especially in the areas that hit together such as the knees. If the exostoses do become malignant, they must be removed.

Further Reading

MHE Research Foundation. 2010. "Hereditary Multiple Exostoses (MHE)." http://www.radix.net/~hogue/mhe.htm. Accessed 5/20/12.

Schmale, G.; W. Wuyte; H. Chansky; and W. Raskind. 2008. "Hereditary Multiple Osteochondromas." *GeneReviews*. http://www.ncbi.nlm.nih.gov/books/NBK1235. Accessed 5/20/12.

Talanow, Roland. 2010. "Hereditary Multiple Exostoses." PedRad (online). http://www.pedrad.info/?search=20080711190908&lang=en. Accessed 5/20/12.

Hereditary Neuralgic Amyotrophy (HNA)

Prevalence Rare; only about 200 families worldwide affected; prevalence unknown

Other Names amyotrophic neuralgia; brachial neuralgia; brachial neuritis; brachial plexus neuritis; familial brachial plexus neuritis; hereditary brachial plexus neuropathy; heredofamilial neuritis with brachial plexus predilection; HNA; NAPB; neuralgic amyotrophy; neuritis with brachial predilection; Parsonage-Turner syndrome (term now obsolete); shoulder girdle neuropathy; three other obsolete names are Feinberg syndrome, Tinel's syndrome, and Kiloh-Nevin syndrome III

Mysterious severe pains in the shoulder and neck area have appeared in the literature for several centuries. As early as 1835, Augustin Nicolas Gendrin and Armand Louis Marie Velpeau, French physicians, described the sequence of pain in the shoulder that would come and go away. Later, M. J. Parsonage and a team described neuralgic amyotrophy as the shoulder-girdle syndrome in an article that appeared in the British journal *Lancet* in 1948. From the many names listed above, it is obvious that interest in the condition expresses its mystery. With the advent of genetics, most of the above names became obsolete, and the condition is referred to as hereditary neuralgic amyotrophy, or HNA.

What Is Hereditary Neuralgic Amyotrophy (HNA)?

Hereditary neuralgic amyotrophy is a condition in which the person experiences severe pain in the shoulder and neck area in episodes. The name gives a clue to what is happening. The word "neuralgic" comes from the Greek words *neuro*, meaning "nerve," and *alg*, meaning "pain." The word "amyotrophy" is made up of three Greek words: *a*, meaning "without"; *myo*, meaning "muscle"; and *trophy*, meaning "nourish or feed." The condition then is characterized by pain to the nerve and wasting away of muscles, which receive little sustenance. Neuralgic pain is felt along a pathway of a network of nerves called the brachial plexus, which control feeling and movement in the shoulders and arms.

HNA is episodic, meaning that it comes and goes in time frames lasting from one to six weeks. During an attack, the immune system appears to attack the nerves of the brachial plexus and cause them to deteriorate. As nerves degenerate, the muscles that the nerves control begin to lose their function and waste away or atrophy. People with HNA usually have the first episode in their 20s, with 28 being the average age of onset. However, there have been cases of children as young as one year of age experiencing the condition. Certain triggers such as stress, exercise, childbirth, surgery, cold, infections, and immunizations may evoke an attack.

The following other characteristics may be seen:

- Severe pain in one or both sides, with the right side being more common
- Pain difficult to control with medications
- Pain that lasts a month on average
- Muscles affected within a couple of weeks begin the wasting process
- Posture change
- Shoulder blades stick out and appear like wings
- Abnormal sensations in skin
- Areas other than shoulder possibly affected

Some families may have other characteristics including short stature, excess skin folds, split uvula, and webbed or fused fingers or toes. They may also have distinctive facial features.

What Is the Genetic Cause of Hereditary Neuralgic Amyotrophy (HNA)?

Mutations in the *SEPT9* gene, officially known as the "septin 9" gene, cause hereditary neuralgic amyotrophy. Normally, *SEPT9* instructs for making the protein septin-9, a member of the family of septins. This family of proteins is very important in a step of cell division called cytokinesis. The septin-9 protein also appears to regulate cell growth by keeping cells from dividing too rapidly or too slowly.

Only a few mutations in *SEPT9* have been related to HNA. One mutation occurs when the amino acid tryptophan replaces arginine. The protein is altered and changes in the function appear to affect the network of the brachial plexus nerves leading to the shoulders and arms. These changes cause the sensation of pain and wasting away of muscle. *SEPT9* is inherited in an autosomal dominant pattern and is located on the long arm (q) of chromosome 17 at position 25.

What Is the Treatment for or Management of Hereditary Neuralgic Amyotrophy (HNA)?

Controlling the pain with anti-inflammatory drugs such as prednisone may be effective in some cases. However, these drugs probably do not slow or stop the nerve degeneration.

Further Reading

Goulder, R., et al. 1994. "Hereditary Neuralgic Amyotrophy and Hereditary Neuroloathy with Liability to Pressure Palsies." *Neurology*. December. 44(12): 2250. http://www.neurology.org/content/44/12/2250.abstract. Accessed 5/20/12.

Hannibal, M.; N. Alfin; P. Chance; and B. Van Engelen. 2008. "Hereditary Neuralgic Amyotrophy." *GeneReviews*. http://www.ncbi.nlm.nih.gov/books/NBK1395. Accessed 5/20/12.

"Hereditary Neuralgic Amyotrophy." 2010. Genetics Home Reference. National Library of Medicine (U.S.). http://ghr.nlm.nih.gov/condition/hereditary-neuralgic-amyotrophy. Accessed 5/20/12.

Hereditary Pancreatitis

Prevalence In the United States, about 1,000 persons affected

Other Names HP; HPC

When the suffix *itis* is used in a word, it means "inflammation of." Pancreatitis is the inflammation of the pancreas, the large gland behind the stomach and the duodenum, the upper part of the small intestine. Very important in digestion, the pancreas secretes digestive juices that join with bile from the liver to digest foods, especially fats. The pancreas also releases the hormones insulin and glucagon, which help the body regulate the glucose from food for energy. Normally, enzymes from the pancreas do not become active until they reach the small intestine; but when the pancreas is inflamed, the enzymes may not be able to move through the pancreatic duct and begin to attack the tissues of the pancreas, causing serious damage.

What Is Hereditary Pancreatitis?

Hereditary pancreatitis is a rare inherited disorder in which the individual has acute attacks of pancreatitis. As the attacks continue, severe scarring of the pancreas may occur that then progresses to chronic pancreatitis. The first attack of hereditary pancreatitis happens in the first two decades of life but can happen at any age.

Symptoms of hereditary pancreatitis are like other forms of the disorder:

- Person looks and feels very ill
- Gradual or sudden pain in upper abdomen and through the back
- Swollen and tender abdomen
- Nausea and vomiting
- Constipation
- Fever
- Rapid pulse

Acute pancreatitis may cause dehydration and low blood pressure, resulting in failure of certain vital organs.

What Are the Genetic Causes of Hereditary Pancreatitis?

Mutations in two genes, *PRSS1* and *SPINK1*, cause hereditary pancreatitis. The presence of the gene can be determined by genetic testing.

PRSS1

Mutations in the *PRSS1* gene, officially known as the "protease, serine, 1 (trypsin 1)" gene, cause hereditary pancreatitis. Normally, *PRSS1* provides instructions for the production of the protein trypsinogen, a member of the trypsin family of serine proteases. This enzyme is made in the pancreas but cleaves to an active form in the small intestine.

Mutations in *PRSS1* cause the enzyme to malfunction and stop the function in the pancreatic ducts, leading to buildup and the severe abdominal pain. *PRSS1* is inherited in an autosomal dominant pattern and located on the long arm (q) of chromosome 32-qter.

SPINK1

Mutations in the *SPINK1* gene, officially known as the "serine peptidase inhibitor, Kazal type 1" gene, cause hereditary pancreatitis. Normally, *SPINK1* instructs for a protein called a trypsin inhibitor, which is secreted from certain cells in the pancreas called the acinar cells into the pancreatic juice. The enzyme prevents the premature activation of pancreatic juices so that it can do the work when released into the small intestine. However, mutations keep this inhibitor from working, leading to the premature release that attacks the tissues of the pancreas. *SPINK1* is inherited in an autosomal dominant pattern and found on the long arm (q) of chromosome 5 at position 32.

What Is the Treatment for Hereditary Pancreatitis?

Alcohol, smoking, and fatty meals are devastating for people with pancreatitis. Treatment for the acute attacks involves a stay in the hospital to give the person fluids, antibiotics, and medication for pain. At this time, there is no cure; relieving the symptoms and prevention are the mainstays of management.

Further Reading

"Fact Sheet—Hereditary Pancreatitis." 2011. Pancreas Foundation. http://pancreasfoundation.org/Docs/patient_info/FactSheetHereditaryPancreatitis.pdf. Accessed 5/20/12.

"Hereditary Pancreatitis." 2009. University of Cincinnati Pancreatic Disease Center. http://www.ucpancreas.org/hereditarypancreatitis.htm. Accessed 5/20/12.

"Pancreatitis." 2011. National Diabetes and Digestive and Kidney Diseases Institute (U.S.). http://digestive.niddk.nih.gov/ddiseases/pubs/pancreatitis. Accessed 5/20/12.

Hereditary Paraganglioma-Pheochromocytoma (Pheochromocytoma)

Prevalence Occurs in about 1in a million people

Other Names familial paraganglioma-pheochromocytoma syndromes; hereditary paraganglioma-pheochromocytoma syndromes; hereditary pheochromocytoma-paraganglioma; paragangliomas 1; paragangliomas 2; paragangliomas 3; paragangliomas 4

In 1886, the first description of a person with pheochromocytoma was documented. Frankel, a German physician, wrote about noncancerous tumors gathered in areas around the nerve groups in the adrenal glands. However, the condition was not named until 1912 when Ludwig Pick coined the term *pheochromocytoma*. In 1926, Cesar Roux in Switzerland and Charles H. Mayo in the United States (one of the founders of the Mayo Clinic) removed the tumors successfully.

Although the condition is rare, pheochromocytoma has generated a lot of interest in television series such as *ER*, *House*, *Private Practice*, and *Mystery Diagnosis*. When President Dwight Eisenhower died of ishemic cardiomyopathy in 1969, the autopsy revealed a 1.5 cm pheochromocytoma on the left adrenal gland. The tumor had not been diagnosed and may have been the source of excessive blood pressure spikes that were documented from 1930 until his death. If this had been located earlier, the president could possibly have been spared severe heart disease.

What Is Hereditary Paraganglioma-Pheochromocytoma (Pheochromocytoma)?

Hereditary paraganglioma-pheochromocytoma is a condition in which noncancerous growths occur in certain body parts or structures outside of the brain and spinal cord. The word "paraganglia" comes from two Greek roots: *para*, meaning "beside," and *ganglia*, meaning "knot." Paraganglia are groups or knots of cells that gather in certain parts of the body, especially in the adrenal gland. The term "pheochromocytoma" literally comes from Greek roots: *phalos*, meaning "dusky"; *chromo*, meaning "color"; *cyto*, meaning "cell"; and *oma*, meaning "tumor of." The pheochromocytoma refers to the yellowish-brown chemical reaction that the cell from the tumor gives when tested with chromic salts. Actually then, paraganglioma-pheochromocytoma is a noncancerous growth that is located near the ganglia, a mass of nerve cells lying outside the brain and spinal cord.

One type of paraganglioma called a pheochromocytoma grows in the adrenal gland. The adrenal glands are located on top of the kidneys and are responsible for stress hormones. Exactly why these pheochromocytomas develop is not exactly clear. Researchers do know that they start in special cells of the adrenal glands, called chromaffin cells. Other paragangliomas with pheochromocytomas are found in the head, neck, and chest.

Pheochromocytomas have another effect. They are associated with the nerves of a system known as the sympathetic nervous system. When a person is threatened with danger, the response is either to flee or to stand up and fight. The alert condition is triggered by a release of hormones from the adrenal glands.

If a pheochromocytoma is present, they also release hormones called catecholamines, such as adrenaline and norepinephrine, in excess to the normal hormones of the adrenal glands. Because of the hyperactivity of the sympathetic nervous system, the following signs and symptoms of pheochromocytoma may develop:

- Skin sensations
- Elevated heart rate
- High blood pressure, which is sporadic and sometimes very difficult to detect
- Heart palpitations
- Anxiety similar to a panic attack
- Excessive sweating
- Headaches
- Weight loss
- Pale appearance
- Pain in the area between the last rib and hip, called the flank area
- Aggressive behavioral problems

The tumors are generally smaller than 10 cm, but may grow larger. Occasionally, the tumor may become cancerous and spread to other parts of the body.

There are four types of hereditary paraganglioma-pheochromocytomas, listed as type 1 through 4. A different gene causes each type. Types 1, 2, and 3 are related to the growths in the neck or head. Type 4 is associated with the adrenal growth and creates the highest risk for cancer. The condition is often diagnosed in the fourth decade.

What Are the Genetic Causes of Hereditary Paraganglioma-Pheochromocytomas?

Several genes are suspected to cause hereditary paraganglioma-pheochromocytomas. They are *SDHD*, *SDHAF2*, *SDHC*, and *SDHB*. All these genes are members of a family of genes called the mitochondrial respiratory chain complex.

SDHD

Mutations in the *SDHD* gene, officially known as the "succinate dehydrogenase complex, subunit D, integral membrane protein" gene, cause type 1 of hereditary paraganglioma- pheochromocytoma. Normally, *SDHD* instructs for an enzyme that becomes one of the four subunits of another enzyme called the succinate dehydrogenase, or the SDH enzyme. SDH is an important player in the mitochondia, the bean-shaped organ that is the powerhouse of the cell. The SDHD protein holds the SDH enzyme to the cell membrane of the mitochrondria so that energy from food can be used.

The mitochondria are essential in the chemistry of cells. Two important activities go on here: the citric acid cycle or Krebs cycle, and oxidative phosphorylation. The SDH enzyme is part of the citric acid cycle converting a compound called succinate into another molecule called fumarate. As this action occurs, negative charged particles are given off which are then transferred through the SDHD protein to the oxidative phosphorylation process. Here in the oxidative phosphorylation pathway, an electrical charge then gives energy for the important source of the body's energy adenosine triphosphate, or ATP. SDH acts upon succinate as an oxygen sensor in the cell, creating pathways that allow cells to grow in a low-oxygen environment. Here it creates the structure to control several genes involved in cell division to form new blood vessels. Thus, SDHD is a tumor suppressor, acting to keep cells from multiplying wildly.

More than 100 mutations in *SDHD* cause hereditary paraganglioma-pheochromocytoma type 1. Individuals with this type may have paraganglions, pheochromocytomas, or both. Type 1 appears more in the head and neck. Type 1 can be hereditary or acquired during one's life. Most of the mutations are the result of a change in only one building block or amino acid, which results in a short protein. Because of the shortened protein, all the pathways that depend on the first reaction cannot take place. This keeps the control mechanism from working and allows uncontrollable growth of the cells. The formation of the cells and new blood vessels leads to the tumors of hereditary paraganglioma-pheochromocytoma. *SDHD* is inherited in an autosomal dominant pattern and is located on the long arm (q) of chromosome 23 at position 11.

SDHAF2

SDHAF2 is a member of a family of genes that affect the process of controlling cell growth and is called a tumor suppressor. Mutations in the *SDHAF2* gene, officially known as the "succinate dehydrogenase complex assembly factor 2" gene, cause type 2 of hereditary paraganglioma-pheochromocytoma. Normally, this gene instructs for a protein that is essential for an important enzyme called the succinate dehydrogenase, or the SDH enzyme, to attach to a molecule called FAD. FAD then acts as a cofactor to carry out the function of SDH. The SDH enzyme works in the mitochondria to link the two pathways of the citric acid cycle or Krebs cycle to the oxidative phosphorylation process. Like the other genes in this family, *SDHAF2* is a tumor suppressor.

One mutation has been implicated in type 2 of hereditary paraganglioma- pheochromocytoma. People with this type have paraganglions, pheochromocytomas, or both. Mutations in just one amino acid building block impair the SDH enzyme and disrupt all the activities that the enzyme begins. Thus the cells divide uncontrollably. *SDHAF2* is inherited in an autosomal dominant pattern and is located on the long arm (q) of chromosome 11 at position 12.2.

SDHC

Mutations in the *SDHC* gene, officially known as the "succinate dehydrogenase complex, subunit C, integral membrane protein, 15kDa" gene, cause type 3 of hereditary paraganglioma- pheochromocytoma. Normally, SDHC instructs for one of the subunits of the SDH enzyme. The SDH enzyme is important in linking the cellular pathways of the citric acid cycle and oxidative phosphorylation in the mitochondria. SDHC is a tumor suppressor gene and is a member of the same family as the other genes in this section.

Over 30 mutations in *SDHC* have been found in people with type 3 of hereditary paraganglioma-pheochromocytoma. Most of the inherited protein mutations are the result of one change in an amino acid building block. This change interrupts the pathway and allows for rapid and uncontrolled cell growth leading to tumors. *SDHC* is inherited in an autosomal dominant pattern and is located on the long arm (q) arm of chromosome 1 at position 23.3.

SDHB

Mutations in the *SDHB* gene, officially known as the "succinate dehydrogenase complex, subunit B, iron sulfur (Ip)" gene, cause type 4 of hereditary paraganglioma-pheochromocytoma. Normally, this gene instructs for one of the four subunits of the succinate dehydrogenase or SDH enzyme. Like the other genes in this family, the *SDHB* and its enzyme SDH plays an important role in linking the citric acid cycle and oxidative phosphorylation. The gene is a tumor suppressor and keeps the cells and blood vessels from growing wildly.

More than 150 mutations cause type 4 of hereditary paraganglioma-pheochromocytoma. The people with this type have paragangliomas, pheochromocytomas, or both. The mutated gene causes only one change in a single amino acid, and because of this exchange, the enzyme does not function properly. This disrupts formation of the proper pathways for controlling the division of cells. *SDHB* is inherited in an autosomal dominant pattern and is located on the short arm (P) of chromosome 1 at position 36-1-p35.

What Is the Treatment for Hereditary Paraganglioma-Pheochromocytoma?

Surgical removal of the tumor is recommended. Early detection is essential to avoid some of the symptoms of the disorder. People with a known tendency are advised to avoid cigarette smoke, which interferes with the proper use of oxygen in the blood.

Further Reading

"Hereditary Paraganglioma-Pheochromocytoma." 2011. Genetics Home Reference. National Library of Medicine (U.S.). http://ghr.nlm.nih.gov/condition/hereditary-paraganglioma-pheochromocytoma. Accessed 5/20/12.

"Pheochromocytoma." 2011. MedlinePlus. National Library of Medicine (U.S.). http://www.nlm.nih.gov/medlineplus/pheochromocytoma.html. Accessed 5/20/12.

Hereditary Spherocytosis

Prevalence 1 in 2,000 people of northern European ancestry; most common form of anemia in that background; can exist in other ethnic groups but less common

Other Names Congenital spherocytic hemolytic anemia; congenital spherocytosis; HS; spherocytic anemia; spherocytosis, type 1

Red blood cells or erythrocytes are normally shaped like biconcave discs that resemble a candy Life Saver or a donut with a filled hole. This shape is ideal for enabling the cells to pass through the blood vessels. The shape also creates more surface space enabling the oxygen–carbon dioxide exchange that takes place in the capillaries, the small vessels between veins and arteries in a most efficient manner. However, if something affects that shape, problems can arise. If the shape is like a ball or sphere, the exchange does not take place efficiently and the red blood cell breaks down. The condition is known as spherocytosis.

What Is Hereditary Spherocytosis?

Hereditary spherocytosis is a condition in which the red blood cells are shaped like a ball. The word comes from three Greek root words: *sphero*, meaning "ball"; *cyt*, meaning "cell"; and *osis*, meaning "condition of." People with this condition have a disorder of the surface layer or outer membrane of the red blood cells that causes a change in shape from the biconcave disc to a ball. This shape causes a premature breakdown of the cells, called hemolysis.

Four forms of HS exist:

- Mild form: These people have very mild symptoms to no symptoms of anemia at all. About 20% to 30% of cases have the mild form.
- Moderate form: These individuals have anemia, jaundice, and enlarged spleen or splenomegaly. They may also develop gallstones. About 60% to 70% of cases have the moderate form, with signs usually appearing in childhood.
- Moderate/severe form: These individuals have all the features of the moderate form, but with more severe anemia. Ten percent of cases have this form.
- Severe form: These individuals have life-threatening anemia and require transfusions to supply red blood cells. They will also have enlarged spleens, jaundice, and high risk for gallstones. This form may also affect other body areas. Some people may be very short, with skeletal abnormalities and delayed sexual development. About 3% to 5% have this type.

What Is the Genetic Cause of Hereditary Spherocytosis?

The following four genes are related to hereditary spherocytosis: *ANK1*, *EPB42*, *SLC4A1*, and *SPTB*.

ANK1

Over half of the conditions are related to the *ANK1* gene, officially known as the "ankyrin 1, erythrocytic" gene. Normally, *ANK1* provides instructions for the ankyrin-1 protein that is active in red blood cells, muscle, and brain cells. The protein is found on the cell membrane at the place where it binds to the other parts of the cell, giving it stability and flexibility. The cell is thus able to pass through arteries, capillaries, and veins without breaking down.

The 55 mutations of *ANK1* interrupt the process by deleting genetic material or exchanging building blocks. A lack of the normal protein leads to a deficiency in another protein called spectrin. The shortage of the two proteins causes the cell membrane not to bind properly to the rest of the red blood cell. The cell membrane becomes flabby and misshapen to give the appearance of the ball instead of a donut shape. The irregular cells are removed by the spleen, causing it to work overtime. The shortage of red blood cells causes anemia and the symptoms of hereditary spherocytosis. *ANK1* is inherited in an autosomal dominant pattern and is located on the short end (p) of chromosome 8 at position 11.1.

EPB42

A second gene is the *EPB42* gene, officially known as the "erythrocyte membrane protein band 4.2" gene. Normally, *EPB42* programs for a special type protein that is associated with production of the protein ankyrin and with determining the erythrocyte shape and its regulation. Mutations in the *EPB42* gene cause the malfunctioning of the protein path, which then results in the round spherical shape of the red blood cells. *EPG42* is inherited in a recessive manner and is located on the long arm (q) of chromosome 15 at position 15-q21.

SLC4A1

The third gene associated with hereditary spherocytosis is the *SLC4A1* gene, officially known as the "solute carrier family 4, anion exchanger, member 1" (erythrocyte membrane protein band 3, Diego blood group) gene. Normally, *SLC4A1* encodes for a protein that is part of a family called the anion exchanger or AE and is found in the plasma membrane of the red blood cell. Mutations in the gene lead to production of a protein that causes the cell membrane to be unstable. The deletion in the gene is common in areas where malaria is endemic. *SLC4A1* is inherited in a dominant pattern and is located on the long arm (q) of chromosome 21-q22.

SPTB

The fourth gene that is related to hereditary spherocytosis is the *SPTB* gene, officially known as the "spectrin, beta, erythrocytic" gene. Normally, *SPTB* encodes for a member of a family of genes that program for spectrin. Spectrin and ankyrin are partners in establishing the flexibility of the cell membrane. Mutations in the gene cause the protein not to function properly and disturb the makeup of the cell membrane of the red blood cell, causing the symptoms of spherocytosis. *SPTB* is inherited in an autosomal dominant pattern and is located on the long arm (q) of chromosome 14 at position 23-q24.2.

What Is the Management of Hereditary Spherocytosis?

Individuals with the mild form may only have to be treated for anemia with nutrition supplements. The other types that cause severe damage to the spleen may require the removal of that organ. Children with spherocytosis may need immunization for pneumonia, influenza, and even protective antibiotic treatment to avoid infection that may risk blood poisoning from the destruction of the blood cells.

Further Reading

"*ANK1*." 2011. Genetics Home Reference. National Library of Medicine (U.S.). http://ghr.nlm.nih.gov/gene/ANK1. Accessed 5/20/12.

"*EPB42*." 2011. Genetics Home Reference. National Library of Medicine (U.S.). http://ghr.nlm.nih.gov/gene/EPB42. Accessed 5/20/12.

"Hereditary Spherocytosis." 2011. Genetics Home Reference. National Library of Medicine (U.S.). http://ghr.nlm.nih.gov/condition/hereditary-spherocytosis. Accessed 5/20/12.

Hermansky-Pudlak Syndrome

Prevalence Rare; 1 in 500,000 to 1 in 1,000,000 worldwide; occurs in 1 in 1800 to 2000 in Puerto Rico

Other Names HPS

Albinism, a condition in which people have little pigment in eyes, hair, and skin, has been identified in both humans and animals for centuries. But in 1959, Frantisek Hermansky and P. Pudlak, Czech internists, were perplexed by two 33-year-olds with albinism, who had two additional health problems: bleeding problems and serious lung congestion. Hermansky and Pudlak wrote about the syndrome in a 1959 article in the journal *Blood*. The syndrome was named for these two doctors.

Groups appeared with the combination of symptoms of Hermansky-Pudlak in other parts of the world. It appeared especially prevalent on the island of Puerto Rico. When the group of disorders were found in a group of people in a village in Switzerland in 1993 and in people of Dutch, Turkish, Pakistani, and Japanese ethnicity, researchers realized Hermansky-Pudlak syndrome was a worldwide syndrome.

What Is Hermansky-Pudlak Syndrome?

Hermansky-Pudlak syndrome (HPS) is a disorder that not only causes albinism, but it also causes bleeding disorders and damage to other organs, such as lungs and kidneys. The following symptoms are part of the syndrome:

- Skin: HPS resembles other types of albinism. The color of skin can range from white or creamy to freckling caused by exposure to the sun.
- Hair: Color varies from yellow to light brown.
- Eyes: Color may vary from blue to light brown. Problems similar to those in other types of albinism may be seen. Eyes may move from side to side or be crossed and sensitive to glare. Vision is also decreased.
- Blood clotting problems: Platelets are the tiny blood vessels that cause the blood to clot. The individual may bruise easily or have nosebleeds. However, the clotting problems may be severe in some types of HPS.
- Lung disease: Tissues in the lungs may scar, a condition known as pulmonary fibrosis. The lungs do not inflate properly, and the condition tends to worsen over time.
- Inflammatory bowel disease: Individuals may have diarrhea or abdominal cramping.
- Kidney failure.

Classified according to their different symptoms and genetic cause, eight different types of HPS are known. Type 1, the type found mostly in Puerto Rico, and type 4 are the most severe; types 3, 5, and 6 are mild types. Types 1, 2, and 4 are the only types that are have the lung disorder. Although types 7 and 8 have been identified, little is known about the complete range of symptoms.

What Are the Genetic Causes of Hermansky-Pudlak Syndrome?

Like the eight different types, eight genes are associated with HPS. These genes all have similar roles. The genes instruct for proteins to make four proteins, which are then involved in specific complexes. In general, these complexes are involved in forming and moving a group of structures within the cells called lysosome-related organelles, or LROs. LROs are highly specialized and found only in certain cells, such as pigment cells that give color to eyes, hair, and skin; blood platelets; and lung cells.

When mutations are present in the genes, the genes do not form the complex of LROs or prevent them from working. The conditions are grouped according to the way the gene may block the function. For example, if the protein does not function in blood platelets, the persons will have a bleeding condition, and blood does not clot. The following genes are associated with *HPS*: *HPS1*, *HPS3*, *HPS4*, *HPS5*, *HPS6*, *AP3B1*, *DTNBP1*, and *BLOC1S3*.

HPS1

Mutations in *HPS1*, officially known as the "Hermansky-Pudlak syndrome 1" gene, cause about 75% of cases of HPS in Puerto Rico. Normally, *HPS1* instructs for making the protein complex called biogenesis of lysosome-related organelles complex-3 or BLOC-3. This complex regulates the lysosome-related organelles within the lysosome, whose function is to destroy unwanted cell parts or recycle them. The protein complexes are found only in certain cells. One type of LRO is found in pigments of skin, hair, and eyes. Another LRO is found in blood platelets, and another in certain lung cells.

At least 27 mutations of *HPS1* appear to block the function of the protein complex BLOC-3, causing it to not function in the cells of the body. People with this genetic defect have very light skin, hair, and eyes. This type of HPS is the most serious. *HPS1* is inherited in an autosomal recessive pattern and is found on the long arm (q) of chromosome 10 at position 23.1-q23.3.

HPS3

A second gene is *HPS3*, known officially as the "Hermansky-Pudlak syndrome 3" gene. This gene instructs for the protein that forms the BLOC2 complex or "biogenesis of lysosome-related organelle complex-2." The function is similar to *HPS1*.

Seven mutations in *HPS3* cause the condition HPS type 3, a mild form of the condition. Mutations of this gene are found on people who live in central Puerto Rico and Ashkenazi Jews of central and eastern Europe. The people may have light hair, eyes, and skin, and also possess the bleeding condition. *HPS3* is inherited in a recessive pattern and is found on the long arm (q) of chromosome 3 at position 24.

Although the other six genes are similar in the way they operate, the variations cause symptoms related to a specific type of condition. Following is a chart of the six genes:

Hermansky-Pudlak syndrome

Gene Symbol	Official Name	Normal Function	Type of HPS	Location on Chromosome
HPS4	Hermansky-Pudlak syndrome 4	Encodes for protein important in lysosome biogenesis	HPS type 4; also a serious type of HPS	Cen of chromosome 22-q12.3
HPS5	Hermansky-Pudlak syndrome 5	Important in lysosome movement and regulation, especially in lungs	HPS type 5	Short arm (p) of chromosome 11 at position 14
HPS6	Hermansky-Pudlak syndrome 6	Synthesis of action of lysosomes in certain cells	HPS type 6	Long arm (q) of chromosome 10 at position 24.32
AP3B1	Adaptor-related protein complex 3, beta 1, subunit	Instructs for protein that is part of AP3 protein complex that interacts with the cell-building protein clathrin	HPS type 2; this type differs from other types in that immunodeficiency is part of the syndrome	Long arm (q) of chromosome 5 at position 14.1
BLOC1S3	Biogenesis of lysosomal organelles complex-1, subunit 3	Instructs for BLOC1 multi-subunit protein complex	HPS type 8; affects eyes and bleeding	Located on long arm (q) of chromosome 13 at position 13.32
DTNBP1	Dystrobrevin binding protein 1	Binds with another complex-the alpha-and beta dystrobrevins	HPS type 7; may be related to premature death in persons with HPS; may also be connected with schizophrenia	Located on the short arm (p) of chromosome 6 at position 22.3

What Is the Treatment for Hermansky-Pudlak Syndrome?

Although there is no cure for HPS, the symptoms can be monitored and controlled. Treatment involves controlling the complications of bleeding, testing the lungs and kidneys for damage, and frequently checking for skin cancers. Important is a periodic CAT scan to test for internal bleeding. There are many aids for vision and other problems. Healthy living such as avoiding smoking can help protect the lungs.

Further Reading

"Hermansky Pudlak Syndrome." 2002. National Organization for Albinism and Hypopigmentation. http://www.albinism.org/publications/HPS.html. Accessed 5/20/12.

"Hermansky-Pudlak Syndrome." 2011. Genetics Home Reference. National Library of Medicine (U.S.). http://ghr.nlm.nih.gov/condition/hermansky-pudlak-syndrome. Accessed 5/20/12.

"Hermansky-Pudlak Syndrome." 2011. Medscape. http://emedicine.medscape.com/article/1200277-overview. Accessed 5/20/12.

Hermansky-Pudlak Syndrome Network. 2011. http://www.hpsnetwork.org. Accessed 5/20/12.

Holocarboxylase Synthetase Deficiency (HCSD)

Prevalence Estimated 1 in 87,000 people

Other Names early-onset biotin-responsive multiple carboxylase deficiency; early-onset combined carboxylase deficiency; HLCS deficiency; infantile multiple carboxylase deficiency; multiple carboxylase deficiency, neonatal form

Knowledge of nutrition and how the body processes food developed slowly throughout the twentieth century. Metabolic syndromes were mysterious facts of nature at the beginning of the century when Archibald Garrod first looked at the black urine from a family and connected the disorder alkaptonuria as an inherited organic disorder. Others emerged as the knowledge of chemical analysis advanced.

In 1976, Karl Roth had two infant patients die with similar unusual symptoms. The newborn children, who had healthy parents, suddenly developed poor appetites, vomiting, irritability, poor muscle tone, and a severe peeling rash. Later when studying the fibroblasts, cells that give rise to connective tissue, he found that these children had a major defect in the way that carbon dioxide was used in the cells. Carbon dioxide is generally thought to be important in plant respiration, but it is also important in animal metabolism, especially related to the function of certain enzymes. He coined the term "multiple carboxylase deficiency." Later he uncovered three gene mutations that involved three different enzymes that did not work properly. The name was changed to holocarboxylase synthetase deficiency, which the medical community uses widely.

What Is Holocarboxylase Synthetase Deficiency (HCSD)?

Holocarboxylase synthetase deficiency is a disorder in which the person cannot use certain foods, such as proteins and carbohydrates, to create energy for the body. The condition arises when an enzyme called "holocarboxylase synthetase," or HCS, is not working or is missing.

Enzymes are necessary for the body to break down food properly. Necessary also for the enzymes to do their work is a cofactor. In this disease, the essential partner is the vitamin biotin, one of the B-complex vitamins. It is essential for the work of certain enzymes to break down food. Biotin is found in egg yolk, milk, liver, and yeast. In addition to HCS, several enzymes use biotin in digestion.

In a normal person, the protein and carbohydrates from body stores and from food are broken down into amino acids and simple sugars. Then the HCS enzyme and biotin use these products for energy and growth. With HCS deficiency, the proteins and carbohydrates are broken down to simple sugar and proteins, but because the HCS enzyme is not there, harmful buildups from the excess amino acids and sugar cause health problems.

When a child is born with HCS deficiency, he or she may appear normal. Biotin from the healthy mother has passed through the placenta and protected the child. The first symptoms, however, may appear hours after birth or several days. The condition is called a metabolic crisis and may exhibit the following signs:

- Excessive sleepiness
- Poor appetite
- Vomiting
- No energy
- Irritability
- Peeling skin
- Low blood sugar or hypoglycemia
- High levels of acid in the blood, called acidosis
- High level of ketones and organic acids in urine
- High levels of ammonia in the blood

If the child is not treated, he may have seizures. Buildup of the toxic chemicals can cause breathing problems, swelling of the brain, coma, and death.

What Is the Genetic Cause of Holocarboxylase Synthetase Deficiency?

Mutations in the *HLCS* gene, officially known as the "holocarboxylase synthetase (biotin proprionyl-Co-A-carboxylase (ATP-hydrolysing) ligase)" gene, cause

HCS deficiency. Normally, *HLCS* instructs for the enzyme holocarboxylase synthetase, or HCS. This enzyme attaches to biotin, one of the vitamin B complex, and aids in the breakdown of proteins and carbohydrates.

About 30 mutations have been identified in *HLCS*. Most of these involve exchange of only one amino acid, but it is enough to disrupt the function of the enzyme. It cannot bind to biotin and thus allow the normal breakdown of amino acids and simple sugars necessary for normal development and growth. Researchers point to the defect in the enzyme as the underlying cause of the signs and symptoms of HCS deficiency. *HLCS* is inherited in an autosomal recessive pattern and is located on the long arm (q) of chromosome 21 at position 22.1.

What Is the Treatment for Holocarboxylase Synthetase Deficiency?

In most states, HCS deficiency is one of the conditions that are tested in newborn screening. If the condition is detected, the child is immediately given biotin. The biotin treatment can prevent the symptoms or can reverse in children who have shown symptoms. A physician who specializes in metabolic disorders works with the primary physician to determine the dosage.

See also Newborn Screening: A Special Topic

Further Reading

"Expanded Newborn Screening." 2011. STAR-G. http://www.newbornscreening.info/Parents/organicaciddisorders/HCSD.html. Accessed 5/20/12.

"Holocarboxylase Synthetase Deficiency. 2011. Medscape. http://emedicine.medscape.com/article/944631-overview. Accessed 5/20/12.

Holt-Oram Syndrome (HOS)

Prevalence 1 out of 100,000 live births; more than 300 cases published showing many different clinical signs

Other Names atriodigital dysplasia; atrio-digital syndrome; cardiac-limb syndrome; heart-hand syndrome, type 1; HOS; ventriculo-radial syndrome

In 1960, British doctors Mary Clayton Holt and Samuel Oram made an unusual connection. Studying four generations of a family that had a specific kind of heart defect, they found that the same people had a congenital birth defect of the thumb. This hand-heart disorder was named for these two doctors—Holt-Oram Syndrome. Later, over 300 published studies showed that the skeletal system disorders and cardiac defects could have several different clinical signs.

What Is Holt-Oram Syndrome (HOS)?

Holt-Oram syndrome is a disorder that affects both the upper skeletal system and the heart. In people with HOS, the skeletal system may display the following conditions:

- Abnormally formed bones in the upper limbs
- At least one problem in the bones of the wrist; this abnormality in the carpal or wrist bones may be seen only in X-ray
- A missing thumb
- Thumb may be very long and appear as a finger
- Partial bones in upper arm or completely missing bones in upper arm
- Abnormal collar bone or shoulder blades
- Left side of the body more often affected

Numerous cardiac abnormalities, which may vary, may be present. In about 75% of people with HOS, the heart defects are so serious as to be life-threatening. The heart is made of four chambers—the upper parts called the atria and the lower parts called ventricles. A tough, muscular wall called the septum separates the left and right sides. In people with HOS, the following cardiac conditions may be present:

- A hole in the septum or muscular wall that separates the two sides of the heart. This defect is the most common. If the hole is in the upper part, it is called an atrial septal defect (ASD); if the hole is between the lower chambers, it is called a ventricular heart defect (VSD).
- Problems in the electrical or conduction system of the heart. The heartbeat can be slow (brachycardia), or rapid (tachycardia), or uncoordinated (fibrillation).

Both heart defects and conduction problems may occur or only one may be present in HOS.

What Is the Genetic Cause of Holt-Oram Syndrome (HOS)?

Mutations in the *TBX5* gene, officially called the "T-box 5" gene, cause Holt-Oram syndrome. Normally, *TBX5* instructs for T-box 5, a protein that is important in forming tissues and organs when the embryo is developing. Called a transcription factor, the T-box 5 protein attaches to special regions of DNA to activate other genes. T-box 5 is very busy turning on genes in the developing embryo. It activates the genes that control the normal development of the hands and arms in the upper limbs and also the genes responsible for the normal development of the heart, especially the septum, the important muscle that divides the two sides. In addition, T-box 5 coordinates the electrical system that controls the heartbeat.

The more than 70 mutations in the *TBX5* gene cause Holt-Oram syndrome. Some of the mutations stop the production of the T-box 5 protein completely, and others have a change in just one of the building blocks of the protein so as to keep it from binding to DNA. Thus the proper genes essential to the development of the limbs and heart are not activated. The individuals with Holt-Oram then have the characteristic symptoms of the disorder. *TBX5* is inherited in an autosomal dominant pattern and is located on the long arm (q) of chromosome 12 at position 24.1.

What Is the Treatment for Holt-Oram Syndrome (HOS)?

The heart defects of the disorder are the most life-threatening. These disorders should be treated first. Surgery can close the hole in the septum. The problems with the electrical system can be managed with surgery or a pacemaker. Some of the disorders of the limbs can be corrected with prostheses. Individuals with these disabilities fall under the Individuals with Disabilities Education Act (IDEA) and can qualify for special educational and occupational training.

Further Reading

"Holt-Oram Syndrome." 2011. Medscape. http://emedicine.medscape.com/article/159911-overview#showall. Accessed 5/20/12.

"Holt-Oram Syndrome." 2011. Orphanet. http://www.orpha.net/data/patho/Pro/en/HoltOramSyndrome-FRenPro1023.pdf. Accessed 5/20/12.

McDermott, D.; J. Fong; and C. Basson. 2011. "Holt-Oram Syndrome." *GeneReviews*. http://www.ncbi.nlm.nih.gov/books/NBK1111. Accessed 5/20/12.

Huntington Disease

Prevalence Affects 3 to 7 per 100,000 people of European ancestry, meaning it affects about 30,000 people in North America with about 100,000 at risk; less common in people of Asian and African descent

Other Names Huntington chorea; Huntington chronic progressive hereditary chorea; Huntington's disease; progressive chorea, chronic hereditary (Huntington)

In 1692, residents of Salem stared at the strange "dancing" movements and violent outburst of tempers of some of their neighbors. Horrified, they concluded that anyone with such bizarre behavior must be flirting with the devil. Two centuries later, New York doctor George Huntington determined that patients with strange twitching behaviors were members of families who had been treated by his father and

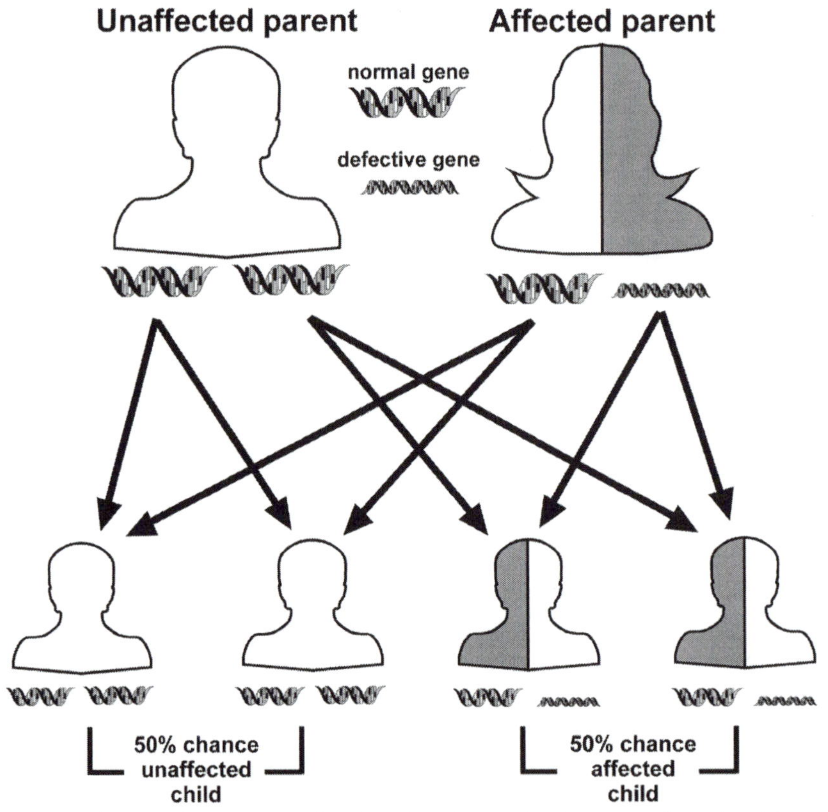

Huntington disease is an example of a dominant single gene disorder. The condition does not usually appear until mid-adulthood. (ABC-CLIO)

grandfather for the same disorder. Huntington traced the condition to a family of immigrants from Bures in Suffolk, who came to Boston in 1630. He noted that if either parent had the disease, one or more offspring suffered from it. In 1872, he fully described the disease that became known as Huntington's chorea, from the Greek word meaning "dance"; it is now just called Huntington disease.

What Is Huntington Disease?

Huntington disease (HD) is a progressive, degenerative brain disorder characterized by uncontrolled movements, emotional outbursts, and mental deterioration. Although HD can begin at any age, the most common age of onset is between 35 and 44. The disease affects about 30,000 people in North America, with about

100,000 at risk. Since symptoms develop in adult life, the person with Huntington disease may pass HD to their offspring without knowing it.

Early symptoms of the disorder include the following:

- Depression
- Irritability
- Lack of coordination
- An unsteady gait
- Small involuntary movements or twitches
- Trouble learning new information
- Trouble making decisions

Then the disease progresses with the following symptoms:

- Jerky body movements more apparent
- Behavior more erratic
- Decline in mental abilities and cognition
- Dementia
- Problems with speaking and swallowing
- Complications such as pneumonia, heart disease, and injury from falls

Eventually, the person is in complete decline and may live 15 to 20 years after the first diagnosis.

A less common early-onset form of Huntington disease begins in early childhood or adolescence. This form is called the juvenile, akinetic-rigid, Westphal variant. The symptoms of abnormal movements, emotional disturbances, and mental disorders are similar. Additional signs of the early-onset form include drooling, slurred speech, and seizures. The early-onset form appears to progress more rapidly, and persons usually live 10 to 15 years after the onset of symptoms.

What Is the Genetic Cause of Huntington Disease?

Changes in the "huntingtin" gene, or *HTT* gene, cause Huntington disease. Normally, this gene encodes for a protein labeled "huntingtin," which appears to play an important role in the embryonic development of nerve cells in the developing brain. The highest level of huntingtin is in the brain, but it is also present in many other cells, with a variety of functions. Huntingtin may be involved in cell chemical signaling, transporting materials, binding proteins, and the extremely important activity of protecting cells from self-destruction, a process called apoptosis. A region of *HTT* gene has a segment of repeats called CAG, or cytosine-adenine-guanosine repeats. The normal number of repeats is 10 to 35 times.

The mutation in huntingtin is a the result of the CAG nucleotide repeat expansion. Instead of repeating the normal 10 to 35 times, people with the disorder may have more than 40 repeats. Individuals with the juvenile form may have between 60 and 120 repeats. The CAG nucleotides instruct the protein synthesis process to include a glutamine amino acid. As the repeats expand the gene, an abnormally long version of the huntingtin protein is produced, having many extra glutamines. This long protein is then cut into smaller, toxic fragments that gather in the neurons. The substances gather especially in the area of the brain called the striatum and cerebral cortex that control movement, thinking, and emotions. The disruption of the neurons and their eventual death lead to the characteristic symptoms of Huntington disease.

As the CAG repeat goes from one generation to another, the size tends to increase. Those who may have 27 to 35 repeats do not develop Huntington disease but can pass the gene to their children, who may develop the disorder. HTT is inherited in a dominant pattern and is located on the short arm (p) of chromosome 4 at 16.3.

What Is the Treatment for Huntington Disease?

Because it is a genetic disorder, no treatment exists. However, some efforts have been made to reduce the severity of the symptoms. A lot of research is focused on help with daily living, in addition to research for new medication. For example, nutritional management may involve adding thickening agents to help swallow and giving very small pieces of food to eat. Physical therapy, occupational therapy, and speech therapy may help some of the cognitive decline.

The Hereditary Disease Foundation and Huntington's Disease Society of America have worked to create public awareness, research, and family support. The death of prominent folk singer Woody Guthrie from complications of Huntington disease and a multitude of efforts by local support groups eventually led to the designation of June 25 as National Huntington's Disease Awareness Day by the U.S. Senate in 2008.

Huntington disease was one of the first diseases diagnosed using the new technique called "genetic testing." This test has caused social, legal, and ethical concerns over access and use of the results. In 1910, Charles Davenport, a leader of the eugenics movement in the United States, proposed forced sterilization for people with Huntington disease. Today, many guidelines have strict procedures over the results; individuals choose how and to whom to reveal the results. For example, testing on a child may result in no medical benefits for that child, but may show that the child has a preexisting condition. The counterargument is that patients and their caregivers need to know about the results so they can make important life decisions and plans.

Huntington disease has been dubbed the "Crown Jewel of Genetic Research," as the story of Nancy Wexler and James Gusella, shows.

Nancy Wexler, James Gusella, and the Crown Jewel of Genetic Research

When Nancy Wexler, a New York psychologist, found out that her mother had Huntington disease and that she had a 50-50 chance of inheriting it, her life was changed. In 1972, at a meeting of the World Federation of Neurological Research on Huntington Disease, a young physician presented video of dozens of people living on Lake Maracaibo in Venezuela who had the writhing "dance" of "Huntington's Chorea." She had talked with several scientists about the idea of finding the causative gene. In 1979, she headed for the two isolated villages of Barranquitas amd Laguenetas and convinced the people there to let her take blood, which yielded 2,000 samples of blood. Through interviews, she constructed a pedigree map of 10,000 people.

She brought the blood samples and pedigree maps to James Gusella and a team of researchers at Harvard who analyzed the DNA samples looking for a marker for the gene that was cross-linked with the traits on the map. They were shocked when they found the marker on the third try. This landmark discovery in 1983 developed other innovations in DNA-marking methods, which became an important step in making the Human Genome Project possible.

In 1993, researchers from the Huntington Collaboration Group found the gene designated IT 15 on chromosome 4 and later called it "huntingtin." All mutations are some form of exaggerated CAG trinucleotide repeats in the coding region of the gene. The research into Huntington disease became known as the "Crown Jewel of Genetic Research" and became a model for research into other neurodegenerative disorders. The federal government has recognized HD as a model for other neurodegenerative diseases, such as Parkinson's, ALS, and Alzheimer disease.

In 2000, Boston researcher Dr. Robert Friedlander reported progress in understanding how inhibiting caspase-1 activity delayed symptoms of HD in mouse models. Using the antibiotic minocycline, his team blocked production of caspases 1 and 3. Although mutant huntingtin protein causes striatal neurons to die, Milan scientist Elena Cattaneo found the normal protein helps regulate the production of BDNF, a protein essential for survival of the striatal cells. Berlin scientist Dr. Erick Wanker has developed a library of 180,000 chemical compounds to screen for prevention of agglutination or clumping of neurons, a characteristic common in HD and other neurodegenerative conditions. He has identified 687 promising substances, concentrating on 100 of these for therapies in cell model cultures.

In March 2001, Dr. Christopher Ross and a Johns Hopkins team described how the HD gene causes the death of cells. A key molecule, CPB, plays a vital role in activating genes needed to reverse cell death. The abnormal gene produces a flawed form of the protein huntingtin that causes clumping in brain cells of HD patients. When the clumped molecules become entangled with the critical protein CBP in the cell nucleus, that regulatory molecule is hijacked, and the pathway essential for cell survival is never activated. They were able to reverse the process *in vitro*, but not in a mouse model. Since finding the abnormal protein huntingtin that causes cell death, researchers are optimistic that here is a needed target for developing new drugs.

Intrastriatal transplantations of fetal striatal neuroblasts have restored motor and cognitive function in experimental animals. Results from two separate studies showed that grafts of human striatal tissues have detectable effects in a limited number of patients with mild to moderate HD. While the use of fetal tissue is limited in the United States and will probably never be used for widespread treatment, researchers believe the transplant principles may apply to other cells such as stem cells that can be grown in the laboratory.

While investigators are working on basic research of understanding mechanisms, several clinical trials are studying drugs and surgical techniques. A disappointing trial (347 patients) found neither the drug remacemide (AstraZeneca) nor coenzyme Q10 (CoQ10) had a statistically significant effect on slowing the progress (*Neurology* 2001 57: 397–404).

Further Reading

"Huntington Disease." 2012. Genetics Home Reference. National Library of Medicine (U.S.). http://ghr.nlm.nih.gov/condition/huntington-disease. Accessed 2/6/12.

"Huntington's Disease." Mayo Clinic. http://www.mayoclinic.com/health/huntingtons-disease/DS00401. Accessed 5/21/12.

Huntington's Disease Society of America. 2010. http://www.hdsa.org. Accessed 5/21/12.

"Nancy Wexler: The Gene Hunter." 2008. Hereditary Disease Foundation. http://www.iwaswondering.org/nancy_homepage.html. Accessed 2/13/12.

"NINDS Huntington's Disease Information Page." 2010. National Institute of Neurological Disorders and Stroke (U.S.). http://www.ninds.nih.gov/disorders/huntington/huntington.htm. Accessed 5/21/12.

Hutchinson-Gilford Progeria Syndrome (HGPS)

Prevalence Rare; about 1 in 4 million newborns; 130 cases described since 1886

Other Names HGPS; Hutchinson-Gilford syndrome; progeria; progeria of childhood

In 1886, a six-year-old boy who had the appearance of an old man was brought into the office of Dr. J. A. Hutchinson. He described the boy in a medical journal, relating how not only his appearance but his organs were like those of a very old person. In 1904, H. Gilford described another boy with similar appearance and published a series of photographs showing how the boy changed over several years. The condition, which is also called progeria, was named after these two doctors—Hutchinson-Gilford progeria syndrome. Sometimes the letters HGPS are used.

The word progeria comes from two Greek terms: *pro*, meaning "in front of," and *geras*, meaning "old age." *Geras* is the same root word used in geriatrics and gerontology, the study of aging.

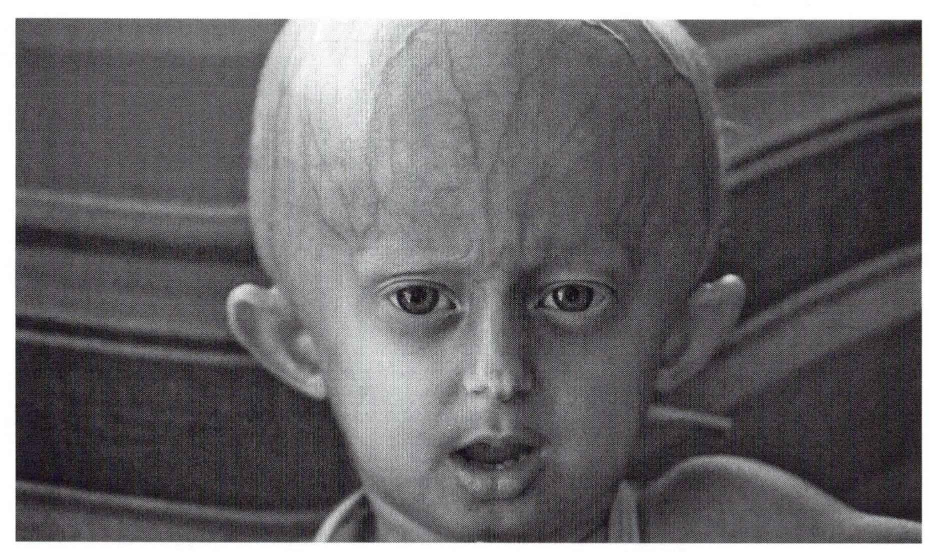

A young girl from Pretoria, South Africa, suffers from the genetic disorder Hutchinson-Gilford progeria syndrome. This child later took part in a U.S. clinical trial for progeria sufferers. (Photo by Foto24/Gallo Images/Getty Images) (Getty Images)

What Is Hutchinson-Gilford Progeria Syndrome (HGPS)?

Hutchinson-Gilford progeria syndrome (HGPS) is a condition in which the child who is normal appearing at birth begins to age rapidly. Over a period of a few months, the child begins to develop many of the conditions, not only in appearance but in the actual physical organs of an older person. The condition affects boys to girls in a ration of 1.5 to 1.

Following are the symptoms of HGPS:

- Aged-looking skin with wrinkles: Brown spots called lipofuscin deposits appear on skin in abundance.
- Loss of fat under the skin: This makes the skin look very thin; veins may show through skin.
- District facial appearance: Eyes appear to protrude and are very prominent. The beaked nose has a very pronounced tip. The child also develops a very small and protruding chin. Ears also protrude.
- Failure to gain weight as a child.
- Balding head, a condition known as alopecia.
- Joint abnormalities and skeletal disorders.
- Osteoporosis.

- Arteriosclerosis of both arteries to the brain and heart. However, the only abnormality appears to be decreased high-density lipoprotein cholesterol levels.
- Affected organs: These include the kidneys, brain, adrenal glands, liver, testes, and heart.

The cause of most deaths is from heart attack or stroke. Life expectancy is no more than about 13 years of age. Surprisingly, the children do not have all the characteristics of aging. They do not develop cataracts, cancer, tumors, or senility. They still possess intellectual functions and skills, such as sitting, standing, and walking.

What Is the Genetic Cause of Hutchinson-Gilford Progeria Syndrome?

Mutations in the *LMNA* gene, officially known as the "lamin A/C" gene, causes HGPS. Normally, *LMNA* instructs for making two different proteins called lamin A and lamin C, both of which are made in more of the body cells. The sequence of the protein is almost identical except that one small difference makes lamin A somewhat longer than lamin C. These two proteins are referred to as intermediate filament proteins that give strength and stability to cells. The two lamins, called lamina, also form a strong mesh-like layer that envelops the nucleus of the cell, permitting molecules to go in an out of the cell. A preparation procedure precedes the insertion of the lamina. Only one of the lamina must undergo the processing in an initial form called prelamina A.

One special mutation in *LMNA* replaces thymine for cytosine in the gene building blocks. The mutation leads to an abnormal version of the lamin A protein, which is called progerin and causes 50 amino acids near the one end of the gene to go missing. When the abnormal version is incorporated into the cells, it interferes with the production of the shape of the nuclear envelope and over time causes damage to the nucleus, causing them to die prematurely. This disruption leads to the signs and symptoms of HGPS. *LMNA* is inherited in an autosomal dominant pattern and is located on long arm (q) of chromosome 1 at position 22.

What Is the Treatment for Hutchinson-Gilford Progeria Syndrome?

Treating the abnormal heart conditions with medications such as statins may extend the lives of the children. Regular diet and exercise is recommended. Hip dislocation and other skeletal disorders may be controlled with physical therapy. Shoe pads may be recommended because the lack of fat on the soles of the feet may cause discomfort. Normal schooling is recommended.

See also Aging and Genetics: A Special Topic

Further Reading

Gordon, L.; T. Brown; and F. Collins. 2011 "Hutchinson-Gilford Progeria Syndrome." *GeneReviews*. http://www.ncbi.nlm.nih.gov/books/NBK1121. Accessed 12/24/11.

"Hutchinson-Gilford Progeria Syndrome." 2011. Genetics Home Reference. National Library of Medicine (U.S.). http://ghr.nlm.nih.gov/condition/hutchinson-gilford-progeria-syndrome. Accessed 12/24/11.

"Hutchinson-Gilford Progeria." 2011. Medscape. http://emedicine.medscape.com/article/1117344-overview#showall. Accessed 12/24/11.

Hypercholesterolemia, Familial

Prevalence More than 34 million Americans affected in some way; inherited form less frequent; about 1 in 500 in some countries; more frequent in Afrikaners in South Africa, French Canadians, Lebanese, and Finns.

Other Names elevated cholesterol; familial hypercholesterolemia; FH; hypercholesterolemic xanthomatosis; low-density lipoprotein receptor mutation; type II hypercholesterolemia

The word "cholesterol" is almost a household name. It is used frequently in television and magazine advertising. According to these ads, cholesterol seems to be a pretty bad thing. Without cholesterol, however, our bodies would not work. It is necessary to make cell membranes, hormones, vitamin D, and certain substances that make digestion possible. Cholesterol is made by the body and is also found in foods that come from animals, such as eggs, red meat, poultry, fish, and dairy products.

Cholesterol is circulated in the bloodstream in little packets called lipoproteins. One can see two root words here: *lipo*, meaning "fat," and protein. Lipoproteins have small bubbles of fat on the inside of the package covered by a protein outer coat. Two kinds of lipoproteins exist: low-density lipoproteins or LDLs, and high density lipoproteins or HDLs. Cholesterol carried by HDLs is designated "good cholesterol." The liver removes the fats from the cells and processes it for body use. Cholesterol carried by LDLs is sometimes called the "bad" cholesterol. The liver does not remove it from the lipoprotein, leaving it to circulate in the bloodstream where it can collect in blood vessels and cause heart disease and other problems. It is the problem of excess cholesterol in the blood that causes hypercholesterolemia.

What Is Hypercholesterolemia?

Hypercholesterolemia is the presence of excess cholesterol in the blood. The long word is easily understood when broken into parts: *hyper* is the Greek root meaning "above or beyond"; cholesterol refers to the fatty chemical; the root *emia* comes from the Greek *heme* meaning "blood." Although all cases are not inherited, the tendency for high levels of LDLs that run in families is called familial hypercholesterolemia or FH.

When cholesterol travels though the blood, it begins to deposit on the walls of the arteries, especially on those leading to the heart. The person develops a form of heart disease called coronary artery disease. This buildup forms plaque, clumps that begin to grow and block the flow of blood. The buildup in the heart arteries is called atherosclerosis. The person may experience angina or pain in the chest and have an increased risk for a heart attack.

As cholesterol is deposited in other organs and tissues, the person with familial hypercholesterolemia experiences other health problems. Following are several areas that can be affected:

- Fatty deposits in tendons cause tendon xanthomas. Most often seen in the Achilles tendon, xanthomas can also occur in tendons of the hands and fingers. The word xanthoma comes from two Greek words: *xanthos*, meaning "yellow," and *oma*, meaning "tumor or growth."
- Xanthomas over the elbows, knees, and buttocks.
- Fatty yellow deposits under the eyelids.
- Cholesterol accumulation on the front surface of the cornea of the eye that appears as a gray ring.

This disorder of LDL begins at birth and can cause problems at an early age. Heart attacks may occur early in life.

What Are the Genetic Causes of Hypercholesterolemia?

The effects of hypercholesterolemia are related to many lifestyle factors and choices of an individual. Diet, exercise, and smoking influence the amount of cholesterol in the blood. Other health factors are involved, such as gender, age, diabetes, and obesity. If a person has a genetic factor for FH, he or she must pay special attention to these lifestyle factors.

A strong family history of early heart attacks and high levels of LDL, in either or both parents, is an indicator of FH. Combination of lifestyle choice with a variety of genes may lead to different degrees of the condition. Mutations in four genes cause FH: *LDLR*, *APOB*, *LDLRAP1*, and *PCSK9*.

LDLR

Mutations in the *LDLR* gene, officially known as the "low density lipoprotein receptor" gene, are the most common causes of FH. Normally, *LDLR* instructs for a protein called the low-density lipoprotein receptor. The receptors sit on the surface of many cells waiting for the low-density lipoproteins that are circulating in the bloodstream to come by. Like a key fitting into a lock, the receptors attach to the LDLs with their cargo of cholesterol and take them into the cells. Inside the cells, three things can happen to the cholesterol. It can be used by the cell for cell functions, stored, or removed from the body. The LDL receptors then go back to the cell surface waiting to pick up more LDLs. The protein made by the *LDLR* gene is critical in controlling the amount of cholesterol circulating in the blood. The proteins are prevalent in the liver, the organ that plays the important role of moving excess cholesterol from the body.

More than 1,000 mutations have been found in *LDLR*. Although some mutations reduce the number of LDL receptors on the cells, others keep the receptors from working properly. The result is that the person will have high cholesterol levels in the blood. The high levels circulate around depositing the waxy substance in the skin, tendons, and especially the arteries that lead to the heart. Cholesterol builds up, causing plaque that interferes with the proper supply of blood and increases the risk of heart attack.

Most people inherit only one copy of the gene from a parent with LH. If in a rare instance, the person inherits the gene from both parents, a very severe form of LH will appear usually in childhood. *LDLR* is inherited in an autosomal dominant pattern and is located on the short arm (p) of chromosome 19 at position 13.3.

APOB

The second gene causing LH is the *APOB* gene, officially known as the "apolipoprotein B (including Ag(x) antigen)" gene. Normally, *APOB* instructs for making two versions of the apolipoprotein B: a short version, B-48, that is produced in the intestines; and a long version, B-100, produced in the liver. Both of these proteins carry fats and cholesterol in the blood. Apolipoprotein B-48 is a building block of a special type of lipoprotein called a chylomicron. When food is digested and broken down, chylomicrons carry the fat and cholesterol into the bloodstream. Apolipoprotein B-100 in the liver makes other kinds of lipoproteins that all transport fats and cholesterol.

Five mutations of *APOB* cause a defective apolipoprotein B-100. Each mutation is the result of a change in only one building block or amino acid. The change in the protein keeps the LDL proteins from binding properly to the receptors on the surface of the cells, and as a result, excess cholesterol circulates in the bloodstream. *APOB* is inherited in an autosomal dominant pattern and is located on the short arm (p) of chromosome 2 at position 24-p23.

LDLRAP1

The third gene, *LDLRAP1*, officially known as the "low density lipoprotein receptor adaptor protein 1" gene, is different from the other genes in that it is inherited in a recessive pattern. Normally, *LDLRAP1* instructs for a protein that removes cholesterol from the bloodstream. This protein works especially in the liver, the organ that takes up excessive cholesterol. The protein interacts with another protein called the LDL receptor, which binds to LDLs that are roaming the blood. The proteins appear to play an important role in moving the receptors from the surface to inside the cells where the LDLs release cholesterol to be used by the cell.

Ten mutations in *LDLRAP1* cause the protein to be small and abnormal. Without this protein, the receptors on the cell do not remove the LDLs from the blood, or they may not be transported properly into cells, especially liver cells. Thus, LDLs remain in the blood. *LDLRAP1* is inherited in a recessive pattern and is located on the short arm (p) of chromosome 1 at position 36-p35.

PCSK9

The fourth gene that causes LH is *PCSK9*, officially known as the "proprotein convertase subkulisin/kexin type 9" gene. Normally, *PCSK9* instructs for a protein that controls cholesterol in the bloodstream. The protein functions by controlling the number of LDL receptors on the surface of the cells. As have been related in other genes, the receptors are very important in binding to the LDLs to carry cholesterol in the bloodstream. Many of the receptors are in the liver.

Several mutations of *PCSK9* cause FH. Most involve changes in a single building block. The mutations disrupt the function of the protein and make it overactive, causing an adjusted workload on the cell receptors. They may break down; fewer receptors are available to remove the LDLs from the blood. *PCSK9* is inherited in an autosomal dominant pattern and is located on the short arm (p) of chromosome 1 at position 32.3.

What Is the Treatment for Hypercholesterolemia?

Hypercholesterolemia is due to a combination of many environmental and genetic factors. Management involves controlling diet, increasing exercise, and not smoking or using alcohol excessively. One gene may be involved in FH or multiple genes. Lifestyle changes in what the person does and eats are critical. Decreasing amounts of beef, chicken, pork, and lamb, substituting low-fat dairy products, and eliminating such oils as coconut and palm oils is necessary. Dietary counseling is available to help people change eating habits. Another management strategy is the use of medications, such as statins. The prognosis is basically good for those that follow their physicians' advice.

Further Reading

"Familial Hypercholesterolemia." 2011. Genetics Home Reference. National Library of Medicine (U.S.). http://www.nlm.nih.gov/medlineplus/ency/article/000392.htm. Accessed 5/21/12.

"Familial Hypercholesterolemia." 2011. Medscape. http://emedicine.medscape.com/article/121298-overview#showall. Accessed 5/21/12.

"Learning about Familial Hypercholesterolemia." 2011. National Human Genome Research Institute (U.S.). http://www.genome.gov/25520184. Accessed 5/21/12.

Hypochondrogenesis

Prevalence Along with achondrogenesis, a similar condition, affects 1 in 40,000 to 60,000 newborns

Other Names None

Several types of dwarfism are inherited. Some conditions are caused by hormone imbalances, metabolic problems, or bone growth problems. Hypochondrogenesis belongs to the latter group, a class of dwarfism called chondrodystrophy or skeletal dysplasia. These disorders are members of a group called collagenopathies in which defects are in the collagen, the connective tissue. The word "hypochondrogenesis" comes from the Greek roots: *hypo*, meaning "under"; *chondro*, meaning "cartilage"; and *genesis*, meaning "giving birth to" or "originating in."

What Is Hypochondrogenesis?

Hypochondrogenesis is a rare disorder of bone growth. Infants born with hypochondrogenesis are abnormally short and have short limbs and unusual development of bones in the spine and pelvis. In addition to the short stature, the child has a flat and oval-shaped face, widely spaced eyes, a small chin, and sometimes a cleft palate.

The condition is noted at birth. The children may have a bloated and distended stomach. This is due to a condition known as hydrops fetalis, a condition in which excess fluid builds up in the abdomen before birth. Compression of the deformed ribs may cause the child to have breathing problems. Because of the serious health problems, many fetuses do not survive to full term. Infants who are born may die at birth or soon after because of respiratory failure. If the child lives past the first year, he or she may be reclassified as having spondyloepiphyseal dysplasia congenita, a similar but milder condition.

What Are the Genetic Causes of Hypochondrogenesis?

Mutations in the *COL2A1* gene, officially known as the "collagen, type II, alpha 1" gene, cause hypochondrogenesis. It is one of the most severe mutations in a group of disorders caused by this gene. Normally, *COL2A1* instructs for making a major ingredient of type II collagen, called the pro-alpha 1(II) chain. Type II collagen is the substance that gives strength and form to the connective tissue. Connective tissue supports muscles, joints, organs, and skin. Type II collagen is like the cartilage found in the nose. It is tough but flexible. During early development, cartilage is found in great abundance in the skeleton, but as one grows, the cartilage becomes bone. It still is found in the ends of the bones to protect them. Collagen II also fills the eyeballs, inner ears, and the vertebrae in the spine.

In order to make Type II collagen, three pro-alpha1(II) chains twist together to form a rope-like molecule. Enzymes in the cells process these molecules and arrange them into long, flexible fibrils that cross-link to make a strong collagen fiber and fill spaces between bone.

About 18 mutations in *COL2A1* cause hypochondrogenesis. Some of the mutations involve a missing part of the gene; others replace a glycine molecule with another amino acid at different places on the chain. Each type of mutation disrupts the proper formation of the collagen chain, resulting in the classic features of hypochondrogenesis. *COL2A1* is inherited in an autosomal dominant pattern and is located on the long arm (q) of chromosome 12 at position 13.11.

What Is the Treatment or Management for Hypochondrogenesis?

Because this is a genetic disorder, there is no cure. For children who live, attacking the most serious problem, assisting in breathing, is primary. If children live past age one, they are usually reclassified and are treated for that disorder. The cleft palate may require corrective surgery.

Further Reading

"*COL2A1*." 2011. Genetics Home Reference. National Library of Medicine (U.S.). http://ghr.nlm.nih.gov/gene/COL2A1. Accessed 5/21/12.

Fergus, Kathleen. 2005. "Hypochondrogenesis." *Gale Encyclopedia of Public Health*. http://www.healthline.com/galecontent/hypochondrogenesis. Accessed 5/21/12.

"Hypochondrogenesis." 2011. Genetics Home Reference. National Library of Medicine (U.S.). http://ghr.nlm.nih.gov/condition/hypochondrogenesis. Accessed 5/21/12.

Hypochondroplasia (HCH)

Prevalence Actual number unknown; more than 200 worldwide diagnoses; more males than females affected

Other Names HCH; hypochondrodysplasia

Dwarfism is caused by several different developmental problems in metabolism, hormones, or bone growth. The word "hypochondroplasia" comes from three Greek words: *hypo*, meaning "under"; *chondro*, meaning "cartilage"; and *plasia*, meaning "form." Children with this condition appear normal at birth, but as they grow, certain bones tend not to develop properly, causing not only short stature but short limbs.

What Is Hypochondroplasia (HCH)?

Hypochondroplasia is a disorder of the long bones of the arms and legs. As the child grows, normally cartilage in the long bones of the arms and legs form strong bones. The process is called ossification. The only cartilage is at the ends of the bones for protection when moving. In people with HCH, the bones, especially the long bones of the arms and legs, do not develop properly. People with HCH have short stature with the average height for men ranging from 4 feet, 6 inches to 5 feet, 5 inches; adult women range from 4 feet, 2 inches to 4 feet, 11 inches.

The signs are similar to those people with achondroplasia but are somewhat less pronounced. Researchers think that this condition is just about as common as achondroplasia, which occurs in 1 in 15,000 to 40,000 newborns.

In addition to short stature and short arms and legs, the following are other features of HCH:

- Stocky build
- Large head, called macrocephaly
- Relatively normal facial appearance
- Short stubby hands and feet
- Limited range of motion of elbows
- Scoliosis less common
- Bow legs, usually mild
- Protruding abdomen
- Adults develop osteoarthritis
- Small percentage with mild to moderate learning disabilities

What Is the Genetic Cause of Hypochondroplasia (HCH)?

Mutations in the *FGFR3* gene, officially known as the "fibroblast growth factor receptor 3" gene, cause hypochondroplasia. Normally, *FGFR3* instructs for the protein fibroblast growth receptor 3. Fibroblast growth factor proteins are crucial for cell process, especially cell growth and division. It is also important for forming blood vessels, wound healing, and the developing embryo. The FGFR3 protein is situated so that part is in the cell and part projects out of the cell surface. The projection can interact with other growth factors to receive all sorts of signals for growth and development. When this occurs, chemical reactions allow the cell to take on certain functions. The protein maintains bone and brain cells and regulates the process of ossification or changing cartilage to bone.

Several mutations in *FGFR3* gene cause hypochondroplasia. Most of the mutations involve changes when lysine replaces asparagine. The changes disrupt the normal function of the protein, causing the receptor to become overactive and lead to disturbances in bone growth. *FGFR3* is inherited in an autosomal dominant pattern is located on the short arm (p) of chromosome 4 at position 16.3.

What Is the Treatment for Hypochondroplasia (HCH)?

The orthopedic surgeon is the mainstay for treatment. Treating the symptoms that would be most disturbing, such as scoliosis or bowed legs, would be primary. Physical therapy may help some people with HCH.

Further Reading

"*FGFR3.*" 2011. Genetics Home Reference. National Library of Medicine (U.S.). http://ghr.nlm.nih.gov/gene/FGFR3. Accessed 5/21/12.

Francomano, Clair. 2005. "Hypochondroplasia." *GeneReviews*. http://www.ncbi.nlm.nih.gov/books/NBK1477. Accessed 5/21/12.

"Hypochondroplasia." Nemours Children's Hospital. http://www.nemours.org/service/medical/orthopedics/dysplasia/hypochondro.html. Accessed 5/21/12.

Hypohidrotic Ectodermal Dysplasia (HED)

Prevalence Affects 1 in 17,000 people worldwide

Other Names andifrotic ectodermal dysplasia; Christ-Siemens-Touraine syndrome; CST syndrome; EDA; HED

What would it be like not to sweat at all? For some people, it may be considered a good attribute. However, not to sweat can be a life-threatening condition. The sweat glands are part of an important temperature control in the body. As the skin perspires, the water evaporates, cooling the body. Absence of sweat glands can lead to dangerous hyperthermia or heatstroke, especially in hot weather.

Ectodermal dysplasias are a large group of disorders that derive from the embryonic ectoderm. Three basic layers form the embryo—the endoderm, mesoderm, and ectoderm. The ectoderm is the layer that eventually becomes skin, hair, nails, sweat glands, and teeth. A first report of this condition was made in 1848 by physician J. Thurman who wrote about two cases in which the skin, hair, and teeth were very imperfectly formed. In 1929, A. A. Weech named the condition in an article in the *American Journal of Disabled Children*, "Hereditary Ectodermal Dysplasia." There are about 150 types of ectodermal dysplasia in humans. The most prevalent kind is hypohidrotic ectodermal dysplasia.

What Is Hypohidrotic Ectodermal Dysplasia (HED)?

Hypohidrotic ectodermal dysplasia (HED) is a condition of the hair, sweat glands, and teeth that is formed before birth when the ectoderm layer of the embryo does not develop properly. If the words are broken down, the name of the condition is not so foreboding. "Hypohidrotic" comes from two Greek stem words: *hypo*, meaning "under," and *hidros* or *hydros*, meaning "water." The ectoderm is the embryonic layer that forms the skin, sweat glands, and teeth, "Dysplasia" literally means "malformed." So the condition HED affects several structures that were formed in the layer of the ectoderm before birth.

Following are the characteristics of HED:

- Absence of sweating: The term "hypohidrosis" is used to describe the person's inability to perspire. The individual has fewer sweat glands (eccrine glands) or the glands do not function properly. This feature can be life-threatening because it interferes with the all-important body cooling mechanism. The sweating mechanism can be tested by bringing an iodine solution near the skin and then raising the room temperature to induce sweating. Iodine turns a color when exposed to sweat and can be used to determine the amount and location of the glands.

- Sparse scalp and body hair: The term "hypotrichosis" is used to describe sparseness of the hair and scalp. The hair may appear light or silver and is very fine and brittle. Sometimes the hair has a twisted or pilled appearance. However, secondary sexual hair such as beard or pubic hair is normal.
- Abnormal teeth: The term "hypodontia" is used to describe the absence of teeth or tiny teeth. Sometimes the teeth may have a sharp or conical appearance.
- Abnormal facial appearance: Many individuals with HED have a distinct facial appearance: a high prominent forehead, a sunken nasal bridge, thick lips, and a long chin.
- Different skin: The skin of people with HED may appear thin, soft, and fragile. There may be a lack of pigment—the element that gives skin color—except around the eyes, which may appear as dark and finely wrinkled. Certain chronic skin problems such as eczema may be present.
- Mucous glands: The glands that secrete mucus can be affected, causing problems with respiration or digestion.

The actor Michael Berryman shows the symptoms of HED.

What Are the Genetic Causes of Hypohidrotic Ectodermal Dysplasia (HED)?

HED can occur via changes in three genes: "ectodysplasin A," or *EDA*; "ectodyspasin A receptor," or *EDAR*; and "*EDAR*-associated death domain," or *EDARADD*. All three of these genes affect activity in the ectoderm:

EDA

Normally *EDA* instructs for a protein called ectodysplasin A that plays an important role before birth. It provides instructions for action between the ectoderm and mesoderm and is important for many body tissues and organs, including the skin, hair, nails, teeth, and eccrine glands. The protein also interacts with a receptor formed by the *EDAR* gene to properly form important structures. Mutations in the gene account for about 95% of HED. *EDA* is inherited in an X-linked recessive pattern and is located on the long arm (q) of the X chromosome at position 12-q13.1.

EDAR

Normally, *EDAR* instructs for a protein called the ectodysplasin A receptor, a protein that is important before birth. Like EDA, it is critical for instructions between the ectoderm and mesoderm embryonic layers. The two proteins together initiate the chemical signals for cell division, growth, and maturation. *EDAR* is inherited in an autosomal recessive pattern and is located on the long arm (q) of chromosome 13.

EDARADD

Normally, *EDARADD* instructs for a protein called the EDAR-associated death domain protein that is part of a signaling pathway before birth. It interacts with the protein produced by the gene *EDAR* to assure the proper formation of teeth, hair, nails, and sweat glands. The mutation in this gene rarely causes HED. *EDARADD* is inherited in an autosomal recessive pattern and is located on the long arm (q) of chromosome 1 at position 42.3.

What Is the Treatment or Management for Hypohidrotic Ectodermal Dysplasia (HED)?

Because this disease is genetic, no cure exists. However, several strategies are used for management. For example, wigs or special hair care formulas may be used to cover missing hair. Dental implants may replace teeth. The most life-threatening symptom—lack of sweating—can be controlled in hot weather by several air-conditioning and cooling vests.

Further Reading

"Ectodermal Dysplasia." 2011. Medscape. http://emedicine.medscape.com/article/1110595-overview. Accessed 5/21/12.

"Hypohidrotic Ectodermal Dysplasia." 2011. RightDiagnosis.com. http://www.rightdiagnosis.com/h/hypohidrotic_ectodermal_dysplasia/intro.htm. Accessed 5/21/12.

Wright, J. T.; D. Grange; and M. Richter. 2011. "Hypohidrotic Ectodermal Dysplasia." *GeneReviews*. http://www.ncbi.nlm.nih.gov/books/NBK1112. Accessed 5/21/12.

Hypophosphatemia

Prevalence X-linked hypophosphatemic rickets most common form; affects 1 in 20,000 newborns; other forms found in only a few families

Other Names hereditary type I hypophosphatemia (HPDR I); hereditary type II hypophosphatemia (HPDR II); hypophosphatemic D-resistant rickets I; hypophosphatemic D-resistant rickets II; phosphate diabetes; X-linked hypophosphatemia (XLH); X-linked vitamin D–resistant rickets

At the end of the nineteenth century, scientists began to uncover the mysteries of nutrition. From 1922 to 1924, British and American scientists concentrated on studying children with rickets. They concluded that their blood lacked something that caused bones to develop properly. After 1900, X-ray photography allowed these investigators to look at the effects of abnormal bone calcification. One of the minerals that was missing in these children was phosphorus. The children did not have the normal amounts of this mineral in the blood. They had a condition called hypophosphatemia.

What Is Hypophosphatemia?

Hypophosphatemia is a condition in which the person has very low levels of the mineral phosphorus in the bloodstream. Phosphorus is essential for forming strong bones and teeth. It is necessary for developing membranes, storing energy, and transporting materials in and out of cells. Thus, phosphorus or the form phosphates play important roles in every cell, tissue, and organ in the body. The word "hypophosphatemia" comes from Greek roots: *hypo*, meaning "under"; phosphate referring to the mineral phosphorus; and *emia*, from *heme*, meaning "blood." A person with hypophosphatemia has deficient phosphorus in the bloodstream.

The symptoms of the disorder begin in early childhood. Following are the signs and symptoms of the disorder:

- Slow growth
- Shorter than peers
- Bowed legs or knock knees
- Problems with walking
- Occasional premature closing of skull bones
- Abnormalities of teeth
- Ligaments not attached to bones properly
- Soft bones in adults, a condition called osteomalacia

Mildly affected people may have few or no symptoms.

What Are the Genetic Causes of Hypophosphatemia?

Two genes are related to the condition: *FGF23* and *PHEX*.

FGF23

Mutations in the *FGFH23* gene, officially known as the "fibroblast growth factor 23" gene, cause hypophosphatemia. Normally, *FGF23* instructs for the fibroblast growth factor 23 protein, which is called upon to help balance phosphate in the body. The kidneys are the main organ for monitoring these levels. Excess phosphate is excreted in urine, and the mineral is reabsorbed for proper body use in the body. The fibroblast growth factor keeps the kidneys from taking inordinate amounts of phosphorus out of the bloodstream. It also monitors the phosphate and vitamin D taken from the intestines as a result of digestion. The protein is normally cut at certain sites when it needs inactivation; this cleavage regulates the amount of the protein that is circulating in the blood.

At least three mutations in *FGFR23* are related to hypophosphatemia. Mutations occur when only a single amino acid is exchanged. That one change is enough to disrupt the cleavage of the protein, causing overactivity of the growth factor in the blood. The kidneys do not function to regulate reabsorption or release into the bloodstream, and thus the amount of phosphates is low. This leads to improper bone development. *FGF23* is inherited in an autosomal dominant pattern and is found on the short arm (p) of chromosome 12 at position 13.3.

PHEX

Officially known as the "phosphate regulating endopeptidase homolog, X-linked" gene. *PHEX* normally instructs for an enzyme that develops bones and teeth. PHEX enzyme cuts other proteins into smaller pieces. PHEX enzyme is possibly involved in regulating phosphate and therefore the formation of growth of bones in childhood and strong bones in adults. The kidneys also play a similar role here, getting rid of excess phosphate but allowing it to be reabsorbed when needed. *PHEX* may also be involved in helping the FGF23 protein to functions properly.

More than 200 mutations have been found in *PHEX*. This is related to the X-linked hypophosphatemic disorders, especially rickets. The mutations affect the normal levels of fibroblast growth factor 23 and leads to the disorder. *PHEX* is inherited in a dominant pattern and is located on the short arm (p) of the X chromosome at position 22.2-p22.1.

What Is the Treatment for Hypophosphatemia?

If patients are malnourished or alcoholics, intravenous use of potassium phosphate is used. Oral supplements are also available. For children with bone disorder, orthopedic surgery and physical therapy may help.

Further Reading

"*FGF23*." 2011. Genetics Home Reference. National Library of Medicine (U.S.). http://ghr.nlm.nih.gov/gene/FGF23. Accessed 5/21/12.

"Hypophosphatemia in Emergency Medicine." 2011. Medscape. http://emedicine.medscape.com/article/767955-overview#showall. Accessed 5/21/12.

"*PHEX.*" 2011. Genetics Home Reference. National Library of Medicine (U.S.). http://ghr.nlm.nih.gov/gene/PHEX. Accessed 5/21/12.

I

Inclusion Body Myopathy-2

Prevalence More than 200 reported; affects about 1 in 1,500 in the Iranian Jewish population and descendants; 15 people in Japanese population identified; 2 in other ethnic groups worldwide

Other Names distal myopathy with rimmed vacuoles; DMRV; hereditary inclusion body myopathy; HIBM; IBM2; inclusion body myopathy, autosomal recessive; inclusion body myopathy, quadriceps-sparing; Nonaka myopathy; QSM; rimmed vacuole myopathy

Inclusion body myopathy-2 is a condition that is part of a member of a group of disorders called hereditary inclusion body myopathies (HIMB). The words of the name give the clue to the symptoms of the disorder. An inclusion body relates to a part of the cell. Bodies that are not a part of the normal cell may be in the cells. Primarily affected are vacuoles or storage tanks of the cell and inclusion of extra fibers. A nonfunctional rim may appear in the vacuole. The word "myopathy" comes from two Greek roots: *myo*, meaning "muscle," and *path*, meaning "disease."

What Is Inclusion Body Myopathy-2?

Inclusion body myopathy-2 is a condition that affects the muscles of the skeletal system. The first sign of the condition is a weakness in the muscle of the lower leg called the tibialis anterior. Other signs and symptoms include:

- Painful movement: The person will have difficulty walking, especially putting weight on the heels and going up stairs. Running is very painful.
- Other muscles: Weakness then develops in the muscles of the upper legs, hip, and shoulders. Interestingly, it does not affect the quadriceps, the large muscle in the front of the thigh.
- Weak fingers: The index finger is weak

- Lost balance: Frequently, the person will lose balance and become likely to have serious falls.
- Muscle biopsy: Typical finding shows the inclusion bodies, rimmed vacuoles, and accumulation of unusual proteins. The accumulations may be similar to the plaques that build up in Alzheimer disease.
- Outcome: The disorder does not affect other muscles such as the eye or heart and does not cause other neurological problems.

The person will probably be confined to a wheelchair about 20 years after diagnosis.

What Is the Genetic Cause of Inclusion Body Myopathy-2?

Changes in the *GNE*, officially known as the "glucosamine (UDP-N-acetyl)-2-epimerase/N-acetylmannosamine kinase" gene, cause inclusion body myopathy-2. Normally *GNE* instructs for making an enzyme found in cells and tissues throughout the body. The enzyme is responsible for the formation of sialic acid in cells. Sialic acid is a simple sugar that attaches to the end of complex molecules on the surface of cell. When it attaches to these molecules, sialic acid assists many cellular functions such as cell movement, cells clinging to other cells, and cell signaling.

The enzyme produced by *GNE* works in two steps. First, it converts a molecule called GleNAc to ManNAc. Second, the enzyme moves a group of oxygen and phosphate atoms to ManNAc to create ManNAc-6-phosphate. Other enzymes convert ManNA-6 phosphate to sialic acid.

The more than 40 mutations in *GNE* disrupt the function of the enzyme. Most mutations involve a change in only a single building block or delete a piece of the enzyme. Different *GNE* mutations cause the conditions in different populations. Researchers are investigating the relationship between changes in the enzyme and the inclusion of the bodies that cause the symptoms of inclusion myopathy-2. *GNE* is inherited in an autosomal recessive pattern and is located on the short arm of chromosome 9 at position 13.3.

What Is the Treatment for Inclusion Body Myopathy-2?

Because this is a genetic disease, there is no cure. Treatment is symptomatic. Some studies have involved the taking of sialic acid as a possible therapeutic agent.

Further Reading

"Inclusion Body Myositis." 2010. Medscape. http://emedicine.medscape.com/article/1172746-overview. Accessed 5/21/12.

"Inclusion Body Myopathy-2, Autosomal Recessive and Nonaka Myopathy via *GNE* Gene Sequencing (Test #367)." 2010. Prevention Genetics. http://www.preventiongenetics.com/ClinicalTesting/TestDescriptions/gne.pdf. Accessed 5/21/12.

O'Ferrall, Erin, and Michael Sinnreich. 2009. "Inclusion Body Myopathy-2." *GeneReviews* .http://www.ncbi.nlm.nih.gov/books/NBK1262. Accessed 5/21/12.

Incontinentia Pigmenti (IP)

Prevalence Rare; 900 to 1,200 cases reported in literature; affects females mostly

Other Names Bloch-Siemens-Sulzberger syndrome; Bloch-Siemens syndrome; Bloch-Sulzberger syndrome; hypomelanosis of Ito; IP; melanoblastosis cutis; naevus pigmentosus systematicus

In 1926, Bruno Bloch, a German dermatologist, wrote about a condition occurring in some female patients in which the children had a rash during infancy, but the conditions worsened with time. He called the condition "incontinentia pigmenti." The essence of the name comes from Latin *continere*, meaning "to stop," and *pigmenti* is the Latinized form referring to the pigment or melanin in the skin. Other systems were also involved. In 1929, Marion Sulzberger, an American dermatologist, published similar results. Sometimes the condition is referred to as Bloch-Sulzberg syndrome but is generally recognized as incontinentia pigmenti as Bloch named it.

What Is Incontinentia Pigmenti (IP)?

Incontinentia pigmenti (IP) is a disorder of the skin, which also can affect many other body systems. It affects females mostly but has been seen in a small number of males. IP is a member of a group of disorders that are known as neurocutaneous conditions, meaning it affects the skin and the nervous system. Following are the symptoms of IP:

- Skin lesions: The lesions go through four stages: blistering from birth to about four months); a wart-like rash for several months; swirling discolored pigment from about six months into adulthood; and wavy or irregular lines of pigment. The problem occurs on trunk and extremities and is slate-gray, blue, or brown.
- Neurological problems: Loss of brain tissue with the formation of small cavities in the small white matter of the brain. This cerebral atrophy is connected with the loss of neurons in the thinking part of the brain. Mental retardation and seizures may occur.
- Eye disorders: The person may develop crossed eyes, cataracts, and severe vision loss.
- Dental problems: Missing or peg-shaped teeth are common; often the person never loses the first set of teeth or milk teeth.

- Motor development: About 10% of children have slow motor development and muscle weakness on one or both sides of the body.
- Loss of hair: The person may lose hair, a condition known as alopecia.
- Breast anomalies.
- Nails: Line and pits may form in the fingernails and toe nails.

What Is the Genetic Cause of Incontinentia Pigmenti (IP)?

Mutations in the *IKBKG* gene, officially known as the "inhibitor of kappa light polypeptide gene enhancer in B-cells, kinase gamma" gene, cause incontinentia pigmenti. Sometimes the gene is called *NEMO*. Normally, *IKBKG* instructs for making a protein that regulates the nuclear factor, kappa-B, a protein complex that binds to DNA and controls the activity of many genes. The IKBKG protein interacts with two enzymes, IKK-alpha and IKK-beta, to activate nuclear factor-kappa-B. It is nuclear factor-kappa-B that regulates the activity of many other genes, including those that control the body's immune and inflammatory reactions. It also keeps the cell from other cell signals that would cause the cell to self-destruct (called apoptosis).

More than 30 mutations cause IP. The most common mutation is a deletion in the *IKBKG* gene producing a nonfunctional version of the protein. Without this protein, nuclear factor-kappa-B cannot be activated, resulting in abnormal cell death and the signs and symptoms of IP. *IKBKG* is inherited in an X–linked dominant pattern and is located on the long arm (q) of the X chromosome at position 28.

What Is the Treatment for Incontinentia Pigmenti (IP)?

Because IP is a genetic disorder, there is no cure. However, treatment and management of the symptoms are possible. Standard management is to control rash and skin infection. The severe skin problems usually disappear by adulthood; however the other manifestations may remain. Diminished vision may be treated with corrective lenses, medication, or surgery. A specialist may treat dental problems. Neurological conditions, such as seizures and muscle spasms, may be controlled by pharmaceuticals or surgery.

Further Reading

"*IKBKG*." 2010. Genetics Home Reference. National Library of Medicine (U.S.). http://ghr.nlm.nih.gov/gene/IKBKG. Accessed 5/21/12.

"Incontinentia Pigmenti." 2010. Genetics Home Reference. National Library of Medicine (U.S.). http://ghr.nlm.nih.gov/condition/incontinentia-pigmenti. Accessed 5/21/12.

"Incontinentia Pigmenti." 2010. National Institute of Neurological Disorders and Stroke (U.S.). http://www.ninds.nih.gov/disorders/incontinentia_pigmenti/incontinentia_pigmenti.htm. Accessed 5/21/12.

Infantile Neuroaxonal Dystrophy

Prevalence Very rare disorder; prevalence unknown

Other Names INAD; NBIA, PLA2G6-related; neuroaxonal dystrophy, juvenile; neuroaxonal dystrophy, late infantile; neurodegeneration with brain iron accumulation, PLA2G6-related; Seitelberger disease; Seitelberger's disease

In 1954, Franz Seitelberger described a condition in infants in which the individual had symptoms of a degenerative brain condition caused by fat storage in the brain. Later physicians noted swelling of the axons, the part of the nerve cell that leads away from the center of the cell. Although Seitelberger is recognized as first writing about the disease, the disorder is named is named for its clinical description, infantile neuroaxonal dystrophy.

What Is Infantile Neuroaxonal Dystrophy?

Infantile neuroaxonal dystrophy is a serious disorder of the nervous system, which affects the axons. The name describes what happens in the disease. The word "infantile" indicates that the disorder appears in infancy between the ages of 6 months and 18 months. It is not observed at birth. However, the disorder can appear later in childhood or during the teenage years. "Neuroaxonal" refers to the axons of the neurons, which are the long arm that carries messages from the center of the nerve cell to other nerve cells. "Dystrophy" comes from two Greek words: *dys*, meaning "with difficulty," and *troph*, meaning "nourish." Thus, in this condition, the axons are not functioning properly because of some malformation. The axons themselves develop swellings called spheroid bodies, and the cerebellum, which is the part of the brain that controls movement, may be damaged. Unusual amounts of iron may accumulate in an area of the brain called the basal ganglia.

Following are the general characteristics of this disorder:

- Distinctive facial features: These features may be present at birth. The child may have a prominent forehead, crossed eyes, a very small nose or jaw, and low-set ears.
- Loss of muscle control: With the damage to the cerebellum and neuronal connections, the child loses the ability to control the head, sit, crawl, or walk. The loss is progressive. The child may just be weak and "floppy" at first but then become very stiff.
- Loss of speech.
- Vision difficulties: The child may have rapid, involuntary eye movements, eyes that do not look in the same direction, and loss of sight because of the deterioration of the optic nerve.
- Breathing difficulties and frequent infections that lead to pneumonia.

- Seizures.
- Hearing loss.
- Loss of cognitive functions, including dementia.

What Is the Genetic Cause of Infantile Neuroaxonal Dystrophy?

Mutations in the *PLA2G6* gene, officially known as the "phospholipase A2, group VI (cytosolic, calcium-independent)" gene, cause infantile neuroaxonal dystrophy. Normally, *PLA2G6* provides instructions for the A2 phospholipase enzyme, which is responsible for breaking down phospholipids, a type of fat. This enzyme is essential in the regulation of a compound called phosphatidylcholine, which is found in the cell membrane.

About 50 mutations in the *PLA2G6* gene cause infantile neuroaxonal dystrophy. The mutations disrupt the work of the enzyme, which must work in the cell membrane in order to properly maintain the cell. Scientists believe that this interruption leads to the development of the spheroid bodies in the axons. In addition, the malfunction of the enzyme may lead to the accumulation of iron, leading to damage to the nerve cells. *PLA2G6* is inherited in an autosomal recessive pattern and is located on the long arm (q) of chromosome 22 at position 13.1.

What Is the Treatment for Infantile Neuroaxonal Dystrophy?

No cure exists for this disorder, and no treatment can stop the progress of the disease. Treatment is symptomatic and supportive. Medications can be given for pain. Physical therapy may be given to help the child be more comfortable. The prognosis of this disease is not good. Death usually occurs between the ages of 5 and 10 years.

Further Reading

"Infantile Neuroaxonal Dystrophy." 2011. Genetics Home Reference. National Library of Medicine (U.S.). http://ghr.nlm.nih.gov/condition/infantile-neuroaxonal-dystrophy. Accessed 1/7/12.

"NINDS Infantile Neuroaxonal Dystrophy Information Page." 2011. National Institute of Neurological Diseases and Stroke (U.S.). http://www.ninds.nih.gov/disorders/neuro axonal_dystrophy/neuroaxonal_dystrophy.htm. Accessed 1/7/12.

"Seitelberger Disease." 2011. RightDiagnosis.com. http://www.rightdiagnosis.com/medical/seitelberger_disease.htm. Accessed 1/7/12.

Infantile Systemic Hyalinosis

Prevalence Unknown; fewer than 20 people reported

Other Names inherited systemic hyalinosis

Some diseases are so rare that they are often misdiagnosed. This misdiagnosis will probably never be the case with infantile systemic hyalinosis. The symptoms are so obvious with the large painful bumps appearing at birth or soon after. The name gives the clues to the symptoms. The disorder is seen in infants. It is systemic meaning that several areas of the body are affected. "Hyalinosis" refers to the word "hyaline" that means clear or glass-like in appearance.

What Is Infantile Systemic Hyalinosis?

Infantile systemic hyalinosis is a painful disorder present at birth that affects many body systems. The organs that are affected with a clear, abnormal substance are the skin, joints, bones, and internal organs. The symptoms are usually present at birth or develop during the first month.

Following are the signs and symptoms of infantile systemic hyalinosis:

- Painful skin bumps that appear on the hands, neck, scalp, ears, creases of the genital area, and nose
- Noncancerous lumps form in the muscles and internal organs
- Bumps can be large or small but increase in number over time
- Diarrhea due to the presence of lumps in the intestines, causing a condition called protein-losing enteropathy
- Failure to gain weight called failure to thrive
- Overgrowth of tissues on the gums
- Bone abnormalities
- Joint abnormalities, making movement difficult and painful
- Recurrent infections

Although this condition has such serious physical conditions, the child has normal mental development. However, because of the severe complications, the child seldom lives past early childhood.

What Is the Genetic Cause of Infantile Systemic Hyalinosis?

Mutations in the *ANTXR2* gene, officially known as the "anthrax toxin receptor 2" gene, cause infantile systemic hyalinosis. The name of the gene does not really give a clue to infantile systemic hyalinosis. The name of the gene refers to the toxin of the condition anthrax, a disease cause by bacteria that produces a deadly toxin. The protein is the same one that allows the toxin produced by the bacteria to attach to the cells and cause death if not treated.

Normally, the gene *ANTXR2* instructs for making the protein anthrax toxin receptor 2 (ANTXR2) that is involved in the formation of the capillaries. The capillaries are the small blood vessels where the arteries and veins meet and the oxygen–carbon dioxide exchange takes place. The protein ANTXR2 also maintains the basement membrane, small, sheet-like linings that separate the cells and tissues.

The more than 10 mutations in the *ANTXR2* gene disrupt the correct formation of the thin delicate layer of basement membrane. Because the membrane is not present or is present in only spotty places, clear hyaline-like fluid leaks into the areas and accumulate as the painful lumps. *ANTXR2* is inherited in an autosomal recessive pattern and is located on the long arm (q) of chromosome 4 at position 21.21.

This condition is similar to a condition called juvenile hyaline fibromatosis, which is also caused by *ANTXR2* and shows some of the same symptoms. However, infantile systemic hyalinosis is much more serious and is always fatal.

What Is the Treatment for Infantile Systemic Hyalinosis?

This condition is a very serious disease with no cure. Physicians will seek to manage the diarrhea caused by the bumps or enteropathy in the intestines. Management is to make the child as comfortable as possible, in spite of the painful lesions.

Further Reading

"*ANTXR2*." 2010. Genetics Home Reference. National Library of Medicine (U.S.). http://ghr.nlm.nih.gov/gene/ANTXR2. Accessed 5/21/12.

"Infantile Systemic Hyalinosis." 2010. Genetics Home Reference. National Library of Medicine (U.S.). http://ghr.nlm.nih.gov/condition/infantile-systemic-hyalinosis. Accessed 5/21/12.

Isovaleric Acidemia (IVA)

Prevalence Affects 1 in 100,000 to 1 in 250,000 people in the United States

Other Names isovaleric acid-CoA dehydrogenase deficiency; isovaleryl-CoA dehydrogenase deficiency; IVA; IVD deficiency

When Archibald Garrod in 1902 speculated on the "inborn error of metabolism" of a condition that he noted in a family with alkaptonuria, he did not realize that he was opening up a whole new area of conditions and disorders caused by faulty metabolism. The condition isovaleric acidemia is one of the conditions referred to as an organic acid disorder. The first patient with the disorder was described in 1966, and the enzyme deficiency that causes the disease was found a few years later.

What Is Isovaleric Acidemia (IVA)?

Isovaleric academia is a rare defect in the body's ability to process the amino acid leucine. Normally, when foods that are high in protein, such as meat, eggs, beans, peanut butter, and nuts, are eaten, the body processes the protein into many different substances. One of the substances is the compound isovaleric acid, which is used by the body to create growth and to produce energy. However, if the balance of protein and isovaleric acid is too high, an enzyme called isovaleryl-CoA dehydrogenase goes to work to help the body get rid of the excess isovaleric acid.

In children with isovaleric acidemia, the enzyme does not work properly. Abnormal levels of the substance build up in the blood, urine, and other tissues. These compounds lead to excess toxic chemicals and can cause health problems that can range from mild to life-threatening. If not treated by dietary restrictions, the following symptoms of isovaleric acidemia may be present:

- Lack of appetite
- Vomiting
- Tiredness
- Brain damage
- Seizures and coma
- Death

A distinctive odor like sweaty feet caused by the buildup of the toxic products may be present. The disorder appears to come and go in some children. Prolonged periods without food, infections, and eating an increased amount of proteins may trigger a flare-up.

What Is the Genetic Cause of Isovaleric Acidemia (IVA)?

Changes in the *IVD* gene, officially known as the "isovaleryl-CoA dehydrogenase" gene, cause isovaleric acidemia. Normally, *IVD* instructs for making the enzyme isovaleryl-CoA dehydrogenase that is essential for processing proteins in the diet. When the body breaks down proteins into smaller proteins called amino acids, the body uses them for energy and growth. In the cells of the body, the enzyme is found within the specialized bean-shaped powerhouses of the cell called mitochondria. In the mitochondria, the chemicals are then converted to use for energy. Isovaleryl-CoA dehydrogenase especially targets an amino acid called leucine for the breakdown for energy.

Of the 25 mutations in IVD, some disrupt the normal function of the enzyme, and others keep the cell from producing the normal enzyme. The child does not process leucine properly, and the signs and symptoms of isovaleric acidemia result when toxic levels are accumulated. *IVD* is inherited in an autosomal recessive pattern and is located on the long arm (q) of chromosome 15 at position 14-q15.

What Is the Treatment for Isovaleric Acidemia (IVA)?

A low-protein food pattern is the only way to keep isovaleric acid at a safe level. The brain then can function normally, and the children will be able to learn and grow.

There are three parts to the successful treatment of IVA:

1. A low-protein food pattern and/or specialized formula. The foods will supply protein without the leucine that is not metabolized.
2. Supplemental carnitine and glycine given orally. Carnitine is essential for muscle energy and binds with isovaleric acid to make it less harmful. Glycine also combines with isovaleric acid and makes it less harmful.
3. Immediate contact with a health provider if illness occurs. Illness and infections can cause the buildup of isovaleric acid. The physician will give the child extra energy foods such as sugar to decrease the amount of protein that is broken down within the body.

Newborns can be screened for isovaleric acidemia as part of newborn screening.

Further Reading

"Isovaleric Acidemia (IVA)." 2010. PerkinElmer Genetics. http://www.perkinelmergenetics.com/IsovalericAcidemia.htm. Accessed 5/20/12.

"Isovaleric Acidemia: A Guide for Parents." 2010. Western States. Genetics. http://www.westernstatesgenetics.org/pacnorgg/PDFs_all-081409/isovaleric_eng.pdf. Accessed 5/20/12.

"*IVD*." 2010. Genetics Home Reference. National Library of Medicine (U.S.). http://ghr.nlm.nih.gov/gene/IVD. Accessed 5/20/12.

J

Jackson-Weiss Syndrome (JWS)

Prevalence Rare genetic disorder; incidence unknown

Other Names acrocephalosyndactyly; JWS; Kuhns toe

In 1976, C. E. Jackson and L. Weiss noted an unusual spectrum of head, face, and foot abnormalities in a large midwestern Amish community in northern Indiana. The great toes were enlarged, and thumb abnormalities were not present. They observed 80 affected individuals with another 50 that were reported to be affected in Ohio and Wisconsin communities. The Amish are an old religious sect that often intermarries and stays within the confines of their compounds.

What Is Jackson-Weiss Syndrome (JWS)?

Jackson-Weiss syndrome has unusual face and foot abnormalities and premature fusion of the bones of the skull. This fusion of the skull bones, called craniosynostosis, prevents normal growth of the skull and affects the shape of the head and face. Following are the symptoms of Jackson-Weiss syndrome:

- Misshapen skull with widely spaced eyes
- Bulging forehead
- Great toes short and wide and bend away from the other toes
- Mid-face flattened
- Some toes fused together
- Hands are always normal

The premature cranial closure is usually not life-threatening, and the person appears to have no consequences for the premature closure except for the cosmetic appearance. People with JWS have normal intelligence as well as a normal life span. Some people with the syndrome have the foot abnormalities and not the head closure.

Jackson-Weiss syndrome is only one of eight syndromes with premature cranial closure. For example, a condition called Pfeiffer syndrome premature had the head closure but no hand abnormalities, and Crouzon disease had the head closures but not the abnormalities of the hands and feet.

What Is the Genetic Cause of Jackson-Weiss Syndrome (JWS)?

Changes in the *FGFR2* gene, officially known as the "fibroblast growth factor receptor 2" gene, cause JWS. Normally, this gene instructs for a protein called fibroblast growth factor receptor 2. The protein is only one of a group of several growth factors that are involved in cell division, cell growth regulation and maturation, formation of blood vessels, wound healing, and embryonic development. FRFG2 protein is found in the cell membrane and is so positioned that a part projects outside the cell and the other end is inside the cell. Thus, the protein can interact with growth signals outside the cell and then trigger a cascade of chemical reactions within the cell that instructs for certain actions. *FGFR2* is especially active in bone cells and in embryonic development.

Mutations of the *FGFR2* gene disrupt the pattern of formation of the protein. The JWS syndrome is caused by a change in a single amino acid protein. The mutations overstimulate signaling of the protein and therefore cause premature closing of the skull and the bone disorders associated with the disease. FRFG2 is inherited in an autosomal dominant pattern and is located on the long arm (q) of chromosome 10 at position 26.

What Is the Treatment for Jackson-Weiss Syndrome (JWS)?

Generally in this type of craniosynotosis, surgery on the head is not essential except for a cosmetic reason. Surgery may help the vision, but generally nothing is done to correct the disorder. Because of the nature of the deformity of the feet, surgery is not usually essential to correct the disorder because the foot malformations cause few functional problems.

Further Reading

"*FGFR2*." 2010. Genetics Home Reference. National Library of Medicine (U.S.). http://ghr.nlm.nih.gov/gene/FGFR2. Accessed 5/21/12.

Jackson, C. E.; L. Weiss; et al. 1976. "Craniosynotosis, Midfacial Hypoplasia and Foot Abnormalities: an Autosomal Dominant Phenotype in a Large Amish Kindred." *Journal of Pediatrics*. June. 88(6): 963–968. doi:10.1016/A00223276(76)81050.PMID 12711196. Accessed 5/21/12.

"Jackson-Weiss Syndrome." 2010. Windows of Hope Project. http://www.wohproject.org/Disorders/a-z/jackson-weiss-syndrome-1. Accessed 5/21/12.

Jacobs Syndrome. *See* 47,XYY Syndrome

Jacobsen Syndrome

Prevalence About 1 out of 100,000 births

Other Names del 11q23.3; del 11qter; distal deletion 11q; distal monosomy 11q; 11q syndrome; JS; JBS; monosomy 11qter; partial deletion 11q; telomeric deletion 11q

In 1973, the Danish physician Petra Jacobsen identified a syndrome that included intellectual disabilities, a distinctive facial appearance, and many physical problems. The condition did not run in families and appeared in most to just happen sporadically; however, he did note that in some rare instances, the syndrome would occur in a family. The condition that he studied was given the name Jacobsen syndrome.

What Is Jacobsen Syndrome?

Jacobsen syndrome is a condition in which the material at the end of chromosome 11 is missing. It is also known as 11q terminal deletion disorder because the missing part happens at the end or terminus of the long arm (q) of chromosome 11. This condition is one of the most studied of all chromosome disorders.

The signs and symptoms of Jacobsen syndrome may vary, but following are the areas of concern:

- Feeding and weight gain: As babies, the child has trouble coordinating the movement necessary for sucking. Some develop reflux when the contents of the stomach move back up into the windpipe.
- Growth: Children are very short compared to others their age.
- Appearance: They may have unusual facial appearances and look more like each other than members of their families. They may have low-set ears, a pointed forehead caused by the early joining of bones in the skull, and wide-set eyes, a condition called hypertelorism. Also present may be a broad bridge to the nose, turned-down corners of the mouth, hooded or drooping eyelids (ptosis), a small lower jaw, and folds of skins on the inside corner of the eyes.
- Delayed development in motor skills such as sitting, standing, and walking.
- Speech: The children talk much later than other children, and some need to learn signing or use pictures to express their needs.

- Learning: Most of these children have learning disabilities and will need special interventions in school.
- Behavior: Many of the children have been diagnosed as having attention deficit hyperactivity disorder (ADHD).
- Bleeding disorder: Over 90% of the children have a serious bleeding disorder called Paris-Trousseau syndrome, which results in abnormal bleeding and bruising. Paris-Trousseau syndrome is caused by abnormality of the blood platelets, a clotting factor in the blood.
- Other medical conditions: Children may have heart defects, frequent ear and sinus infections, bone abnormalities, digestive system disorder, kidney problems, and abnormal genitalia.

Many children have lived to adulthood, but the exact life expectancy is unknown because of possible complications of the medical disorders.

What Is the Genetic Cause of Jacobsen Syndrome?

Jacobsen syndrome is caused by missing parts of chromosome 11, and the more parts that are missing, the more serious the condition. The size of the deletion varies from about 5 million to 16 million building blocks. This deletion includes from about 170 to 340 genes. The gene region here appears to contain critical genes for normal development and for many body parts, including the face, brain, and heart.

Most cases are not passed from one generation to the other, probably a result in meiosis during the formation of the egg and sperm or in early embryonic development. These people have no history of the syndrome, although they can pass it to the next generation. There is an inheritance pattern among about 10% of people with Jacobsen. In this situation, the parent carries a unique chromosome arrangement called balanced translation. In this case, part of chromosome 11 trades places with another chromosome. No material is actually deleted, just misplaced. The people with the balanced translocation do not have any of the symptoms or health problems, but they may pass an unbalanced chromosome to their children. These children have extra chromosomes in place of the regular chromosome and display the health problems characteristic of Jacobsen syndrome.

What Is the Treatment for Jacobsen Syndrome?

Treatment of the life-threatening disorders is given top priority. Cosmetic surgery can treat some of the unique facial disorder such as ptosis or drooping eyelids. Children with this disorder will qualify for special education to assist in learning and behavior.

Further Reading

"Jacobsen Syndrome." 2010. Genetics and Rare Diseases Information Center (GARD). National Institutes of Health (U.S.). http://rarediseases.info.nih.gov/GARD/QnASelected.aspx?diseaseID=307. Accessed 5/21/12.

"Jacobsen Syndrome." 2010. Genetics Home Reference. National Library of Medicine (U.S.). http://www.ghr.nlm.nih.gov/condition/jacobsen-syndrome. Accessed 5/21/12.

Mattina, T.; C. Perotta; and P. Grossfield. 2009. "Jacobsen Syndrome." *Orphanet Journal of Rare Diseases*. http://www.ojrd.com/content/4/1/9. Accessed 5/21/12.

Jervell and Lange-Nielsen Syndrome (JLNS)

Prevalence Uncommon; affects 1.6 in 1 million people worldwide; higher prevalence in Denmark and Norway, where it affects at least 1 in 200,000 people

Other Names cardio-auditory-syncope syndrome; cardioauditory syndrome of Jervell and Lange-Nielsen; JLNS; sudo-cardiac syndrome

In 1957, Jervell and Lange-Nielsen described a condition in which children had profound deafness as well as heart abnormalities in several Scandinavian families. Later in 2008, a researcher traced the prevalence to four Norwegian founder mutations. The syndrome appears to be more common in cultures where intermarriage among relatives is common.

What Is Jervell and Lange-Nielsen Syndrome (JLNS)?

Jervell and Lange-Nielsen syndrome is a condition is which the person has profound deafness from birth and an abnormal heart function. The heart disorder is related to an arrhythmia or disruption in the normal heartbeat, called a long QT syndrome. The QT interval is an expression on the electrocardiogram that shows the relationship at the beginning of a beat (Q) to the end of the T wave. The QT represents the duration of time of the electrical activity of the ventricles, the pumping quadrant of the heart. People with this disorder have a long QT, which indicates that the heart takes a longer time to recharge between beats. The irregular heartbeat may cause dizziness, fainting, or even sudden death.

What Are the Genetic Causes of Jervell and Lange-Nielsen Syndrome (JLNS)?

Changes in two genes, *KCNE1* and *KCNQ1*, cause JNLS.

KCNE1

The official name of the *KCNE1* gene is "potassium voltage-gated channel, Isk-related family, member 1." The K in the symbol is the chemical symbol for the element potassium. Normally, *KCNE1* instructs for making a protein that controls the activity of potassium channels. These important channels carry positively charged potassium atoms in and out of the cells and are critical to the cell's ability to generate and transmit electrical signals. The channels are especially active in the inner ear and in cardiac muscle, where they carry the potassium ions out of the cell. In the inner ear, having the proper balance of potassium is essential for normal hearing. In the heart, the channels are necessary to recharge the muscle after each contraction, thereby maintaining a regular heartbeat.

Mutations in the *KCNE1* gene may result from a change in a single amino acid and cause an altered protein. This altered protein then disrupts the regulation of the flow of potassium ions through channels in the inner ear and cardiac muscle. This loss of the function leads to the profound hearing loss and cardiac arrhythmias of JLNS. About 90% of the cases of this disorder are traced to mutations in the *KCNE1* gene. *KCNE1* is inherited in an autosomal recessive pattern and is located on the long arm (q) of chromosome 21 at position 22.1-q22.2.

KCNQ1

The official name of the *KCNQ1* gene is "potassium voltage channel, KQT-like subfamily, member 1." Normally, this gene has a similar function as that of *KCNE1*: transporting positively charged atoms of potassium in and out of cells and working in the cell to generate and transmit electrical signals. The KCNQ1 protein is active in the inner eat and heart muscle. The protein interacts with proteins in the KCNE1 family to form these working potassium channels. A molecule called PIP2 must bind to the channels made with KCNQ1 protein for the proper functioning. PIP2 starts the action of the ion channel and keeps it open, allowing the ions to flow out of the cell. At least 12 mutations of *KCNQ1* cause JLNS. These mutations cause the protein to not function properly and to disrupt the normal flow of potassium ions across the cell membrane, causing hearing loss and abnormal heartbeat. Only about 10% of the cases are attributed to mutations in this gene. *KNCQ1* is inherited in an autosomal recessive pattern and is located on the short arm (q) of chromosome 11 at position 15.5.

What Is the Treatment for Jervell and Lange-Nielsen Syndrome (JLNS)?

Treatment for JLNS is symptomatic. A cochlear implant, as well as educational interventions, may assist in helping the child with profound hearing loss. Medication such as beta blockers may help regulate the heartbeat. If there is a history of cardiac arrest, other treatments may used such as a pacemaker.

Further Reading

"*KCNE1*." 2010. Genetics Home Reference. National Library of Medicine (U.S.). http://ghr.nlm.nih.gov/gene/KCNE1. Accessed 5/21/12.

Tranebjaerg, L.; R. Samson; and G. Green. 2010. "Jervell and Lange-Nielsen Syndrome." *GeneReviews*. http://www.ncbi.nlm.nih.gov/books/NBK1405. Accessed 5/21/12.

Jeune Syndrome (Asphyxiating Thoracic Dystrophy)

Prevalence Affects about 1 in 100,000 to 130,000 people

Other Names asphyxiating thoracic chondrodystrophy; asphyxiating thoracic dysplasia; asphyxiating thoracic dystrophy (ATD); chondroectodermal dysplasia-like syndrome; infantile thoracic dystrophy; Jeune thoracic dysplasia; Jeune thoracic dystrophy; thoracic asphyxiant dystrophy

In 1955, M. Jeune and a team of French doctors published an article in the Archives of French Pediatrics that described a potentially lethal form of dwarfism. Several other conditions were part of the syndrome. The name, asphyxiating thoracic dystrophy, was given to the disorder because the children had very narrow thoraxes. Both names are used for the syndrome—asphyxiating thoracic dystrophy and Jeune syndrome.

What Is Jeune Syndrome (Asphyxiating Thoracic Dystrophy)?

Jeune syndrome is a serious form of dwarfism, in which the infant has a very small chest, kidney disorders, and other problems. The term asphyxiating thoracic dystrophy is also used and is a clinical description of what happens in the disease. The child smothers or is asphyxiated because the chest or thorax is deformed and does not allow the infant to breathe properly. The following symptoms characterize Jeune syndrome:

- A long narrow, small chest and short ribs that reduces lung capacity
- Very short arms and legs when compared to the trunk
- Very short overall
- Usually shaped pelvic bones
- Extra fingers and toes, a condition known as polydactyly
- Kidney lesions, which can lead to lethal kidney failure
- Liver problems
- Heart and circulatory issues

- Intestinal malabsorption
- Retinal degeneration
- Pancreatic cysts
- Dental abnormalities

Some people with asphyxiating thoracic dystrophy experience only mild breathing difficulties, such as rapid breathing or shortness of breath. These individuals may live into adolescence or adulthood.

What Is the Genetic Cause of Jeune Syndrome (Asphyxiating Thoracic Dystrophy)?

Mutations in the *IFT80* gene, officially known as the "intraflagellar transport 80 homolog (Chlamydomonas)" gene, causes Jeune syndrome. Normally, *IFT80* provides instructions for making proteins that are found in the cilia, the hairlike projections that stick out of the surface of the cells. The cilia have many important functions in cells, ranging from cell movement, to directing signaling pathways, to playing a role in the senses. The role of movement in the cells is especially important because here it helps assemble and maintain cell structure.

About three mutations in *IFT80* cause Jeune syndrome. In two of the mutations, only one amino acid building block is exchanged and the third deletes an amino acid in the protein. The changes in the protein disrupt the intracellular movement function of the cilia and lead to the many abnormalities of the disorder. *IFT80* is inherited in an autosomal recessive pattern and is located on the long arm (q) of chromosome 3 at position 25.33.

What Is the Treatment for Jeune Syndrome (Asphyxiating Thoracic Dystrophy)?

The life-threatening symptom of the syndrome is the respiratory distress and recurrent respiratory conditions. Many children die with the syndrome before suitable treatment can be found. There is a possibility of surgically repairing the rib cage with bone grafts or artificial plates so that the lungs can expand. Treating the recurrent respiratory infections is essential. Individuals who survive infancy may begin to have normal chest development. If other conditions, such as heart or kidney disorders, appear, they must be treated according to conventional procedures.

Further Reading

"Asphyxiating Thoracic Dystrophy." 2011. Genetics Home Reference. National Library of Medicine (U.S.). http://ghr.nlm.nih.gov/condition/asphyxiating-thoracic-dystrophy. Accessed 1/7/12.

"*IFT80*." 2011. Genetics Home Reference. National Library of Medicine (U.S.). http://ghr.nlm.nih.gov/gene/IFT80. Accessed 1/7/12.

"Jeune Syndrome." 2011. Genetics and Rare Diseases Information Center (GARD). National Institutes of Health (U.S.). http://rarediseases.info.nih.gov/GARD/Condition/3049/Jeune_syndrome.aspx. Accessed 1/7/12.

Job Syndrome

Prevalence Rare; affects 1 per million; only about 250 people with Job syndrome in medical literature

Other Names Buckley syndrome; HIES; HIE syndrome; hyper-IgE syndrome; hyperimmunoglobulin E-recurrent infection syndrome; Job-Buckley syndrome; Job's syndrome

In the Old Testament Book of Job, the story is told of Job, a godly man who was very wealthy and happy with his status in life. Satan states to God that if this man loses his wealth, family, and health, he will curse God and die. God allows Job to be tested. In one of his tests, he is covered with boils and sores. The condition "Job's syndrome" or "Job syndrome" was named after this affliction of Job.

In 1966 S. Davis and a team described in the British medical journal *Lancet* cases of two girls with red hair, chronic dermatitis, and recurrent boils and abscesses caused by staphylococcal infections. They dubbed the condition Job syndrome after the one of the trials of the biblical character. Later, in 1972, R. Buckley wrote about two boys who had similar symptoms but had elevated blood serum levels of immunoglobulin E. This condition they called "hyperimmunoglobulin E-recurrent infection syndrome" or HIES. The two observations are not thought to be the same syndrome.

What Is Job Syndrome?

Not only does Job syndrome affect the skin, but the condition affects several other systems, especially the immune system. The disease usually manifests itself early in childhood, but because it is so rare, it may take years for it to be diagnosed properly. Following are the symptoms of Job syndrome:

- Persistent skin abscesses, open sores, blisters, rashes, and collections of pus, and scaling: The infections may be caused by a number of organisms, including staphylococcus and the yeast candida.
- Recurrent sinus infections.
- Frequent bouts of pneumonia: Several kinds of bacteria invade the lungs, causing serious damage.

- High levels of immunoglobulin E: People with this condition have very high levels of this blood protein, which is essential in responding against foreign invaders. The relationship between the high levels of IgE and Job syndrome has not been determined.
- Skeletal abnormalities: The person may display curvature of the spine (scoliosis), hyperextendability of joints, reduced bone density, and fractures.
- Dental abnormalities: The individual may have structural problems with teeth. The baby teeth may not fall out so that the person may have two sets of teeth.
- Distinctive facial features: Face is asymmetrical, prominent forehead, deep-set eyes, and a broad nasal bridge, with a wide fleshy nose tip.
- Facial skin: Skin on face is coarse with large pores.
- Chronic dry eye.

What Is the Genetic Cause of Job syndrome?

Changes in the *STAT3* gene, officially known as the "signal transducer and activator or transcription 3" gene, causes Job syndrome. *STAT3* is a member of a large family of genes known as *STAT* genes. Normally, these genes instruct for a protein that guides chemical signaling pathways within the cell. The proteins are activated by chemical signals to move into the nucleus and bind to areas of DNA. In this process, the STAT proteins turn off certain genes or turn them on. Thus, they are called transcription factors.

The STAT3 protein has a long list of functions. In addition to regulating cell transcription, the protein is essential to several systems, especially the immune system. It mediates the response to foreign invaders and regulates the process of inflammation, the way that the body responds to outside organisms. Active also in the skeletal system, it works to maintain normal development as well as break down bone tissue.

About 16 mutations have been located in the *STAT3* gene. The mutations then affect the production of the protein, making it ineffective in doing the work for the immune system. Thus, people with the abnormality have the compromised immune function and are susceptible to infections and other symptoms.

STAT3 is inherited in two different patterns. One form is autosomal recessive and is less common. Autosomal recessive means that each parent carries a copy of the normal gene and passes it to the offspring without having symptoms of the disease. The recessive form is associated with a different pattern of symptoms, such as fewer bacterial infections of the skin and lungs, and no skeletal or dental disorders. The autosomal dominant pattern is the more serious of the two and is associated with many of the symptoms outlined above. *STAT3* is located on the long arm (q) of chromosome 17 at position 21.31.

What Is the Treatment for Job Syndrome?

Most patients with Job syndrome are treated with high-powered antibiotics to protect from bacterial infections. It is absolutely imperative to have good skin care and pay immediate attention to an outbreak. Some patients with severe eczema infections may take doses of intravenous gamma globulin.

Further Reading

"Job Syndrome." 2008. Job Syndrome Support Group. http://jobsyndrome.com. Accessed 5/21/12.

"Job Syndrome." 2010. Genetics Home Reference. National Library of Medicine (U.S.). http://ghr.nlm.nih.gov/condition/job-syndrome. Accessed 5/21/12.

"Job Syndrome." 2011. RightDiagnosis.com. http://www.rightdiagnosis.com/j/job_syndrome/intro.htm. Accessed 5/21/12.

Joubert Syndrome

Prevalence Affects between 1 in 80,000 and 1 in 100,000 newborns; incidence may be more because of misdiagnosis

Other Names AH1-related Joubert syndrome; CEP290-related Joubert syndrome; CORS2-related Joubert syndrome; JBTS1-related Joubert syndrome; NPHP1-related Joubert syndrome; TMEM67-related Joubert syndrome

In 1969, pioneering pediatric neurologist Marie Joubert at McGill University in Montreal, Canada, observed some patients with some serious brain conditions along with many other malformations and disorders. She was the first to publish work on the syndrome in the journal *Neurology*. For her pioneer work on the disorder, the name was called Joubert syndrome in her honor.

What Is Joubert Syndrome?

Joubert syndrome is a condition characterized by an underdeveloped area of the brain called the cerebellum vermis and a malformed brain stem. The cerebellum is the area of the brain that controls balance and coordination; the brain stem controls all the automatic processes such as breathing. Magnetic resonance imaging (MRI) reveals a malformed area of the section of the cerebellum and abnormalities or absence of the brain stem. The picture on the MRI appeals like a molar tooth; hence, the definitive diagnosis sign is called the "molar tooth sign."

Following are the symptoms of the disorder:

- Very weak muscle tone in infancy
- Abnormal breathing pattern

- Abnormal sleep pattern including sleep apnea
- Unusual eye movements and tongue movements; retinitis pigmentosa, which may cause blindness
- Seizures
- Delayed development
- Skeletal deformities
- Polydactyly
- Impaired intellectual function
- Distinct facial characteristics, with broad forehead, arched eyebrows, droopy eyelid, widely spaced eyes, and a triangular shaped mouth
- Possible kidney and liver disorders

What Are the Genetic Causes of Joubert Syndrome?

The inheritance of Joubert syndrome is very complex and may involve mutations in at least 10 genes. Some researchers have shown that a number of genetic disorders may have a common root cause. An emerging class of genetic disorders are called ciliopathies. These conditions are related to the cilia, tiny hair or projections from the surface of cells. Cilia are especially important for cells of the brain, cells of the kidney and liver, and sensory function.

Mutation in the following 10 genes have been connected to Joubert Syndrome:

- *AHI1*
- *ARL13B*
- *CC2D2A*
- *CEP290*
- *INPP5E*
- *NPHP1*
- *OFD1*
- *RPGRIP1L*
- *TMEM216*
- *TMEM67*

Changes in one of these genes can lead to defects that will disrupt chemical signaling and pathways. However, mutations in these genes account for only about half of the known cases. Generally the genes are inherited in an autosomal recessive pattern.. For families that are known carriers, prenatal diagnosis for *AH11*, *TMEM67*, *CEP290*, and *NPHP1* is conducted. A few rare cases have been traced to the X chromosome and affects only males.

What Is the Treatment for Joubert Syndrome?

Treatment is symptomatic and supportive. Stimulation of the infant and occupational and speech therapy may help. If the infant has an abnormal breathing pattern, he or she must be monitored consistently.

Further Reading

Joubert Foundation. 2011. http://www.joubertfoundation.com. Accessed 5/21/12.

"Joubert Syndrome." 2011. Cleveland Clinic. http://my.clevelandclinic.org/disorders/joubert_syndrome/hic_joubert_syndrome.aspx. Accessed 5/21/12.

Parisi, M., and I. Glass. 2007. "Joubert Syndrome and Related Disorders." *GeneReviews*. http://www.ncbi.nlm.nih.gov/books/NBK1325. Accessed 5/21/12.

Juvenile Polyposis Syndrome (JPS)

Prevalence Occurs in about 1 in 100,000 persons worldwide

Other Names *BMPR1A*-related juvenile polyposis; juvenile polyposis coli; *SMDA4*-related juvenile polyposis

In naming genetic conditions, occasionally the name will give general clues to the nature of the condition, but not exactly. Such is the case with juvenile polyposis syndrome. The word "juvenile" does not refer to an age but to a type of growth. However, most people with the growths do have at least a few before the age of 20. The word "polyposis" comes from two Greek roots: *polyp*, meaning "growth," and *osis*, meaning "condition of." A polyp is a growth arising form the lumen or space within the colon or stomach.

What Is Juvenile Polyposis Syndrome (JPS)?

Juvenile polyposis syndrome is a condition in which the person has bleeding growths called polyps in the gastrointestinal (GI) tract, specifically in the stomach, small intestine, large intestine or colon, and rectum. The polyps are usually noncancerous, but because they occasionally may become cancerous, the person must be consistently under a physician's observation.

Most people have the at least four or five polyps by the age of 20. Some with the condition may have only four or five during a lifetime; others may have more than 100. According to the World Health Organization, the diagnosis of juvenile polyposis is having one of the following conditions:

1. More than five of the type of polyps in the colon or rectum

2. Polyps generalized through the GI tract
3. Any number of polyps if the person has a family history of JPS

Juvenile polyps are hematomas, meaning a growth that bleeds; they are not the same as adenomas, which are growths in the glands of the epithelium lining of the GI.

Following are some of the symptoms and signs of JPS:

- Gastrointestinal bleeding, causing a shortage of red blood cells and anemia
- Abdominal pain
- Diarrhea
- Twisting of the intestines, called malrotation
- Heart or brain disorders
- Cleft palate
- Extra fingers and toes, a condition called polydactyly
- Abnormal sex organs
- Urinary tract infections

Three distinct types have been described based of a combination of symptoms:

1. If the condition is present in infants, it is the most severe and has the poorest outcome possibilities. Children will have severe diarrhea and develop a condition called protein-losing enteropathy, which causes them to fail to thrive.
2. If polyps develop through the entire GI tract, it is called generalized polyposis.
3. If the polyps are localized in the colon, the condition is called juvenile polyposis coli.

Most of the juvenile polyps are noncancerous, but there is a 10% to 50% risk of developing a cancer in the GI tract. However, the most common cancer is cancer of the colon.

What Are the Genetic Causes of Juvenile Polyposis Syndrome (JPS)?

Mutations in two genes cause JPS. About 20% of persons with JPS have mutations in the *BMPR1A* gene, officially known as the "bone morphogenetic protein receptor, type IA" gene. About 20% of persons have mutations in the *SMAD4* gene, officially known as the "SMAD family member 4" gene. Genetic testing of both genes is available for clinical diagnosis.

BMPR1A

Normally, *BMPR1A* instructs for a protein called bone morphogenetic protein receptor 1A. This protein has a specific receptor site that locks into other proteins.

The proteins are called ligands and lock together like pieces of a puzzle. The BMPR1A protein binds to factors in the proteins in the transforming growth factor beta (TGF-β) pathway, which signals to the cell instructions for producing other proteins. *BMPR1A* works with *SMDA4* to regulate cell growth and division. More than 60 mutations of *BMPR1A* have been found to cause JPS. Most mutations produce an abnormally short, nonfunctional protein that disrupts the binding to ligands in the TGF-β pathway and activation of *SMAD4*. Cell growth is not regulated or controlled leading to development of polyps. *BMPR1A* is inherited in an autosomal dominant pattern and is located on the long arm (q) of chromosome 10 at position 22.3.

SMAD4

Normally, *SMAD4* instructs for making a protein that is involved in transmitting chemical signals from the cell surface to the nucleus. Like its companion gene *BMPR1A*, this gene works in the transforming growth factor beta (TGF-β) signaling pathway that controls activity of particular genes and regulates cell growth and division. More than 60 mutations of *SMAD4* have been related to JPS. One mutation that deletes four DNA building blocks in a region of the gene is related to a more aggressive form of JPS that has many generalized polyps and a greater risk for cancer. *SMAD4* is inherited in an autosomal dominant pattern and is located on the long arm (q) of chromosome 18 at position 21.1.

What Is the Treatment for Juvenile Polyposis Syndrome (JPS)?

The treatment is management of the various manifestations. A routine colonoscopy to remove bleeding polyps is essential. If many polyps are present, removal of part of the stomach or intestine may be necessary. Prevention of cancer is a main goal. Other symptoms may be treated with medications or surgery.

Further Reading

"*BMPR1A*." 2010. Genetics Home Reference. National Library of Medicine (U.S.). http://ghr.nlm.nih.gov/gene/BMPR1A. Accessed 5/21/12.

Haidle, Joy Larsen, and James Howe. 2003. "Juvenile Polyposis Syndrome." *GeneReviews*. http://www.ncbi.nlm.nih.gov/books/NBK1469. Accessed 5/21/12.

"Juvenile Polyposis Syndrome." 2010. Genetics Home Reference. National Library of Medicine (U.S.). http://ghr.nlm.nih.gov/condition/juvenile-polyposis-syndrome. Accessed 5/21/12.

"*SMAD4*." 2010. Genetics Home Reference. National Library of Medicine (U.S.). http://ghr.nlm.nih.gov/gene/SMAD4. Accessed 5/21/12.

Juvenile Primary Lateral Sclerosis

Prevalence Rare; only a small number of reported cases

Other Names JPLS; juvenile PLS; PLSJ; primary lateral sclerosis, juvenile

The famous baseball player Lou Gehrig made known a condition called amyotrophic lateral sclerosis (ALS). Gehrig's muscles in his arms and legs became so weak that over a number of years he could not function. When he announced to the world that he had this disease with a long name, people dubbed it "Lou Gehrig's disease." Juvenile primary lateral sclerosis is the juvenile form of this disease and is caused by the same gene.

What Is Juvenile Primary Lateral Sclerosis?

Juvenile primary lateral sclerosis is a condition caused by damage to the motor neurons. Motor neurons control movements to arms, legs, and voluntary muscles. They are located in the brain and spinal cord. Symptoms of this disorder begin in early childhood and progress over a period of 15 to 20 years. The following symptoms progress as the child ages:

- Clumsiness
- Muscle spasms
- Progressive leg weakness
- Difficulty with balance
- Progressive arm stiffness and weakness
- Slurred speech
- Difficulty swallowing
- Drooling
- Inability to walk

The person will probably be confined to a wheelchair and dependent upon other persons for care. However, it can begin at any age and differs from ALS in severity. Primary juvenile lateral sclerosis is not fatal.

What Are the Genetic Causes of Juvenile Primary Lateral Sclerosis?

Changes in the *ALS2* gene, officially known as the "amyotrophic lateral sclerosis 2 (juvenile)" gene, cause juvenile lateral sclerosis. Normally, *ALS2* instructs for a protein called alsin, which is found in many tissues and organs, especially the brain. Alsin is prevalent in motor neurons, the long specialized nerve cells in the brain and spinal cord that are related to voluntary movements. The protein may also control cell membrane organization and movement of molecules inside cells, as well as the development of axons and dendrites.

Alsin is the same protein that is related to Lou Gehrig disease. Three mutations in the gene appear to cause the juvenile form. Two mutations delete DNA building blocks and a third one replaces a building block with the wrong one. When the alterations of the building blocks occur, the protein action is disrupted and motor neurons do not function properly. *ALS2* is inherited in an autosomal recessive pattern and is located on the long arm (q) of chromosome 2 at position 33.1.

What Is the Treatment for Juvenile Primary Lateral Sclerosis?

No treatments can prevent or stop the condition, so most strategies include preserving function. Medications are available to relieve spasms, or a medication pump may be used to deliver medication if the person cannot take the medication orally. Other medications may treat cramps or spasticity. Physical and speech therapy can help compensate for weak muscles or help in the ability to form facial speaking patterns. Other assistive devices, such as a cane, walker, or wheelchair may be used.

Further Reading

"*ALS2*." 2010. Genetics Home Reference. National Library of Medicine (U.S.). http://ghr.nlm.nih.gov/gene/ALS2. Accessed 5/21/12.

"Juvenile Primary Lateral Sclerosis." 2010. Genetics Home Reference. National Library of Medicine (U.S.). http://ghr.nlm.nih.gov/condition/juvenile-primary-lateral-sclerosis. Accessed 5/21/12.

"Symptoms of Juvenile Primary Lateral Sclerosis." 2010. RightDiagnosis.com. http://www.rightdiagnosis.com/j/juvenile_primary_lateral_sclerosis/symptoms.htm. Accessed 5/21/12.

K

Kabuki Syndrome

Prevalence Rare; 1 in 32,000 births

Other Names Kabuki makeup syndrome; KMS; Nikawa-Kuroki syndrome

Kabuki is an ancient Japanese theater art in which performers and the makeup that they are wearing gives them a special look. High arched eyebrows, heavy mascara on the eyelashes, painted fissures from the corner of the eye, and white makeup are the distinct features worn by the actors. So in 1981, when Japanese doctors Nikawa and Kuroki observed a condition in which the children had those distinctive features, they chose the name "Kabuki make-up syndrome." The term "make-up," however, was later dropped because families of people with Kabuki syndrome were offended. So the condition is now called Kabuki syndrome.

What Is Kabuki Syndrome?

Kabuki syndrome is a condition in which the person has a wide range of congenital disorders and obvious intellectual disabilities. Like most genetic disorders, there is a wide variety in the symptoms. The condition is difficult to diagnose in the first year and may resemble several other disorders, especially CHARGE syndrome. Following are some of the many problems related to the disorder:

- Facial features: The distinctive appearance includes long eyelids with turning up at the side of the lower eyelid, prominent earlobes, depressed nasal tip, arched eyebrows, and fissures or creases on the outer edge of the eyes.
- Skeletal deformities: The person may have short fingers, abnormally short bones, rib anomalies, turned-in fifth finger, and vertebral problems such as scoliosis.
- Skin problems, including persistent finger fetal pads
- Very short stature.

- Mild to moderate intellectual disability.
- Behavioral changes.
- Hypotonia or muscle weakness.
- Hyperextensible joints.
- Feeding difficulties.
- Immune system disorder with recurring respiratory or ear infections in early childhood years.
- Urinary tract infections.
- Blood disorders: These include anemia and polycythemia, a condition where many red blood cells are produced.
- Gastrointestinal problems: These include both diarrhea and constipation.

Kabuki syndrome affects so many different systems, but most of the above symptoms are ones that can be corrected with good medical care.

What Are the Genetic Causes of Kabuki Syndrome?

Changes in the *MLL2* gene, officially known as the "myeloid/lymphoid or mixed-lineage leukemia 2" gene, cause Kabuki syndrome. Normally, *MLL2* instructs for the productions of a protein that is a histone methyltransferase and part of a larger protein complex called ASCOM. This complex is a transcription factor that regulates many other genes.

Mutations in *MLL2* disrupt the function of this important complex. This affects many pathways and causes the multiple symptoms. The discovery of the gene announced August 15, 2010, by University of Washington researchers made quite a stir in the genetic world because of a newly developed technique. Using a new strategy, instead of sequencing the entire genome looking for the markers for the gene, the scientists sequenced just the exome, the 1%–2% of the human genome that contains protein-coding genes. This strategy, called "second generation DNA sequencing," is less expensive and very rapid. The *MLL2* gene is located on the long arm (q) of chromosome 12 at position 12-q14.

What Is the Treatment for Kabuki Syndrome?

Treatment for this disorder is symptomatic. Present data does not point to a shortened life span for those with the disorder. Most of the medical issues relate to heart, kidney, or GI symptoms and are usually resolved with medical intervention early in life.

Further Reading

"Discovered Gene Causes Kabuki Syndrome." 2010. PhysOrg. http://www.physorg.com/news201094837.html. Accessed 5/21/12.

"Facts about Kabuki." 2010. http://kabukisyndrome.com/kabuki.html. Accessed 5/21/12.

"Kabuki Syndrome." 2010. Genetics Home Reference. National Library of Medicine (U.S.). http://ghr.nlm.nih.gov/condition/kabuki-syndrome. Accessed 5/21/12.

Kallmann Syndrome

Prevalence Occurrence is 1 in 10,000 male births and 1 in 50,000 female births

Other Names familial hypogonadism with anosmia; hypogonadotropic hypogonadism; hypothalamic hypogonadism

In 1856, Spanish doctor Aurelaina Maestre de San Juan noted that a condition that affected the development of the sex organs was also connected to the inability to smell. In 1944, Josef Kallmann, a German American geneticist, described the syndrome connecting the two. He and a team of researchers reported their findings in the *American Journal of Mental Deficiencies*.

Several well-known people have had Kallmann syndrome. One, jazz singer "Little Jimmy" Scott, had an unusual high contralto voice because of his inherited condition. Also, Brian Brett, a Canadian writer, had not entered puberty by the age of 20 when his endocrine condition called Kallmann syndrome was discovered. He began taking testosterone and lived to have a normal life. He recounted his growing up with Kallmann syndrome in a 2004 memoir, *Uproar's Your Only Music*.

What Is Kallmann Syndrome?

Kallmann syndrome is a condition with two seemingly different symptoms: lack of a sense of smell, called anosmia; and decreased functioning of the glands that produce the sex hormones, called hypogonadism. With Kallmann syndrome, there is a deficiency in the gonadotropin-releasing hormone, or GnRH. These hormones direct sexual development. Men with the condition often have small sex organs; undescended testes, which is a condition called chryptorchidism; and lack of secondary sex characteristics such as facial hair and lower male voice. Women may not have a monthly menstrual period and may have little or no breast development. However, the distinguishing factor between Kallmann and the other forms of hypogonadism is the diminished or complete absence of the sense of smell.

There are several other signs of the disorder:

- Failure of one kidney to develop
- Cleft lip and cleft palate
- Abnormal eye movements

- Hearing loss
- Abnormal tooth development
- A unique condition called bimanual synkinesis in which the person cannot move hands separately; certain tasks such as playing a musical instrument require this skill

What Are the Genetic Causes of Kallmann Syndrome?

Mutations in several genes cause the four different types of Kallmann syndrome:

1. Type 1: Mutations in "Kallmann syndrome 1 sequence," or *KAL1*, cause type 1. *KAL1* instructs for a protein called anosmin-1, which is found in many parts of the developing embryo. Anosmin-1 helps form the spaces between cells. Researchers have found that the protein aids the movement of nerve cells that transmit nerve impulses, especially related to the sense of smell. At least 60 mutations in *KAL1* delete part or all of this gene. The action of the protein is disrupted during embryonic development causing possibly the loss of the sense of smell and the migration of GnRH-producing nerve cells to their normal locations. *KAL1* is inherited in an X-linked recessive pattern and is found on the short arm (p) of the X chromosome at position 22.32.
2. Type 2: Mutations in the "fibroblast growth factor receptor 1" gene, or *FGFR1*, causes type 2 Kallmann syndrome. Normally, *FRFG1* instructs for the protein called fibroblast growth factor receptor 1, one of several growth factor receptors. The protein is important in regulating cell growth, forming blood vessels, healing wounds, and developing the embryo. It plays an important part in the development of the nervous system. More than 40 mutations of *FGFR1* result in small or nonfunctional versions of the proteins. Because of the mutations, the FGFR1 protein cannot function properly, causing the problems with the sense of smell and sexual development. This type causes about 10% of the cases. *FRFG1* is an autosomal dominant gene found on the short arm (p) of chromosome 8 at 11.2-11.1.
3. Types 3 and 4: Mutations in the "prokinectin receptor-2" gene, or *PROKR2* and *PROK2* genes, cause type 3 and 4. Normally, these genes play a role in the development of certain areas of the brain before birth. The areas are those that relate to the production of the GnRH and the olfactory sense of smell. *PROKR2* is located on the short arm (p) of chromosome 20 at position 13.

What Is the Treatment for Kallmann Syndrome?

Hormone therapy has known to help restore deficient hormones. Doses of gonadal steroids such as testosterone or hCG injections in males and estrogen or progestin in females have been found to be effective.

Further Reading

Kallmann Support Group (UK). 2008. http://www.kallmanns.org. Accessed 5/21/12.

"Kallmann Syndrome." 2010. Genetics Home Reference. National Library of Medicine (U.S.). http://ghr.nlm.nih.gov/condition/kallmann-syndrome. Accessed 5/21/12.

"Kallmann Syndrome." *GeneReviews*. http://www.ncbi.nlm.nih.gov/books/NBK1334. Accessed 5/21/12.

Kartagener Syndrome

Prevalence	One case per 32,000 live births; some physicians believe it is underdiagnosed and may be as many as 1 in 15,000 births
Other Names	immobile cilia syndrome; Loeffler's disease; primary cilia dyskinesia; Zivert's syndrome

The 39-year-old woman smiled with her eyes as she peered over her mask and patiently waited to see her doctor. She had a double lung transplant and had to wear the mask any time she was in public because she had no resistance to infection. Her condition started at birth when she had difficulty clearing the fetal fluid from her lungs. Colds, infections, and even pneumonia plagued her throughout her life. She had been born with a condition in which the cleaning mechanism in her lungs did not work and over the years had caused such damage that she would have died without the transplant. Her condition was a rare hereditary syndrome called Kartagener syndrome.

Although Dr. A. K. Zivert first described a combination of abnormal body symptoms in 1904, Manes Kartagener, a Swiss physician, published a report in which he recognized a distinct congenital syndrome with many serious characteristics in 1933. Other doctors described the problem as a movement of cilia, the cleansing mechanism in the lungs, and added the term primary cilia dyskinesia in 1981; however, the syndrome had acquired the name of the doctor who published the first study of this condition—Kartagener syndrome.

What Is Kartagener Syndrome?

Kartagener syndrome is also known as primary ciliary dyskinesia (PCD). The cilia are microscopic hairlike projections that line the respiratory tract and act as a cleansing mechanism to get rid of foreign particles or a clearing process for mucus. In normal people, cilia beat from 7 to 22 times per second and work together in the respiratory system to sweep out foreign particles that the person may breath and move out of the lungs mucus, bacteria, or other particles that may collect. Kartagener disease is called a ciliopathic disease from two Greek words: *cilia*, meaning "hair," and *path*, meaning "disease." Cilia are also found in other organs such as the reproductive system. The tail-like flagella of sperm are similar to cilia in that they move the sperm forward.

Several unique features of Kartagener syndrome may occur. People with Kartagener syndrome may have a chronic cough that lasts year-round and develop bronchiectasis, a condition in which the bronchi, the passages leading to the lungs, are damaged.

As a result of abnormal embryonic development, the person may have organs that are the mirror image of each other. This condition is called *situs inversus totalis*. For example, the heart points to the right side of the body instead of to the left. The viscera or intestines may be transposed, and the frontal sinuses may be in an abnormal position. In fact, any place or organ that has ciliary movement may be affected. The ear has cilia that assist in cleaning; people with Kartagener syndrome may have serious ear infections because the normal mechanism is not working. Males may not be able to have children because the sperm depend on flagella to propel themselves toward the egg. Females may also be infertile because the ova depend on cilia for propulsion through the fallopian tubes.

What Is the Genetic Cause of Kartagener Syndrome?

Two genes, *DNAI1* and *DNAH5*, cause Kartagener syndrome. Normally, these genes do several things. They instruct for the proteins that form the inner structure of cilia and then produce the force necessary for these cilia to be flexible. The coordination of this back-and-forth movement is essential for the formation of many organs such as the ears, nose, respiratory system, and reproductive system. During embryonic development, cilia are responsible for the left-right axis and making sure the organs are on the proper side of the body.

What happens in Kartagener syndrome? Mutations in the genes result in defective cilia or cilia that cannot move (immotile). Because cilia are not functioning properly, the many symptoms of the syndrome appear. The condition is inherited in an autosomal recessive pattern. Parents each carry a mutated gene but show no symptoms of the disorder. *DNAI1* is located on the short arm of chromosome 9 at position 13.3; *DNAH5* is located on the long arm of chromosome 5 at position 15.2.

What Is the Treatment for Kartagener Syndrome?

Basically, the symptoms must be treated because there is no cure for the disorder. If infections are in the ear, bronchi, or lungs, they may be treated with antibiotics. Severe cases may involve lung transplants.

Further Reading

American Lung Association. 2010. "Primary Ciliary Dyskinesia: Symptoms, Diagnosis, and Treatment." http://www.lungusa.org/lung-disease/primary-ciliary-dyskinesia/symptoms-diagnosis-and.html. Accessed 5/21/12.

"Kartagener Syndrome." 2010. Medscape. http://emedicine.medscape.com/article/299299-overview. Accessed 5/21/12.

"Primary Ciliary Dyskinesia." 2010. Genetics Home Reference. National Library of Medicine (U.S.). http://ghr.nlm.nih.gov/condition/primary-ciliary-dyskinesia. Accessed 5/21/12.

Karyotyping: A Special Topic

Other Names chromosome analysis

In 1842, when Wilhelm von Nageli looked at plant cells using his crude microscope, he noted bodies that responded to staining. Walther Flemming in 1882 first discovered mitosis or cell division in salamanders; he described the role of some unique bodies in the process. However, another German doctor, Wilhem von Waldeyer-Hartz, in the late 1880s coined the term "chromosomes" from two Greek words: *chromo*, meaning "color," and *some*, meaning "body." Chromosomes were colored bodies.

In the early twentieth century, several scientists studied chromosomes, and it took many years for the scientists to decide exactly how many chromosomes were

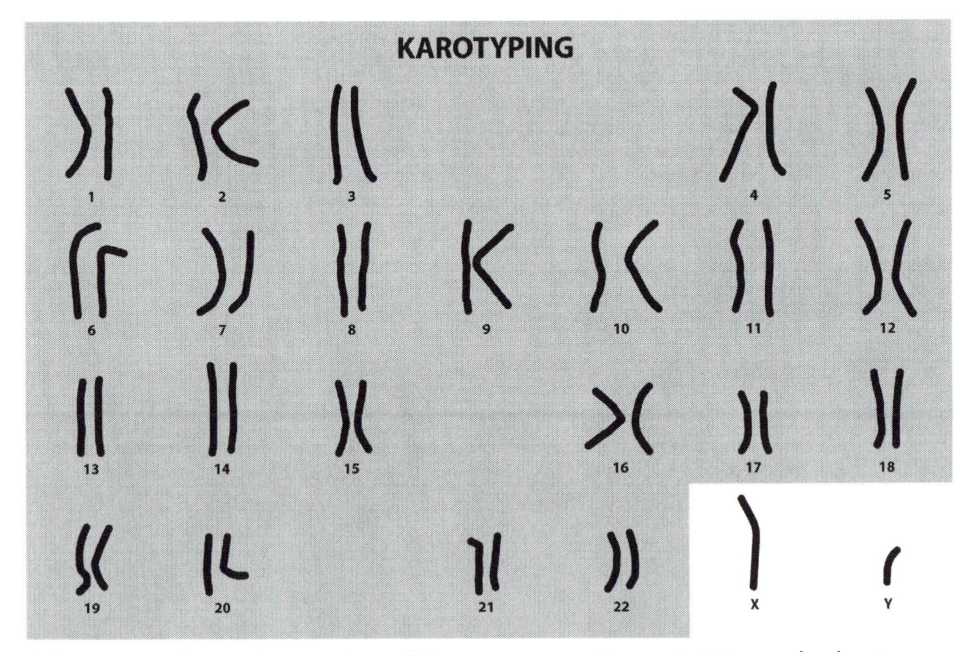

A karyotype shows the number of chromosomes. Here are 22 matched pairs and then one unmatched pair. This karyotype is typical of a male. (ABC-CLIO)

present in the human being. In 1922, Painter finally settled the question by deciding that the human had 44 somatic chromosomes and that the sex chromosomes had an XX/XY system, making a total of 46.

What Is Karyotyping?

Karyotyping is a test that examines the chromosome in cells to determine if a person has a specific genetic disease or disorder. The karyotype of the human shows 22 pairs of body or somatic chromosomes pictured in order of 1 to 22, and the last pair consists of the sex chromosomes X and Y.

The karyotype reveals the following six characteristics:

- Size of the chromosome
- Position of the centromeres
- Bands on the chromosomes; these reflect light and dark bands
- Number of the chromosomes
- Possible small bodies called satellites that may be attached by a thin thread
- Darker stains, which indicate tighter packing

Making the Karyotype for Observation

The process of making the karyotype follows:

1. Cells obtained from blood, skin, or other body tissues are put in a culture to grow. White blood cells may be the preferred cells for humans because they grow more easily in culture.
2. Cells are pretreated in a hypotonic solution, making them swell.
3. A solution called colchicine stops the cells in the metaphase stage of mitosis.
4. Cells are stained with a suitable dye such as Giemsa in the metaphase stage.
5. The preparation is forced onto the slide, making the chromosomes appear in a single plane.
6. A photomicrograph is taken.
7. Parts are cut apart and arranged from the largest somatic pairs to the smallest, numbering from 1 to 22. The XX/XY sex chromosomes are put last.

Karyotypes are arranged with the short arm of the chromosome on top, and the long arm on the bottom. The short arm is called p and the long arm q.

Recently, more advanced techniques have been used.

- Spectral karyotype or the SKY technique: This type enables the reader to see all pairs of chromosomes in different colors. Fluorescent labels can generate different colors, which are read by an interferometer using an image-processing software program.

- Digital stereotyping: This technique uses short sequences of DNA from specific loci to isolate and enumerate chromosomes. This is also called virtual karyotyping.

What the Karyotype May Reveal

The karyotype may reveal certain abnormalities. Some of the following disorders may show up in a karyotype:

- Turner syndrome, showing only one X chromosome.
- Klinefelter syndrome, showing an 47,XXY and an extra X chromosome. This disorder is the most common male chromosome disease.
- Trisomy 18, or Edwards syndrome, showing three copies of chromosome 18.
- Down syndrome, showing three copies of chromosome 21.
- Trisomy 13 or Patau syndrome, showing three copies of chromosome 13.
- Cri-du-Chat (cry of the cat) syndrome, showing a missing piece of the short arm of chromosome 5.
- Angelman syndrome, showing a segment of the long arm of chromosome 15 is missing.
- Prader-Willi syndrome, showing a segment of the long arm of chromosome 15 is missing.

A doctor may order some other tests that will amplify the karyotype, such as telomere studies to look at the ends of chromosomes, microarray to look at small changes in chromosomes, and fluorescent in situ hybridization or FISH to look for small mistakes such as deletions.

Further Reading

"Karyotyping." 2010. MedlinePlus. National Library of Medicine (U.S.). http://www.nlm.nih.gov/medlineplus/ency/article/003935.htm. Accessed 12/26/11.

"Karyotyping—Overview." 2011. University of Maryland Medical Center. http://www.umm.edu/ency/article/003935.htm. Accessed 2/14/12.

Kawasaki Disease

Prevalence In Western countries and the United States, affects about 1 in 10,000 children under five each year; 10 to 20 times more common in Asia, especially Japan, Korea, and Taiwan and people of Asian descent

Other Names Kawasaki syndrome; KD; mucocutaneous lymph node syndrome

In 1967, T. Kawasaki, a Japanese doctor, described a condition in which young children developed a high fever and had a tongue that he called "strawberry tongue." The fever lasted for several days and soon showed a red skin rash all over the body and swollen lymph glands in the neck. The condition, which is seen in populations worldwide, especially of people of Japanese and Korean ancestry, was named for the doctor—Kawasaki disease.

What Is Kawasaki Disease?

Kawasaki disease is a perplexing disorder that affects the blood vessels, lining of the mucous membranes, and skin. It appears mostly in infants and young children. Following are the symptoms of the disorder:

- Prolonged high fever that lasts for at least five days. The fever does not respond to antibiotics.
- Red, swollen lips, tongue, and lining of the mouth. The characteristic "strawberry tongue" is present.
- Rash, also known as erythema, all over the body.
- Redness in the whites of the eyes.
- Redness in the palms of the hands and soles of feet.
- Swollen lymph glands in the neck, which may be located on one side.
- Irritability and lethargy. Children may be fussier than normal and complain of stomach pain, headache, or joint pain.

The disorder inflames the blood vessels and may affect the arteries of the heart. The inflammation causes the walls of the blood vessels to be very weak and thin and may lead to aneurysms in some children. This development can be life-threatening. However, the blood vessels of most children return to normal after a few months.

What Is the Genetic Cause of Kawasaki Disease?

The exact cause of Kawasaki disease is unknown. Although much research has been done on the disorders, it still remains a perplexing disease. Several theories and ideas abound about its cause:

- *Some problem in the immune system*: Perhaps something causes an abnormal response in the immune system. However, the triggers for such a response have never been found and are unknown.
- *Infection*: Some bacteria or virus may be involved, but the causative agent has never been found. They suspect that there may be an infection because outbreaks occur in communities and in certain seasons (spring). This idea is only a theory because even with sophisticated microbiological techniques, no bacteria or virus has been isolated.

- *Genetic*: There is probably a genetic connection with one variation that may increase risk.

ITPKC

A variation of the *ITPKC* gene, officially known as the "inositol-trisphosphate 3-kinase C" gene, may play a factor in Kawasaki disease. Normally, *ITPKC* provides instructions for making a version of the enzyme inositol 1,4,5-trisphosphate 3-kinase (ITPK). The enzyme is responsible for adding oxygen and phosphorus atoms to a molecule that regulates the amount of calcium in the cells. The ITPKC enzyme is essential in a pathway that calcium atoms affect. This pathway acts to control the activity of certain immune cells known as T cells. When T cells are overproduced, infection can occur that causes damage to body tissues.

A variation in the *ITPKC* gene has been found to increase the risk of Kawasaki disease. The change involves only one amino acid building block. The change disrupts the making of the enzyme and thus interferes with the body's ability to limit T-cell activity. Inflammation can occur that could lead to damage to blood vessels and other symptoms of the disorder. Although the disorder appears to be passed though families, the exact pattern of inheritance is unknown. Children of parents who have had Kawasaki disease have twice the risk of developing the disorder; those with affected siblings have a tenfold risk. *ITPKC* is located on the long arm (q) of chromosome 19 at position 13.1.

What Is the Treatment for Kawasaki Disease?

The condition cannot be prevented. Early treatment is essential to reduce the risk of serious heart defects. The condition does not appear to respond to antibiotics but will respond to intravenous gamma globulin. Aspirin, given in high doses, keeps the blood from forming a dangerous blood clot in the vessels. Researchers continue to look for the cause and better ways to treat this perplexing disorder.

Further Reading

"*ITPKC*." 2011. Genetics Home Reference. National Library of Medicine (U.S.). http://ghr.nlm.nih.gov/gene/ITPKC. Accessed 1/5/12.

"Kawasaki Disease." 2011. Genetics Home Reference. National Library of Medicine (U.S.). http://ghr.nlm.nih.gov/condition/kawasaki-disease. Accessed 1/5/12.

"Kawasaki Disease." 2012. Healthy Kids. American Academy of Pediatrics. http://www.healthychildren.org/English/health-issues/conditions/heart/pages/Kawasaki-Disease.aspx. Accessed 1/5/12.

"What Is Kawasaki Disease?" 2011. National Heart Lung and Blood Institute (U.S.). http://www.nhlbi.nih.gov/health/health-topics/topics/kd. Accessed 1/5/12.

Kearns-Sayre Syndrome (KSS)

Prevalence About 1 to 3 per 100,000 people

Other Names chronic progressive external ophthalmoplegia and myopathy; chronic progressive external ophthalmoplegia with ragged red fibers; CPEO with myopathy; CPEO with ragged red fibers; Kearns-Sayre mitochondrial cytopathy; KSS; mitochondrial cytopathy; oculocraniosomatic syndrome (obsolete); ophthalmoplegia, pigmentary degeneration of the retina and cardiomyopathy; ophthalmoplegia plus syndrome

In 1958, Dr. Thomas P. Kearns and Dr. George Sayre wrote about two patients who had three types of symptoms: eye weakness, eye pigment irregularities, and heart abnormalities. Two years later in 1960, Jager and coauthors found the same symptoms in a 13-year-old boy. Kearns in 1965 then published studies of nine other cases with this triad of symptoms. The condition was connected to deletions in the mitochondrial DNA in 1988.

What Is Kearns-Sayre Syndrome (KSS)?

Kearns-Sayre syndrome (KSS) is a disorder that affects two seemingly unrelated body parts—the eyes and heart. The symptoms appear before the age of 20 and can be described as the following:

- Progressive weakness or paralysis of the eye muscles: This condition is often referred to as CPEO, which stands for called chronic progressive external ophthalmoplegia.
- Accumulation of pigment on the nerves of the eye: This condition is called atypical retinitis pigmentosa and shows an abnormal collection of colored material on the membrane lining the eyes. This pigment disrupts the normal functioning of the eye, leading to inflammation and progressive degeneration of the retina. The light-sensing photoreceptors on the retina may be affected and cause loss of vision.
- Heart disorders: The heart muscles develop a condition called cardiomyopathy.
- Short stature.
- Muscle weakness: When individual muscle cells are viewed under a microscope, there is an abundance of mitochondria called ragged-red fibers.
- Hearing loss.
- Elevated spinal fluid.
- Progressive disorder of the cerebellum of the brain: Loss of balance may cause unsteadiness.
- Other disorders such as kidney disorders, loss of cognitive functions, diabetes mellitus, and Addison disease.

What Are the Genetic Causes of Kearns-Sayre Syndrome (KSS)?

Defects in the mitochondria cause KSS. Normally, mitochondria are the powerhouses of the cells, which use oxygen to make energy from food that the cells can use, a process called phosphorylation. Mitochondria have their own DNA, which is notated as mtDNA. These genes make the mitochondria do their work properly.

Individuals with KSS are missing a large part of the mtDNA. The missing parts can range from 1,000 to 10,000 amino acid building blocks, but these deletions disrupt the process of oxidative phosphorylation. The most common change takes away 4,997 amino acid blocks affecting 12 genes. Scientists surmise that the loss of these energy genes especially affects the eyes, which have a high demand for energy.

KSS is not inherited but is a somatic mutation that happens after conception and is present in certain cells. The egg cells from the mother contribute the mitochondria to the offspring. Females only pass along mitochondrial conditions, and they can appear in every family and affect both make and female.

What Is the Treatment for Kearns-Sayre Syndrome (KSS)?

This rare genetic disorder has no cure. Most of the treatment is symptomatic, treating the disorders of the heart, which may be life-threatening, and the many secondary disorders of the endocrine gland. The person can receive aids to assist hearing and vision.

Further Reading

"Kearns-Sayre Syndrome." 2011. Genetics and Rare Diseases Information (GARD). National Institutes of Health (U.S.). http://rarediseases.info.nih.gov/GARD/Condition/6817/Kearns_Sayre_syndrome.aspx. Accessed 12/23/11.

"Kearns-Sayre Syndrome." 2011. Medscape. http://emedicine.medscape.com/article/950897-overview#showall. Accessed 12/23/11.

"Kearns Sayre Syndrome: A Mitochondrial Disorder." 2011. About.com Rare Diseases. http://rarediseases.about.com/cs/kearnssayresynd/a/012404.htm. Accessed 12/23/11.

Kennedy Disease (KD)

Prevalence Estimated 1 in 40,000 men

Other Names KD; spinobulbar muscular atrophy; X-linked bulbo-spinal atrophy; X-linked spinal and bulbar muscular atrophy (SMBA)

Three months after completing his residency at the Mayo Clinic in 1964, Dr. William Kennedy saw a 57-year-old patient of French and Native American ancestry who had experienced muscle weakness for over 20 years. He found other family members had the same condition. His task was to develop an extensive pedigree of relatives with a similar disorder.

Two months later, a 68-year-old man from Iowa was referred to him, and he noted the similarity. He loaded all his equipment into his car and then drove to Iowa to contact family members to build their pedigrees. Kennedy had identified a new disease. Dr. Paul Delwalde, a Belgium neurologist, first used the name Kennedy disease in a paper in 1979. The name for the disorder has stuck and is Kennedy disease.

What Is Kennedy Disease (KD)?

Kennedy disease (KD) is a disorder that affects both sensory and motor neurons. It involves the degeneration of the lower neurons and bulbar musculature. The disease affects only males and usually appears between the ages of 20 and 40, although the ages of men with the disorder have ranged from teenagers to 70. The disease is a member of a group of disorders called spinal muscular atrophy (SMA).

The following symptoms are common in this disease:

- Early symptoms include tremor of the outstretched hands
- Muscle cramps with exertion
- Muscle twitches under the skin
- Limb weakness in the pelvic or shoulder regions
- Weakness of face and tongue muscles
- Difficulty swallowing
- Difficulty speaking
- Recurring aspiration pneumonia
- Enlargement of the male breast, a condition called gynecomastia
- Low sperm count and infertility
- Type II diabetes

What Is the Genetic Cause of Kennedy Disease (KD)?

Changes in the *AR* gene, officially known as the "androgen receptor" gene, cause Kennedy disease. Normally, *AR* provides instructions for proteins called androgen receptors, such as testosterone. The receptors, which are present before birth, are important for normal male sexual development. A region of *AR* has CAG repeats. For a normal person, the repeats range from fewer than 10 to about 36. In Kennedy disease, the CAG repeat is from 38 to 60 times. Researchers believe these fragments collect within cells and disrupt the normal functions. *AR* is an X-linked recessive condition amd is located on the long arm (q) of the X chromosome at position 12.

What Is the Treatment for Kennedy Disease (KD)?

No treatment is currently known. Treatment must target the symptoms and be supportive with physical therapy and rehabilitation to slow muscle weakness. The disease is progressive, and the individual may keep walking and active until late in the disease. The life span is usually normal. The National Institute of Neurological Disease (NINDS) is supporting research on the broad spectrum of motor neuron disease.

Further Reading

"Kennedy Disease." 2010. Medscape. http://emedicine.medscape.com/article/1172604-overview. Accessed 5/21/12.

Kennedy's Disease Association. 2010. http://www.kennedysdisease.org. Accessed 5/21/12.

"NINDS Kennedy's Disease Information Page." 2010. National Institute of Neurological Disorders and Stroke. http://www.ninds.nih.gov/disorders/kennedys/kennedys.htm. Accessed 5/21/12.

Kleefstra Syndrome

Prevalence Unknown; genetic testing to distinguish from disorders with similar features

Other Names 9q34.3 deletion syndrome; 9q34.3 microdeletion syndrome; 9q subtelomeric deletion syndrome; 9q syndrome

Dr. Tjitske Kleefstra, a clinical geneticist working at the Department of Human Genetics, Radboud University, Nijmegen Medical Centre in the Netherlands, came across a female who appeared to have an exchange of chromosome parts between chromosomes 9 and X. Dr. Kleefstra tells on the Kleefstra disorder website how she became especially interested in genetic disorders that cause intellectual disabilities. She and her colleagues found how the breakage was on chromosome 9 at a place where the *EHMT1* gene was located. For her diligent work on this disorder, the syndrome was named Kleefstra syndrome.

What Is Kleefstra Syndrome?

Kleefstra syndrome is a condition that affects muscles, speech, and intellectual ability. Several body parts are involved. Following are the characteristics of the syndrome:

- Developmental delay and severe mental retardation: Structural brain defects affect intellectual development.
- Small head with a wide short skull

- Distinct facial appearance: The person has a flat face with eyebrows that grow together in the middle, wide-spaced eyes, nostrils that are turned to the front rather than downward, lower lip that is turned out, a protruding jaw, and a sunken appearance in the center of the face. Some describe the mouth as "fish-like" with a very large tongue.
- Speech: The person may have severely limited speech or may not be able to speak at all.
- Weak muscle tone, a condition called hypotonia.
- Congenital heart defects.
- Genitourinary disorders.
- Respiratory infections.
- Seizures.
- High birth weight.
- Childhood obesity.

Some of the developmental disorders may manifest as autism or disorders of communication and social interaction.

What Is the Genetic Cause of Kleefstra Syndrome?

Mutations in the *EHMT1* gene, officially known as the "euchromatic histone-lysine N-methyltransferase 1" gene, cause Kleefstra syndrome. Normally, *EHMT1* provides instructions for the euchromatic histone methyltransferase 1 enzyme. In the body, histones are proteins that bind to certain parts of DNA to give chromosomes shape. The enzymes that are part of the histone family add a methyl group and work to control the activity certain genes. The process is essential for normal development.

Loss of the *EHMT1* gene in a translocation process interrupts the complete action of the methyltransferase 1 enzyme. About a million amino acid building blocks of DNA are lost on chromosome 9. Also, about 25 percent of the people have mutations in the *EHMT1* gene that have the same effect. Without the enzyme, the body does not function properly, causing the symptoms of Kleefstra syndrome. *EHMT1* is inherited in an autosomal dominant pattern and is located on the long arm (q) of chromosome 9 at position 34.3

What Is the Treatment for Kleefstra Syndrome?

Kleefstra syndrome is a very serious genetic disorder that affects so many body parts. It is essential that a multidisciplinary team that cares for children with intellectual disabilities be assembled. Treating any of the life-threatening symptoms such as heart, kidney, or other medical issues is of prime importance. The person qualifies for early childhood intervention programs and special education services.

Further Reading

Kleefstra, Tjitske; Willy M. Nillesen; and Helger G. Yntema. 2010. "Kleefstra Syndrome." *GeneReviews*. http://www.ncbi.nlm.nih.gov/books/NBK47079. Accessed 1/6/12.

"Kleefstra Syndrome." 2011. Genetics Home Reference. National Library of Medicine (U.S.). http://ghr.nlm.nih.gov/condition/kleefstra-syndrome. Accessed 1/6/12.

Kleefstrasyndrome.org. 2011. http://www.kleefstrasyndrome.org. Accessed 1/6/12.

Klinefelter Syndrome

Prevalence	Occurs only in males; 1 in 500 to 1,000 born with an extra X chromosome; more than 3,000 affected males born yearly; about 250,000 males born in United States each year; 5 to 29 times higher in a population with mental retardation than in the general population; most common sex chromosome disorder in males; second-most common condition caused by extra chromosome; one of the 10 most researched genetic conditions listed on the National Institutes of Health websites
Other Names	47,XXY; Klinefelter's syndrome; XXY syndrome; XXY trisomy

In 1942, Dr. Harry Klinefelter of the Massachusetts General Hospital published a report of nine men who had enlarged breasts, sparse facial and body hair, small testes, and infertility. However, it was not until 1959 that Drs. Patricia Jacobs and J. A. Strong in Edinburgh reported the karyotype of a man with Klinefelter syndrome. The karyotype showed the man had an extra X chromosome with the genotype XXY, rather than the normal male XY. Klinefelter syndrome is defined clinically as 47,XXY. A normal person has 46 chromosomes; those with Klinefelter syndrome have 47.

What Is Klinefelter Syndrome?

Klinefelter syndrome is a condition in which the person has an extra X chromosome that affects male sexual development. Normally, females have an XX chromosome pattern and males have an XY pattern. In Klinefelter syndrome, there is an extra X chromosome. Although not all males with XXY show the symptoms of the disorder, those who have the extra X chromosome are usually referred to as "XXY males" or "47,XXY males."

Symptoms of XXY males vary depending on the number of cells that have the XXY pattern, how much testosterone (male hormone) the person has, and the age of diagnosis. Following are the main symptoms of XXY males:

- Physical development: A condition called hypogonadism describes decreased development of the hormone/endocrine function that controls testosterone production and results in small testicles and small penis. As babies, XXY males have weak muscles and reduced strength; however, as they enter puberty, they become tall but less muscular. They may not develop secondary sex characteristics, and they have less facial hair, broader hips, and higher pitched voice than other boys. Between 96% and 99% of males with XXY are infertile.

- Language development: About 25% of XXY males have trouble with language. They may be slow to speak, have trouble reading, and show difficulty communicating thoughts and ideas. However, unlike Down syndrome (trisomy 21) and other multiple chromosomal disorders, intellectual development may be in the normal range. Some may have minor developmental and learning disabilities.
- Social development: As babies, these males may be quiet and not demanding. When they go to school, again they may call little attention to themselves and appear as shy. They may struggle to fit in, especially when physical activity and sports are involved.
- Body image: The individual may develop some female body characteristics. Because of the low serum level of testosterone and possible presence of female hormones, the individual may have increased breast tissue, a condition called gynecomastia. About one-third of XXY males have this condition; some may have breasts large enough to require cosmetic surgery.
- Other conditions: People at the severe end of the spectrum can develop germ cell tumors, male breast cancer, and osteoporosis. Other conditions may include an increased risk for diabetes, varicose veins, pulmonary disease, lupus erythematosis, and rheumatoid arthritis. The chances of getting these disorders are similar to females.

What Is the Genetic Cause of Klinefelter Syndrome?

Kleinfelter syndrome or 47XYY is related to the X and Y chromosomes. The extra copy of the X chromosome in the cells disrupts the normal male pattern by preventing the testes from functioning normally and reducing the amount of testosterone. Some makes have the extra X chromosome only in some of their cells. This condition is called mosaic Klinefelter syndrome (46,XXY/47XXY). The signs and symptoms may be milder depending on how many cells have the extra chromosome. Mental retardation is common in those with multiple X chromosomes.

Several variants of Klinefelter syndrome may have other configurations. Studies have shown that the more X chromosomes that the person has, the more severe the symptoms. For example, people with 49,XXXXY may have more severe symptoms than one with 48,XXXY. There may also be the variants 48,XXYY and 49, XXXYY with extra Y chromosomes.

What Is the Treatment for Klinefelter Syndrome?

Many males with XXY go undiagnosed, and some do not receive the diagnosis until they are adults. Infertility and gynecomastia are the two most common symptoms that lead one to suspect XXY.

Klinefelter syndrome is not inherited. It happens as a random occurrence during the formation or meiosis of sperm and egg. An error in cell division that results in an abnormal number of chromosomes is called nondisjunction.

Several types of treatments are available to those with Klinefelter syndrome

- Educational treatments: Children in school can qualify for special services. Special education teachers help by breaking down complex materials into smaller chunks for comprehension.
- Therapy treatments: Individuals can be helped with physical therapy to improve muscle tone. Speech therapy, occupational therapy, behavioral and mental health counseling, and family counseling are also available. The therapy treatments may reduce some of the social problems the person may encounter and assist in building self-esteem.
- Medical treatments: Testosterone replacement therapy (TRT) may help males get testosterone levels back to a normal range. This treatment may overcome some problems such as muscle weakness, high-pitched voice, and sparse facial and body hair. The androgen receptor gene (AR) found on the X chromosome has a trinucleotide repeat sequence (CAG). Individuals with short AR CAG repeats have been shown to respond to male hormones. Continued research is ongoing on treatments for Klinefelter syndrome. The important linchpin is early identification.

Further Reading

"Klinefelter Syndrome." 2010. National Center for Biotechnology Information (U.S.). http://www.ncbi.nlm.nih.gov/books/NBK22183. Accessed 5/21/12.

"Klinefelter Syndrome." 2010. Eunice Kennedy Shriver National Institute of Child Health and Human Development. http://www.nichd.nih.gov/health/topics/klinefelter_syndrome.cfm. Accessed 5/21/12.

"Klinefelter Syndrome." 2010. Genetics Home Reference. National Library of Medicine (U.S.). http://ghr.nlm.nih.gov/condition/klinefelter-syndrome. Accessed 5/21/12.

"Klinefelter Syndrome." 2010. Medscape. http://emedicine.medscape.com/article/945649-overview. Accessed 5/21/12.

Klippel-Feil Syndrome

Prevalence Incidence about 1 in 42,000 births; more cases in boys than girls

Other Names KFS

In 1912, Maurice Klippel and Andre Feil, French doctors, described patients who had a short neck, low hairline, and decreased range of motion. Their names were given to this rare disorder. Since then, several archeological radiographs have turned up some famous people who might have had the condition. The 18th Dynasty Egyptian pharaoh Tutankhamen might have suffered from Klippel-Feil. In 2009, an archeological dig in Vietnam unearthed a young man with KFS. The article published in *Discovery News* headlined him as the "oldest known paralyzed human."

What Is Klippel-Feil Syndrome?

Klippel-Feil syndrome (KFS) is a skeletal disorder. There are three basic types of the disorders that relate to malformations in the vertebrae or backbone. Following are the three types:

- Type I: Patients have a single massive fusion of the cervical vertebrae.
- Type II: Patients have fusion of one or two vertebrae.
- Type III: In addition to the fusions of Types I or II, the person will have thoracic (chest) and lumber anomalies.

In addition to the fusion of the neck bones, the patient may have some other abnormalities:

- Scoliosis
- Spina bifida, a birth defect
- Cleft palate
- Defects of the ribs
- Problems with the urinary tract, kidney, muscles, brain, and skeleton
- Facial defects
- Problems hearing and breathing.

What Are the Genetic Causes of Klippel-Feil Syndrome?

KFS is believed to occur in the early stage of fetal development when the spine is formed. For some reason, the cervical vertebrae do not segment properly. Some of the cases occur sporadically, but there is some indication that the condition may appear in families. Some researchers suspect that the condition is related to maternal use of alcohol, causing fetal alcohol syndrome.

Human pedigrees have been developed to find a genetic locus for KFS. Autosomal dominant inheritance has been associated with C2-C3 fusion. Autosomal recessive inheritance is associated with C5-C6. A dominant form has been mapped to a gene called *GDF6*, or the "growth differentiation factor 6." This gene is a member of the bone morphogenetic protein (BMP) and the TGF-beta superfamily of signaling molecules. Mutations in the gene cause a failure in the normal segmentation of the vertebrae during the early weeks of fetal development, causing the clinical triad of short neck, low posterior hairline, and limited neck movement. *GDF6* is located on the long arm (q) of chromosome 8 at position 22.1.

What Is the Treatment for Klippel-Feil Syndrome?

Treatment of KFS is symptomatic and may include surgery to relieve cervical and constriction of the spinal cord. Surgery may also correct scoliosis.

Further Reading

"Klippel Feil Syndrome." Dr. Reddy's Laboratories Case Report. http://www.bhj.org/journal/1999_4103_july99/case_586.htm. Accessed 5/21/12.

"Klippel-Feil Syndrome." 2010. Medscape. http://emedicine.medscape.com/article/1264848-overview. Accessed 5/21/12.

Kniest Dysplasia

Prevalence One in 1,000,000 live births; affects males and females alike

Other Names None

German pediatrician Wilhelm Kniest described patients with a very rare condition that includes short stature, malformed bones and joints, and other sensory problems. The condition was named for Kniest in 1952—Kniest dysplasia.

What Is Kniest Dysplasia?

Kniest dysplasia is a form of dwarfism in which the person has short stature, but also has a short trunk, shortened limbs, and large joints. The word "dysplasia" comes from two Greek roots: *dys*, meaning "with difficulty," and *plas*, meaning "form." The person may be from 42 inches to 58 inches in height.

The first symptoms of Kniest dysplasia occur at birth and may include the following:

- Short limbs, barrel chest
- Joints do not extend all the way and may be swollen

- Eye problems that can lead to blindness
- Facial features include a broad round face and wide-set eyes
- Cleft palate in some infants
- Flattened bones of the spine
- Dumbbell-shaped bones in the arms and legs
- Long and knobby fingers
- Inguinal hernias
- Spinal curvature that may worsen over time

Intelligence is usually not affected.

What Are the Genetic Causes of Kniest Dysplasia?

Mutations in the *COL2A1* gene, officially known as the "collagen type II, alpha 1" gene, cause Kniest dysplasia. Normally, *COL2A1* instructs for a protein that makes type II collagen. This collagen is found in cartilage and the clear gel that fills the eye. Type II collagen is necessary for normal development of bones and other connective tissues that from the body's supportive framework.

Mutations in the *COL2A1* gene interfere with the making of the collagen molecules. The abnormal collagen leads to the signs and symptoms of Kniest dysplasia. Most of the mutations cause an abnormally short collagen chain to be produced. *COL2A1* is inherited in an autosomal dominant pattern and is located on the short arm (q) of chromosome 12 at position 13.11.

What Is the Treatment for Kniest Dysplasia?

Medical care should be consistent in patients with Kniest dysplasia. If surgery is necessary, anesthesia must be used with care because of reactions caused by cervical spine instability, lung capacity, and small airways. Orthopedic care should be used to evaluate hip, spine, and knee complications. Because of the poorly developed type II collagen, people with Kniest may develop arthritis. Exercise should be undertaken with caution.

Further Reading

"*COL2A1*." 2010. Genetics Home Reference. National Library of Medicine (U.S.). http://ghr.nlm.nih.gov/gene/COL2A1. Accessed 5/21/12.

"Kniest Dysplasia." 2010. Genetics Home Reference. National Library of Medicine (U.S.). http://ghr.nlm.nih.gov/condition/kniest-dysplasia. Accessed 5/21/12.

Kniest SED Group. http://www.ksginfo.org/kniest.html. Accessed 5/21/12.

Knobloch Syndrome

Prevalence Rare; exact prevalence unknown

Other Names retinal detachment and occipital encephalocele

In 1971, W. H. Knobloch and J. M. Layer published an article that told about families of 5 to 10 siblings who had severe nearsightedness, a detached retina, and an encephalocele. The parents of the siblings did not have the disorder, nor were they related to each other. The condition was later named for Dr. Knobloch.

What Is Knobloch Syndrome?

Knobloch syndrome is a disorder that affects the eyes and is also characterized by an encephalocele, a protrusion of the brain through an opening in the skull. Following are the symptoms of the condition:

- Vision disorder: The person has severe nearsightedness, a condition called myopia. The myopia develops during the first year of life. The person may also be born with cataracts.
- Retinal detachment: The retina, the back part of the eye, is detached, causing serious vision problems.
- Encephalocele.
- Other disorders.

Intelligence does not appear to be affected by this disorder.

What Is the Genetic Cause of Knobloch Syndrome?

Mutations in the *COL18A1* gene, officially known as the "collagen, type XVIII, alpha 1" gene, causes Knobloch syndrome. Normally, *COL18A1* provides instructions for making three proteins that are called alpha 1 subunits. These units attach to form collagen XVIII, a major protein found in the basement membranes of cells. The basement membranes are the thin layers that line and separate cells from each other. Three versions of the alpha 1 subunit make three different lengths of the collagen XVIII protein. The short version is located in the basement cells of body tissues, which include the eye. Longer versions are found in the liver. If a piece of the longer form is cut, it forms the protein endostatin, which blocks formation of blood vessels. It appears that all three forms are important for the normal development of the eye.

About a dozen mutations in *COL18A1* cause Knobloch syndrome. Mutations result in a very short version of the proteins or replacement of a single amino acid building block that disrupts the formation of the alpha subunit of collagen XVIII. Even the loss of one of the forms may cause the signs and symptoms of the disorder. *COL18A1* is inherited in an autosomal recessive pattern and is located on the long arm (q) of chromosome 21 at position at position 22.3.

What Is the Treatment for Knobloch Syndrome?

Glasses may correct the nearsightedness. Most individuals with retinal detachment may require surgery. New techniques have been developed for eye surgery. If there are tears in the retina, laser surgery may sear the tear before a complete detachment can occur. If the detachment is small, a gas bubble, called a pneumatic retinoplexy, may help the retina float back in place. Severe detachment requires more complex surgery.

Further Reading

"Knobloch Syndrome." 2011. Genetics and Rare Diseases Information Center (GARD). National Institutes of Health (U.S.). http://rarediseases.info.nih.gov/GARD/Condition/380/QnA/25961/Knobloch_syndrome.aspx. Accessed 1/6/12.

"Knobloch Syndrome." 2011. Genetics Home Reference. National Library of Medicine (U.S.). http://ghr.nlm.nih.gov/condition/knobloch-syndrome. Accessed 1/6/12.

"Retinal Detachment." 2011. MedlinePlus. National Library of Medicine (U.S.). http://www.nlm.nih.gov/medlineplus/ency/article/001027.htm. Accessed 1/6/12.

Krabbe Disease

Prevalence Rare; about 1 in 100,000; 1 in 6,000 in some Arab communities in Israel; Scandinavian rates 1 in 50,000 births

Other Names galactosylceramide lipidosis; globoid cell leukodystrophy; Krabbe's leukodystrophy; Krabbe's syndrome 1

Knud Haraldsen Krabbe, a gifted Danish neurologist, was very interested in conditions that affected the nervous system. He observed a disorder that is occasionally seen in Scandinavia. The child was normal at birth but then developed a fatal degenerative disorder. For his research on the disorder, he was honored with the name—Krabbe disease.

Jim Kelly, longtime quarterback for the Buffalo Bills, became a strong advocate for the recognition of Krabbe disease. His son Hunter died from Krabbe disease in 2005 at the age of eight. Another note of interest relates to the animal world. Cats and dogs—especially Westies and Cairn Terriers—may have Krabbe disease.

What Is Krabbe Disease?

Krabbe disease affects the myelin sheath and leads to a serious degenerative disorder. Krabbe disease is a type of leukodystrophy, which affects the white matter covering of the neurons or nerve cells.

Two forms of Krabbe disease exist: an early-onset form that appears the first few months of life; the child usually dies before the age of two. The second form, late-onset, begins in late childhood or early adolescence.

Some of the symptoms of Krabbe disease include the following:

- Changing muscle tone from floppy to rigid
- Irritability
- Hearing loss that leads to deafness
- Vision loss that may lead to blindness
- Feeding difficulties
- Failure to thrive
- Severe seizures that may begin at a very early age
- Fevers without inflammation
- Vomiting
- Slowing of mental and motor development

What Is the Genetic Cause of Krabbe Disease?

Changes in the *GALC* gene, officially called the "galactosylceramidase" gene, causes Krabbe disease. Normally, *GALC* provides instructions for making an enzyme called galactosylceramidase. Through a process called hydrolysis, this enzyme uses water molecules to break down certain fats called galactolipids, which are found primarily in the brain and kidneys. The enzyme is also found within the lysosome of the cells where it processes galactosylceramide and psychosine. Galactosylceramide is an important component of myelin, the fatty sheath that covers certain nerve cells. The myelin is important for the rapid transmission of nerve impulses.

The more than 70 mutations of *GALC* cause Krabbe disease. The mutations impede that activity of the galactosylceraminidase enzymes so that the materials that make up myelin cannot be broken down. These products then accumulate in cells that damage the myelin-producing cells. Without myelin, neurons do not function properly, leading to the fatal signs of Krabbe disease. *GALC* is inherited in an autosomal recessive pattern and is located on the long arm (q) of chromosome 14 at position 31.

What Is the Treatment for Krabbe Disease?

There is no treatment for Krabbe disease. Some people have had bone marrow transplants, but there are risks involved with this procedure. A 2005 article in the *New England Journal of Medicine* told of successful cord blood transplants, as long as they were given before symptoms appeared.

Further Reading

"*GALC*." 2010. Genetics Home Reference. National Library of Medicine (U.S.). http://ghr.nlm.nih.gov/gene/GALC. Accessed 5/21/12.

"Krabbe Disease." 2010. Genetics Home Reference. National Library of Medicine (U.S.). http://ghr.nlm.nih.gov/condition/krabbe-disease. Accessed 5/21/12.

"Krabbe Disease." 2010. MedlinePlus. National Library of Medicine (U.S.). http://www.nlm.nih.gov/medlineplus/ency/article/001198.htm. Accessed 5/21/12.

Kufs Disease

Prevalence Affects 2 to 4 per 100,000

Other Names adult neuronal ceroid lipofuscinosis; CLN4A; Kuf's disease; Kufs type neuronal ceroid lipofuscinosis

A condition that appeared in young adults and caused normal-appearing individuals to progressively lose their memories and ability to walk and move baffled Dr. Hugo Friedrich Kufs, a German neuropathologist. He described three people with similar conditions in 1925, 1929, and 1931. Although other physicians studied the disease, the medical profession has called the disorder Kufs disease.

What Is Kufs Disease?

Kufs disease is a progressive disorder of the nervous system, which mimics mental illness and affects movement and sight. It is one of the diseases that are classified in a family of disorders called neuronal ceroid lipofuscinoses or NCL. These diseases are characterized by deposits of lipofuscins, yellowish-brown pigments that are dissolved in fat and located throughout the nervous system. The disorder is very difficult to distinguish from other progressive nervous system disorders. The symptoms usually begin about the age of 30 but can occur any time from teenage years to later life. The person does not live more than 10 years after diagnosis.

Two types of the disorder exist, which have some overlapping signs: type A and type B.

Type A

The following symptoms occur over time:

- Loss of intellectual functions—the person begins to lose memory, cognition, and logical thinking
- Impaired muscle coordination
- Involuntary tics and tremors

- Speech difficulty
- Seizure and uncontrollable muscle jerking, a condition of myoclonic epilepsy

Type B

- Has characteristic dementia, muscle coordination issues, and tics
- Not associated with epilepsy or speech disorders
- Notable changes in personality

What Is the Genetic Cause of Kufs Disease?

Mutations in the *PPT1* gene, officially known as the "palmitoyl-protein thioesterase 1" gene, cause Kufs disease. Normally, *PPT1* provides instructions for the palmitoyl-protein thioesterase 1, an enzyme found in the lysosomes. The lysosomes are structures in the cells that dispose of unwanted cell material and repackage different types of molecules. One of the functions of palmitoyl-protein thioesterase 1 is to take certain fats out of long-chain fatty acids from other proteins for use in creating energy. The enzyme has other important roles: transporting molecules form the cell surface and into the cell; moving small sacs called vesicles in cell, helping the self-destruction of cells, and helping in the functioning of the synapses, the gaps between neurons.

About three mutations in the *PPT1* gene are related to Kufs disease. The mutations change only one amino acid building block or create premature stop signals in the instructions for palmitoyl-protein thioesterase 1. The mutations do allow some function of the enzyme, which accounts for the appearance of the symptoms later in life. The disruption of the enzyme causes the symptoms of Kufs disease. *PPT1* is inherited in an autosomal recessive pattern and is located on the short arm (p) of chromosome 1 at position 32.

What Is the Treatment for Kufs Disease?

Treatment of this multisymptom disorder is symptomatic. Certain symptoms such as seizures, depression, movement disorders, and behavior problems can be treated with drugs. Physical and occupational therapy may also help with movement disorders.

Further Reading

"Kufs Disease." Genetics Home Reference. National Library of Medicine (U.S.). http://ghr.nlm.nih.gov/condition/kufs-disease. Accessed 12/24/11.

"Kufs Disease." 2010. Whonamedit? http://www.whonamedit.com/synd.cfm/1914.html. Accessed 12/24/11.